W9-AAK-466

ACS SYMPOSIUM SERIES **932**

Carbohydrate Drug Design

Anatole A. Klyosov, Editor
Pro-Pharmaceuticals, Inc.

Zbigniew J. Witczak, Editor
Wilkes University

David Platt, Editor
Pro-Pharmaceuticals, Inc.

Sponsored by the
ACS Division of Carbohydrate Chemistry

American Chemical Society, Washington, DC

Library of Congress Cataloging-in-Publication Data

Carbohydrate drug design / Anatole A. Klyosov, Zbigniew J. Witczak, David Platt, editors.

p. cm.—(ACS symposium series ; 932)

Includes bibliographical references and indexes.

ISBN 13: 978–0–8412–3946–3 (alk. paper)

1. Carbohydrate drugs—Congresses.

I. Klyosov, Anatole A., 1946- II. Witczak, Zbigniew J., 1947- III. Platt, David, 1953- IV. Series.

RS431.C23C367 2006
615.3´—dc22 2005057228

The paper used in this publication meets the minimum requirements of American National Standard for Information Sciences—Permanence of Paper for Printed Library Materials, ANSI Z39.48–1984.

Distributed by Oxford University Press

ISBN 10: 0–8412–3946–0

PRINTED IN THE UNITED STATES OF AMERICA

Foreword

The ACS Symposium Series was first published in 1974 to provide a mechanism for publishing symposia quickly in book form. The purpose of the series is to publish timely, comprehensive books developed from ACS sponsored symposia based on current scientific research. Occasionally, books are developed from symposia sponsored by other organizations when the topic is of keen interest to the chemistry audience.

Before agreeing to publish a book, the proposed table of contents is reviewed for appropriate and comprehensive coverage and for interest to the audience. Some papers may be excluded to better focus the book; others may be added to provide comprehensiveness. When appropriate, overview or introductory chapters are added. Drafts of chapters are peer-reviewed prior to final acceptance or rejection, and manuscripts are prepared in camera-ready format.

As a rule, only original research papers and original review papers are included in the volumes. Verbatim reproductions of previously published papers are not accepted.

ACS Books Department

Contents

Pathogen Management

New Approaches in Synthesis and Computational Studies

Indexes

Preface

During the past few years, it has become increasingly clear that carbohydrates, which can be targeted to specific diseases, represent a whole new dimension in drug design. Characterized by a variety of terms—*specific recognition, lectins,* and *molecular diversity*, just to cite a few—this new dimension based on carbohydrates, has essentially introduced a new *language* into chemistry, biochemistry, and related disciplines.

Vocabulary has been building so fast, in fact, that most professional chemists (i.e., practitioners who are among the few who work in the area of carbohydrates) may sometimes feel a bit illiterate. This book will help them and many other scientists to better understand the current state of the art and the challenges that remain in successfully consummating matches of carbohydrate-based drugs and the deadly diseases they target.

Only three books have previously been published in this area: *Carbohydrates in Drug Design;* Z. J. Witczak and K. Nieforth, editors, 1997; *Complex Carbohydrates in Drug Research: Structural and Functional Aspects;* Klaus Bock, Henrik Clausen, and Dennis Boch, editors, 1998; and *Carbohydrate-Based Drug Discovery;* C.-H. Wong, editor, 2003. This current book complements and extends these predecessor volumes in several critical ways. It includes preclinical studies and clinical trials of carbohydrate-based drugs as well as analyzing their delivery, biocompatibility, clearance, and metabolic pathways. Further, this book explores a number of other features of carbohydrate drugs and their targets, such as the structure of antibodies with unusually high-affinity for carbohydrates, and protein–glucan interactions and their inhibitors. Galactomannans and thio-, imino, nitro, and aminosugars are particularly considered with respect to their structural and functional impacts.

It is our hope that this volume will not only update existing publications on carbohydrate-based drug design but also further shape the emerging data and thinking in this new area. In that spirit, we have systematized the important information presented at a recent symposium, *Carbohydrate Drug Design,* that ran as a part of the American Chemical Society meeting from March 28–April 1, 2004, in Anaheim, California. Beyond documenting much of the content of that seminal meeting, we have included several additional chapters on carbohydrate-based vaccines and carbohydrate-based treatment of some infections, along with new data on the chemistry of carbohydrates and the computation of carbohydrate structures.

Last, but not least, we must thank our wives—Gail, Wanda, and Naomi—who had to tolerate our spending so many eveenings and weekends cloistered with computers, books, and papers in order to prepare this book.

Anatole A. Klyosov
Pro-Pharmaceuticals, Inc.
189 Well Avenue
Newton, MA 02459

Zbigniew J. Witczak
Department of Pharmaceutical Sciences
Nesbitt School of Pharmacy
Wilkes University
137 South Franklin Street
Wilkes-Barre, PA 18766

David Platt
Pro-Pharmaceuticals, Inc.
189 Well Avenue
Newton, MA 02459

Carbohydrate Drug Design

Overview

Chapter 1

Carbohydrates and Drug Design: What Is New in This Book

Anatole A. Klyosov

Pro-Pharmaceuticals, 189 Well Avenue, Newton, MA 02459

A good drug is a target-specific drug; its users can expect high efficiency and few, if any, side effects. Target specificity also means recognition, and this is where carbohydrates come in. While many drugs contain carbohydrates as part of their molecules, other drugs – lacking carbohydrates covalently bound to their molecules – can be guided by them. Carbohydrates' value is that they provide a guidance mechanism for sick cells, enabling drugs to arrive there with precision and act properly. On the other hand, carbohydrates can provide a defense mechanism to sick or deadly cells, preventing a drug to act properly.

This book covers, or, in some places, touches on – all these aspects of carbohydrates in drug design. It emerged from topics discussed at the symposium "Carbohydrate Drug Design" which was a part of 227th American Chemical Society meeting that ran from March 28 to April 1, 2004, in Anaheim, California. The reasons for organizing this symposium were to establish a new role for carbohydrates in concert with known drugs, taking into account newly acquired knowledge in the field, and to outline innovative ways of designing new drugs based on that knowledge.

This chapter introduces the book's content in accordance with the following six categories, described below:

1. Carbohydrate drugs
2. Cancer and a combination carbohydrate-assisted chemotherapy
3. Carbohydrate-based HIV-1 vaccines
4. Polysaccharides and infections
5. Aminosugars
6. Computational studies

1. Carbohydrate Drugs

A few dozen FDA-approved prescription drugs contain carbohydrate moieties as part of their structures. Typically, removal of the sugar eliminates the therapeutic value of the drug. These drugs can be divided into five categories, as follows:

- Monosaccharide conjugates
- Disaccharides and disaccharide conjugates
- Trisaccharides
- Oligosaccharides and polysaccharides
- Macrolides

1.1 Monosaccharide conjugates

Monosaccharide conjugates include, in turn, four groups of prescription drugs:

- Anthracycline antibiotics and agents
  - Doxorubicin
  - Daunorubicin
  - Epirubicin
  - Idarubicin

- Nucleotides and nucleosides and their analogs
  - Fludarabine Phosphate
  - Stavudine
  - Adenosine
  - Gemcitabine
  - Ribavirin
  - Acadesine

- Polyenes
  - Amphotericin B

- Other agents
  - Etoposide
  - Lincomycin
  - Clindamycin
  - Pentostatin

1.1.1 Anthracycline antibiotics and agents

The first group is represented by cytotoxic anthracycline antibiotics of microbial origin (**Doxorubicin** and **Daunorubicin**) or their semi-synthetic derivatives (**Epirubicin** and **Idarubicin**).

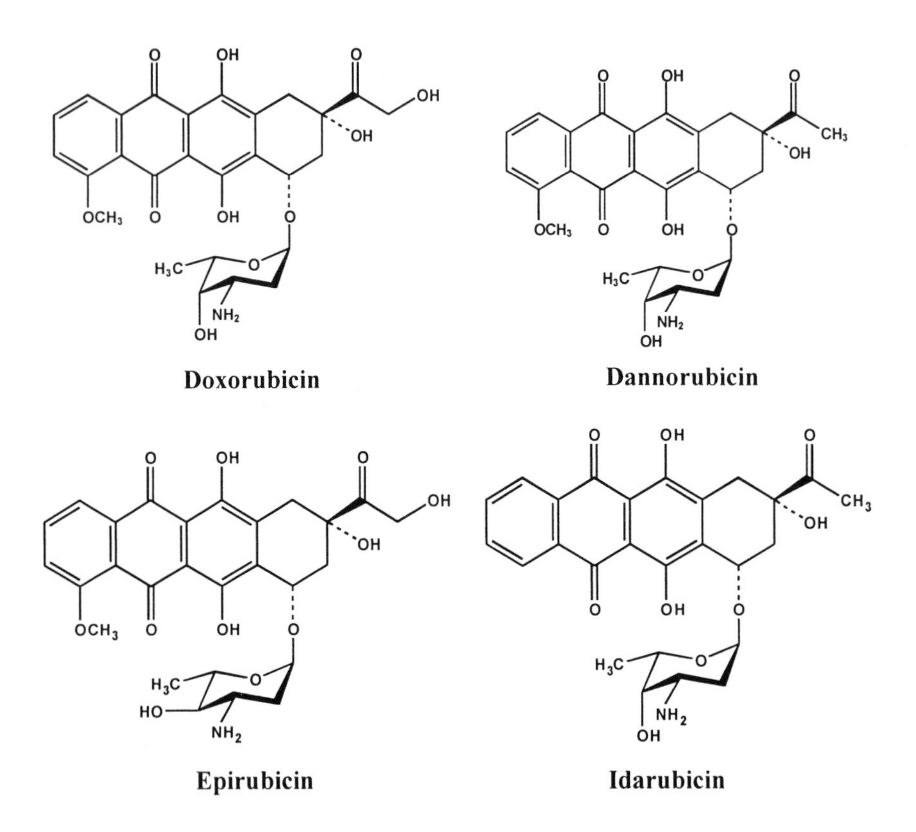

All of these drugs are potent neoplastic agents consisting of a naphthacenequinone nucleus linked through a glycosidic bond at ring atom 7 to an amine sugar, daunosamine. All of them bind to nucleic acid, presumably by specific intercalation of the planar anthracycline nucleus with the DNA double helix, between nucleotide base pairs, with consequent inhibition of nucleic acids (DNA and RNA) and protein synthesis. They all inhibit topoisomerase II activity by stabilizing the DNA-topoisomerase II complex, blocking the ligation-religation reaction. All of these drugs show the cytotoxic effect on malignant cells and – as side effects – on various organs. Intercalation inhibits nucleotide replication and action of DNA and RNA polymerases. All of them induce apoptosis, which may be an integral component of the cellular action related to antitumor therapeutic effects as well as toxicities.

1.1.2 Nucleotides and nucleosides and their analogs

The second group of monosaccharide drugs is represented by an assortment of nucleotides and nucleosides and their synthetic analogs. Among them are:

- Potent neoplastic agents, such as **Fludarabine Phosphate** (fluorinated arabinofuranosyladenine 5'-monophosphate), whose metabolic products inhibit DNA synthesis. This drug is indicated for the treatment of patients with B-cell chronic lymphocytic leukemia, while another such agent **Gemcitabine** (2'-deoxy-2',2'-difluorocytidine), is a nucleoside analogue that inhibits DNA synthesis and exhibits antitumor activity.

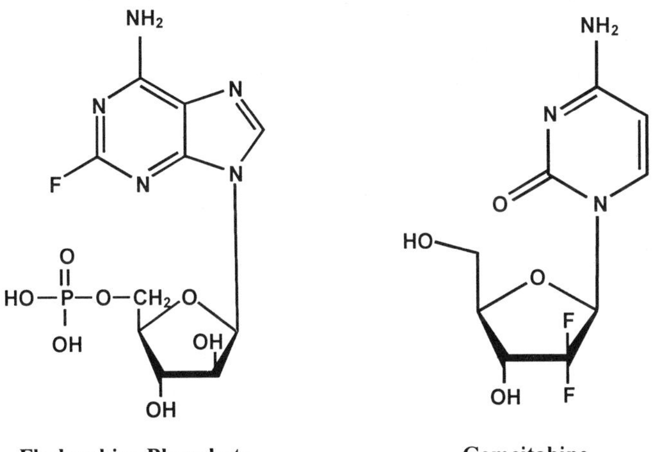

Fludarabine Phosphate **Gemcitabine**

- Drugs active against the human immunodeficiency virus (HIV) such as **Stavudine**, a synthetic thymidine nucleoside analog. This drug is a derivative of deoxythymidine, which inhibits the replication of HIV in human cells.

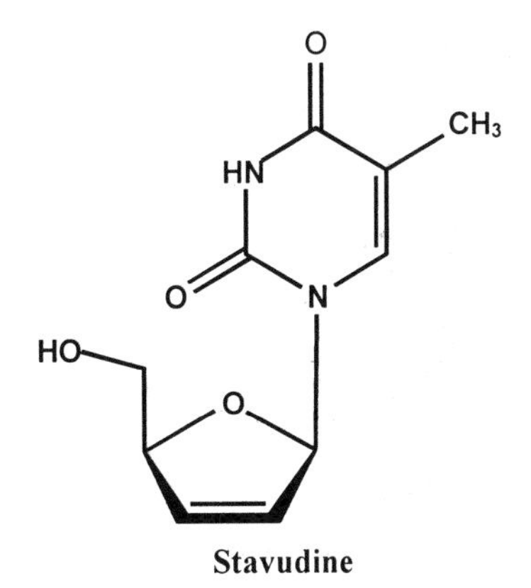

Stavudine

- An antiarrhythmic drug **Adenosine** (6-amino-9-β-D-ribofuranosyl-9-H-purine), which presents in all cells of the body and apparently activates purine receptors (cell-surface adenosine receptors). These molecules in turn activate relaxation of vascular smooth muscle through a number of biochemical events, and they are therefore indicated in patients with paroxysmal supraventricular tachycardia.

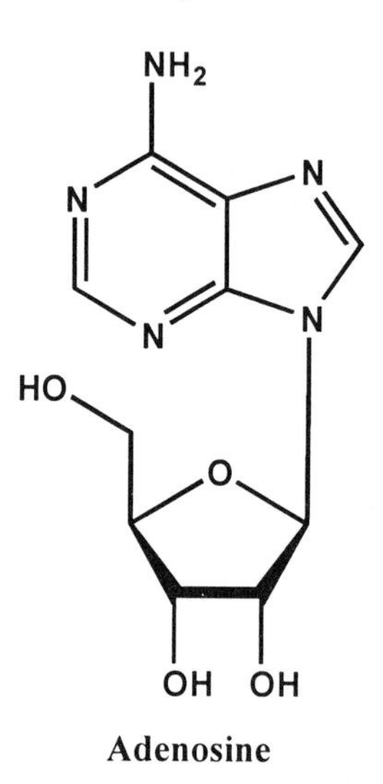

Adenosine

- The first synthetic, non-interferon type antiviral drug **Ribavirin** (ribofuranosyl-triazole derivative), a nucleoside analog, which is particularly active against respiratory syncytial virus (RSV)

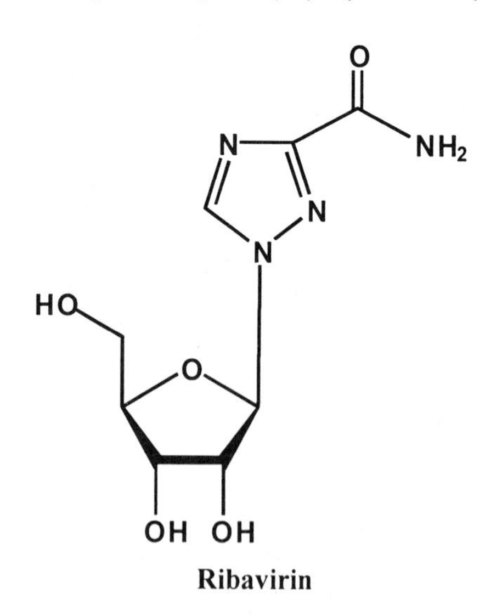

Ribavirin

- A cardioprotective agent **Acadesine**, a ribofuranosyl-imidazole derivative and a purine nucleoside analog, which is employed in particular in coronary artery bypass graft surgery.

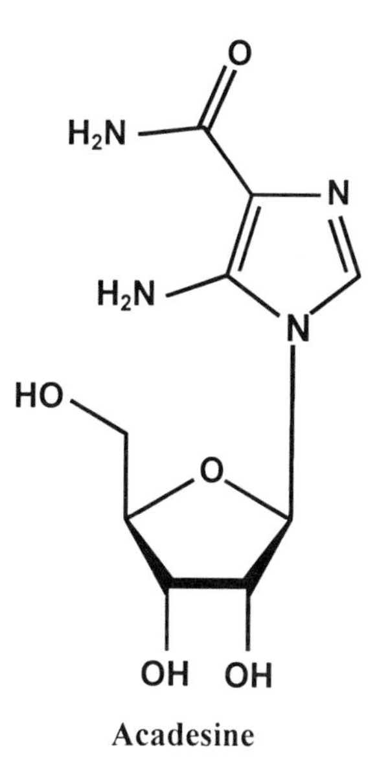

Acadesine

1.1.3 Polyenes

The third group, polyenes, is exemplified by **Amphotericin B**, which is an antifungal antibiotic of microbial origin. Amphotericin B is a 3-Amino-3,6-dideoxy-β-D-mannopyranosyl derivative of an octahydroxypolyene containing seven carbon-carbon double bonds in a macrocyclic 38-member ring. The drug changes the permeability of the cell membrane of susceptible fungi by binding to sterols in the membrane. This binding causes leakage of intracellular content – and, as a consequence, cell death.

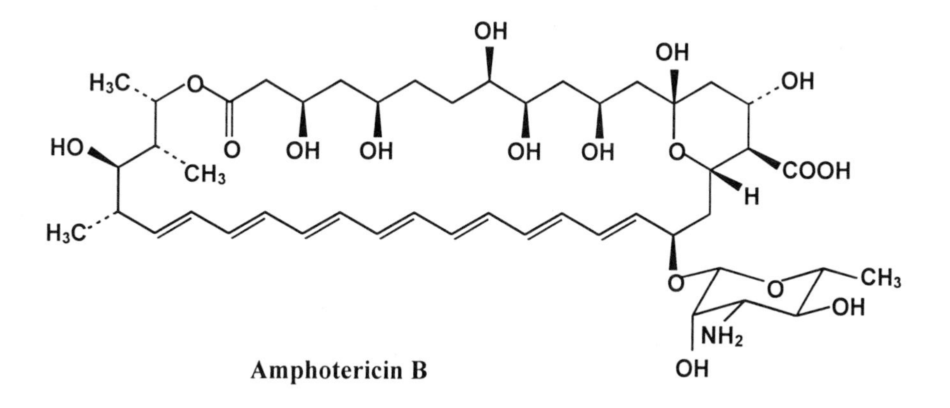

Amphotericin B

1.1.4 Other agents

The forth group of monosaccharide drugs contains a number of assorted compounds, such as:

- The cancer chemotherapeutic agent **Etoposide**, a semi-synthetic β-D-glucopyranoside derivative of podophyllotoxin.

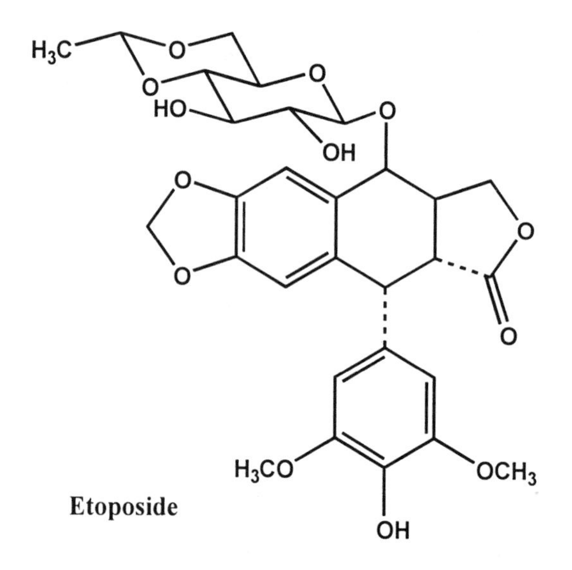

Etoposide

- An antibacterial antibiotic of microbial origin **Lincomycin**, which is a derivative of 1-thio-D-*erythro*-α-D-*galacto*-octopyranoside,

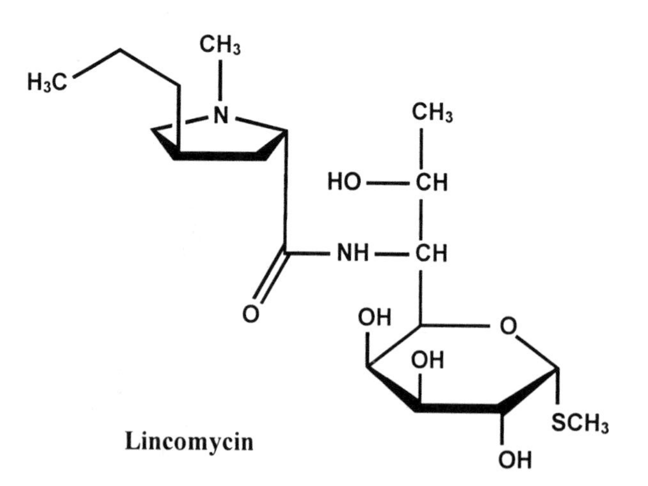

Lincomycin

- A semisynthetic antibiotic, **Clindamycin**, which is a derivative of 1-thio-L-*threo*-α-D-*galacto*-octopyranoside and produced from Lincomycin. Clindamycin is indicated in the treatment of infections caused by susceptible anaerobic bacteria, streptococci, pneumococci, and staphylococci.

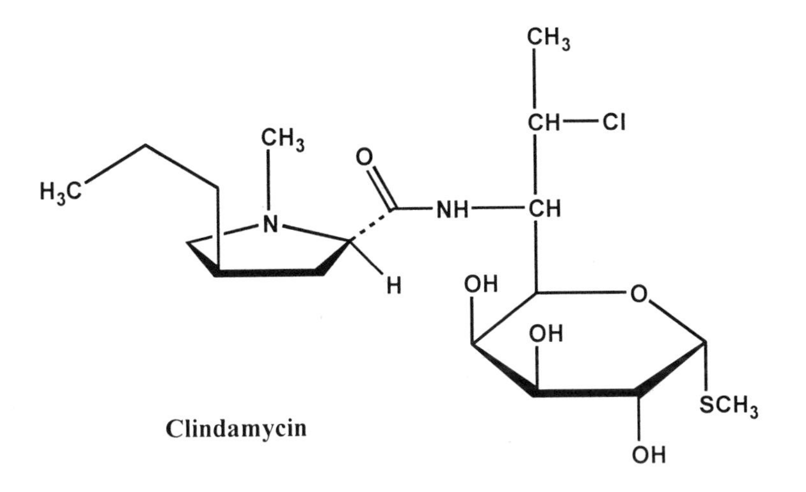

Clindamycin

- An antitumor drug, **Pentostatin,** that inhibits RNA and DNA synthesis by being a direct inhibitor of enzymes adenosine deaminase and ribonucleotide reductase, particularly in cells of the lymphoid system.

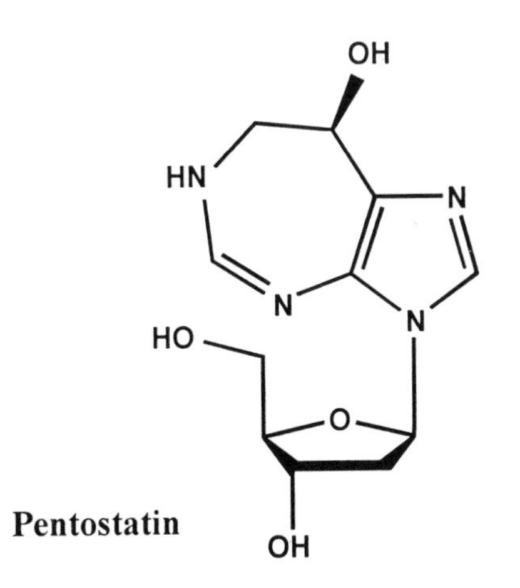

Pentostatin

1.2 Disaccharides and disaccharide conjugates

The next subcategory of prescription carbohydrate drugs, disaccharides and their conjugates, is represented by the following medications:

- An antipeptic and antiulcerative drug, **Sucralfate**, which is a β-D-fructofuranosyl-α-D-glucopyranoside basic aluminum sucrose sulfate complex. It accelerates healing of duodenal ulcers, in part by inhibiting pepsin activity in gastric juice.

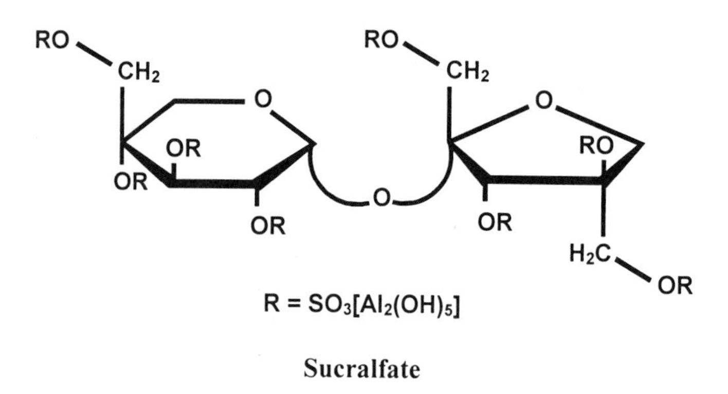

Sucralfate

- A synthetic colonic acidifier **Lactulose**, 4-O-β-D-galactosyl-D-fructose, which promotes laxation.

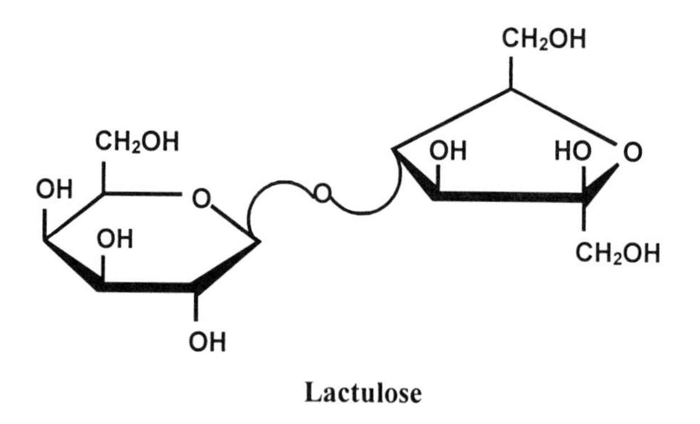

Lactulose

- A microbial amphoteric tricyclic glycopeptide antibiotic, **Vancomycin**, which inhibits bacterial cell-wall biosynthesis. Vancomycin is active against staphylococci, streptococci, enterococci, and diphtheroids, and it is indicated for treatment of systemic infections.

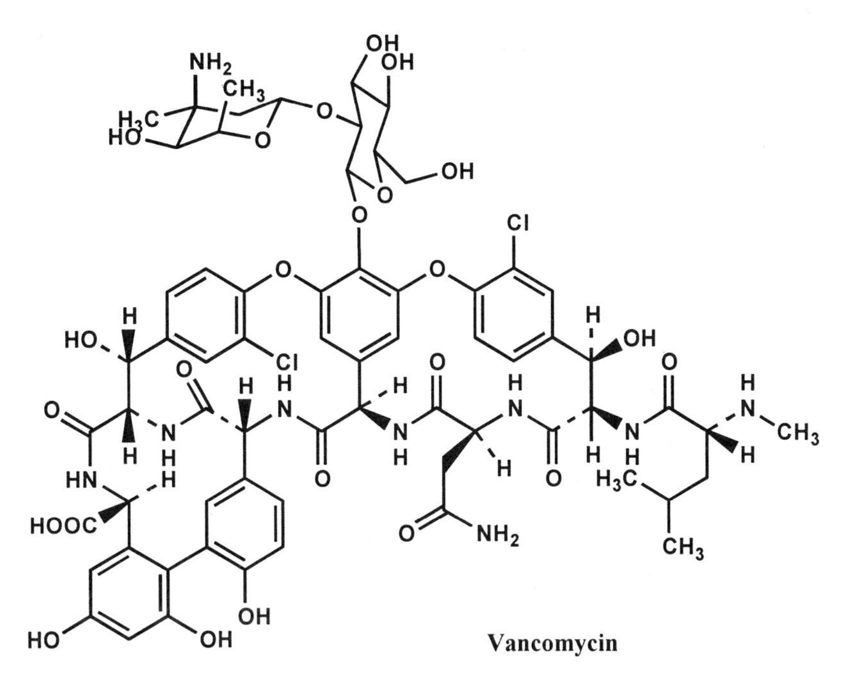

Vancomycin

1.3 Trisaccharides

The third category, trisaccharides and their conjugates, is represented by the following two prescription drugs:

An antibacterial aminoglycoside antibiotic of microbial origin, **Tobramycin**, which is a derivative of an aminoglucopyranosyl-ribohexopyranosyl-L-streptamine. The drug acts primarily by disrupting protein synthesis through altering cell membrane permeability, thereby breaching the cell envelope and causing eventual cell death. It is indicated for the management of cystic fibrosis patients,

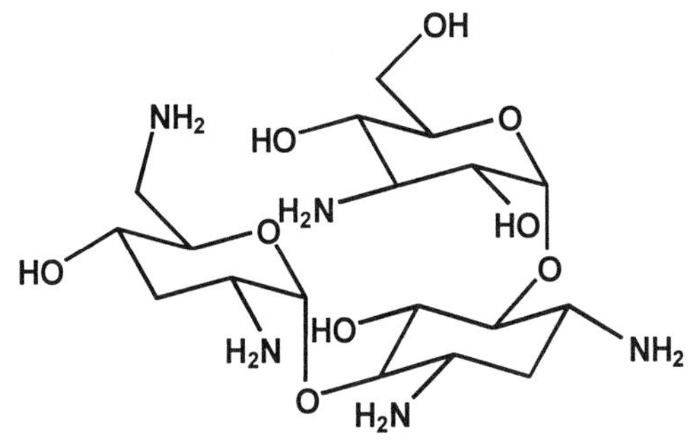

Tobramycin

- A cardiac glycoside, **Digoxin**, that belongs to a closely related group of drugs of plant origin and that contains a sugar and a cardenolide; the sugar part consists of (O-2,6-dideoxy-β-D-*ribo*-hexapyranosyl)$_3$. Digoxin inhibits sodium-potassium ATPase, that in turn leads to an increase in the intracellular concentration of sodium and calcium. This results in a chain of biochemical events that have multiple effects on cardiac muscle and the cardiovascular system in general.

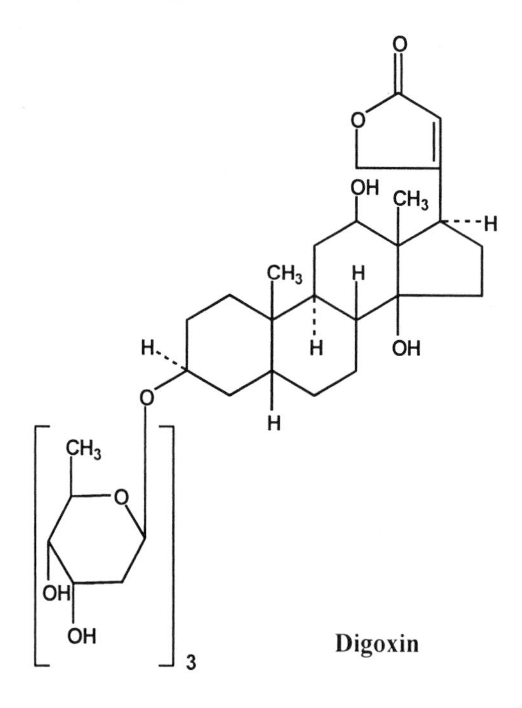

Digoxin

1.4 Oligosaccharides and polysaccarides

Prescription drugs made of oligosaccharides and polysaccharides include two principal groups:

- Heparin and heparin-like saccharides
  - Heparin
  - Enoxaparin
  - Tinzaparin
  - Dalteparin
  - Danaparoid
  - Pentosan polysulfate

- Complex oligosaccharides
  - Streptomycin
  - Neomycin
  - Acarbose

1.4.1 Heparin and heparin-like saccharides

The first group is represented by heparin and a series of its low-molecular weight fragments and analogs, all of them being antithrombotic agents.

Heparin is a heterogeneous group of glycosaminoglycans, straight-chain anionic mucopolysaccharides that have anticoagulant activity; in particular they inhibit formation of fibrin clots in blood. These drugs' variably sulfated polysaccharide chains are composed of repeating units of D-glucosamine and L-iduronic or D-glucuronic acids.

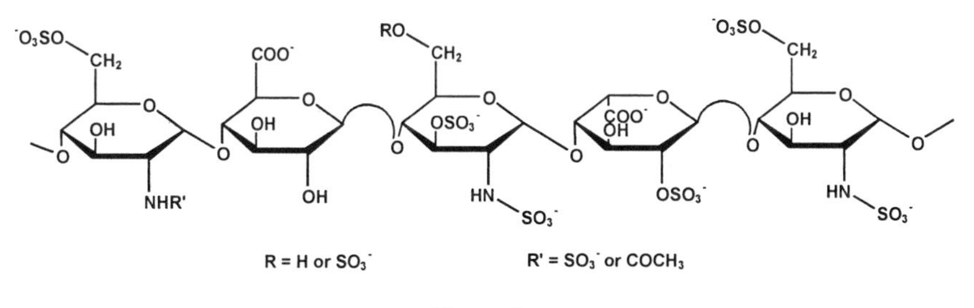

Heparin

Enoxaparin, **Tinzaparin** and **Dalteparin** are all prepared by controlled depolymerization of Heparin or its derivatives. This is accomplished by alkaline degradation, enzymatic hydrolysis, and nitrous acid fragmentation, respectively. **Danaparoid** is a complex glycosaminoglycuronan whose active components are heparan sulfate, dermatan sulfate, and chondroitin sulfate. Finally, **Pentosan Polysulfate** is a semi-synthetic sulfated heparin-like oligomer. Composed of β-D-xylopyranose residues, it shows anticoagulant and fibrinolytic effects.

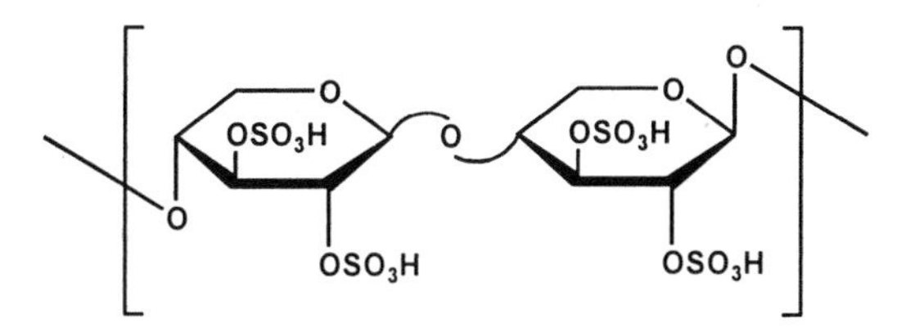

Pentosan Polysulfate

1.4.2 Complex oligosaccharides

The group of complex oligosaccharides contains two fundamentally different kinds of prescription drugs. The first are bactericidal aminoglycoside antibiotics of microbial origin, **Streptomycin** and **Neomycin**, which act by interfering with normal protein synthesis. Streptomycin is usually available as the sulfate (2:3) salt,

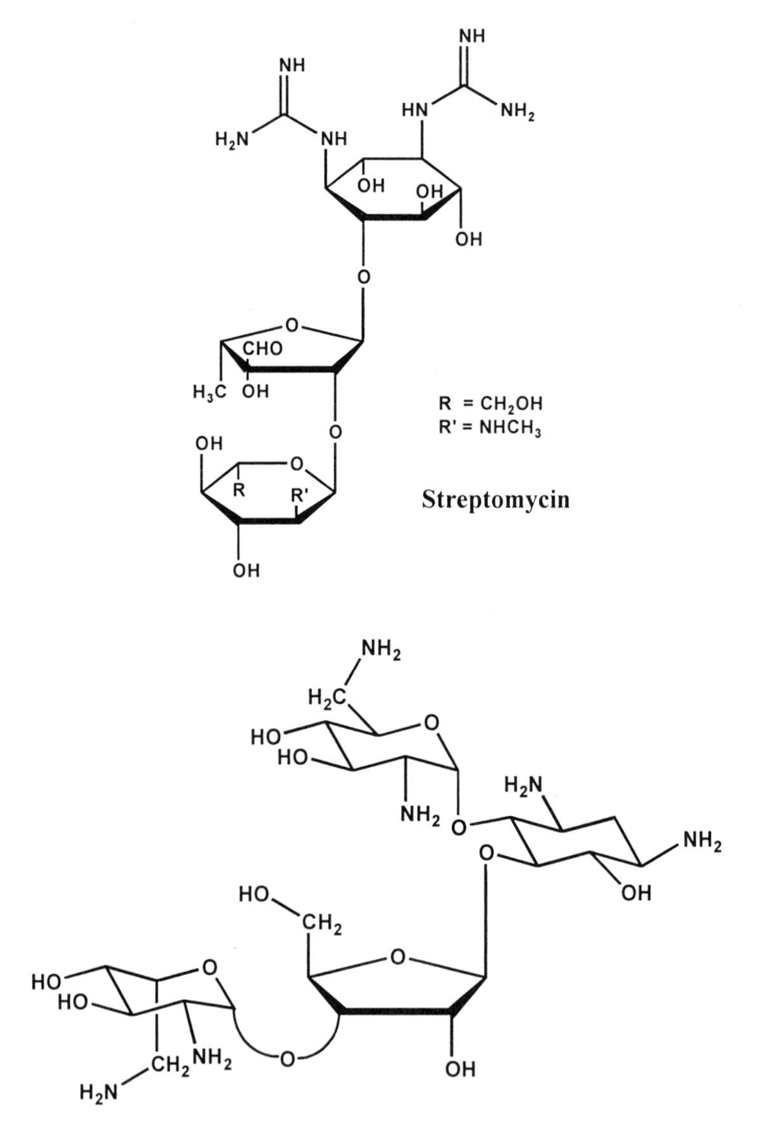

Streptomycin

Neomycin

The second kind of complex oligosaccharide **Acarbose**, also of microbial origin, inhibits α-glucosidase and delays the digestion of ingested carbohydrates, making the drug beneficial for the management of type 2 diabetes mellitus.

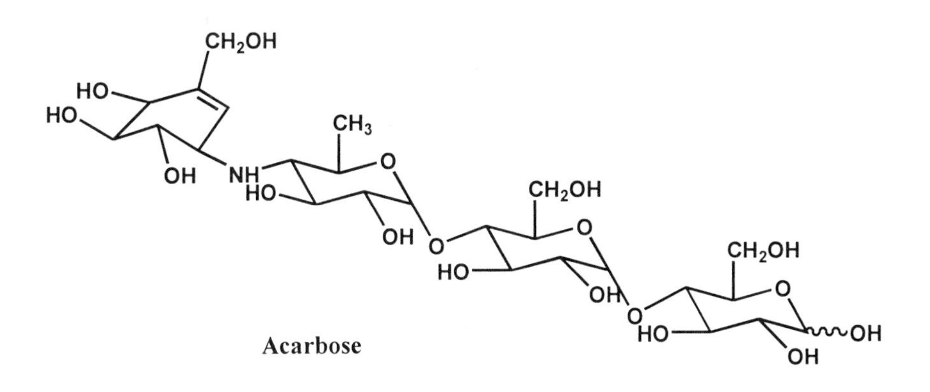

Acarbose

1.5 Macrolides

The fifth and final subcategory of prescription carbohydrate drugs is represented by macrolide group of antibiotics, of which there are four. The first, **Erythromycin**, is of microbial origin; it appears to inhibit protein synthesis in susceptible organisms by binding to ribosomal subunits and thereby inhibiting translocation of aminoacyl transfer-RNA. The other three – **Dirithromycin**, **Clarithromycin** and **Azithromycin** – are semi-synthetic macrolide antibiotics derived of Erythromycin. Dirithromycin is a pro-drug that is transformed during intestinal absorption into an anti-bacterial active form, Erythromycylamine. Clarithromycin is 6-O-methylerythromycin. Azithromycin is N-methyl-11-aza-10-deoxo-10-dihydroerythromycin.

Among these near-forty carbohydrate drugs are some of the most widely used prescription drugs in the United States. Based on a total of more than three billion U.S. prescriptions (for all drugs), the following carbohydrate drugs make it into the top 200:

- Digoxin (Lanoxin, **Glaxo SmithKline**; Digitek, **Bertek**)
- Azithromycin (Zithromax, **Pfizer**)
- Clarithromycin (Biaxin, **Abbott**)
- Neomycin (**Baush & Lomb**)
- Erythromycin (Ery-Tab, **Abbott**)
- Tobramycin (Tobradex, **Alcon**)

An essential component of these drugs (five of which are antibiotics), as well of all other carbohydrate drugs described in this chapter, is their sugar moiety.

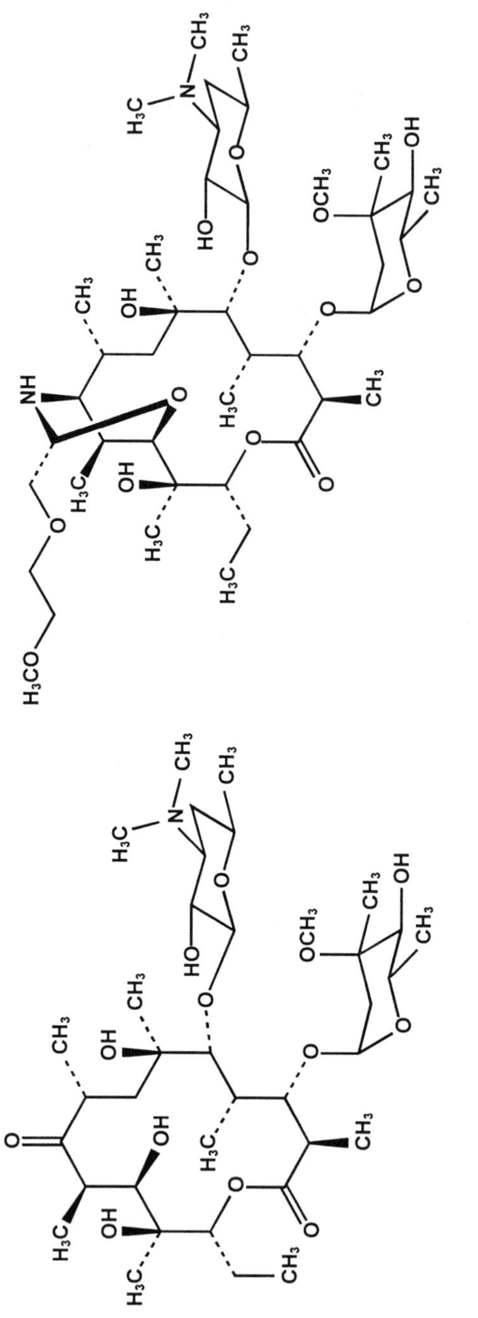
Dirithromycin

Erythromycin

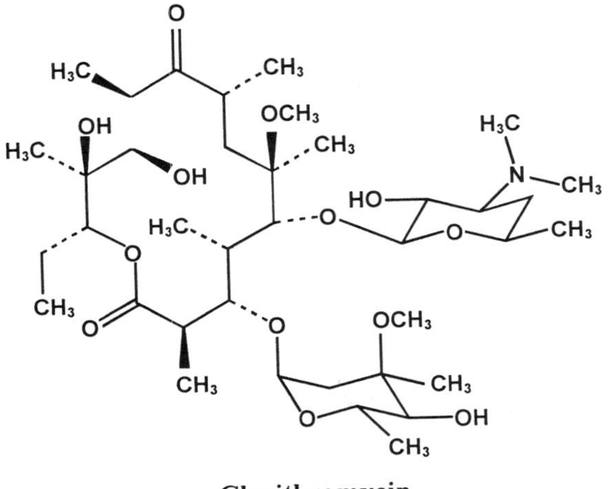

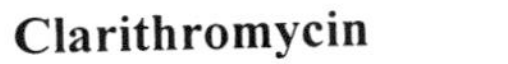

Clarithromycin

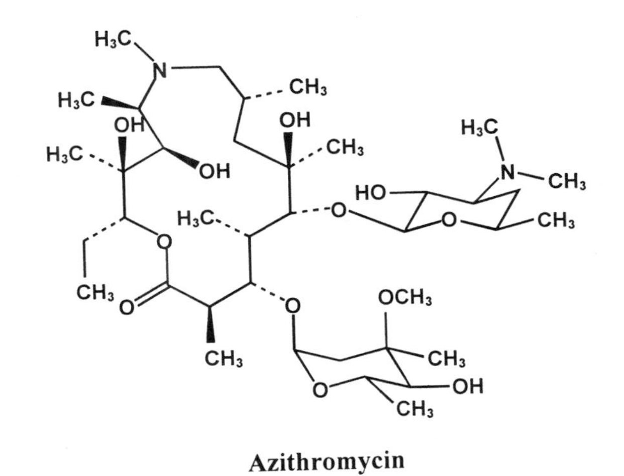

Azithromycin

Removal of the sugar residue typically leads to elimination of the drug's therapeutic properties. On the other hand, addition of a certain sugar moiety sometimes enhances the recognized potential of the drug at the target level. This approach is exemplified in the section "Cancer" of this book – particularly in the chapter "Synthesis and Biological Activity of Doxo-Galactose and Doxo-Galactomannan, New Conjugates of Doxorubicin with D-galactose and 1,4-β-D-galactomannan".

2. Cancer and a combination carbohydrate-assisted chemotherapy

Doxorubicin consists of an aglycone, named adriamycinone, and the aminosugar residue daunosamine attached via α(1→4) linkage to 7–hydroxy group of the aglycone. The drug is considered one of the most active antitumor agents. Unfortunately, by virtue of its mechanism of action, which is directed at replication, transcription, and recombination of DNA, the drug shows severe side effects. This has stimulated several studies aimed at chemical modification of doxorubicin sugar moiety in order to reduce side effects, or to increase the drug efficacy, or both.

Authors of the above-mentioned chapter have elongated doxorubicin at its sugar part by linking it to a stable D-galactose residue, which was chosen because of the important role that galactose-specific receptors – called galectins – play in tumor development. The best of six synthesized derivatives, in which D-galactose had the α-anomeric configuration and the two sugar residues were linked via 1→6 bond was, tentatively named Doxo-Galactose. It had a lower toxicity and a higher efficacy compared to the parent doxorubicin when tested on mice bearing lymphocite leukemia P-388 at single and multiple i.v. injection regimens. While doxorubicin showed the maximal antitumor effect of 70 percent (increase of lifespan) at the MTD (maximal tolerated dose) of 7 mg/kg, Doxo-Galactose showed a 118-percent increase of lifespan and no toxicity at 40 mg/kg. Moreover, Doxo-Galactose did not show any cumulative toxic effect, unlike the parent doxorubicin, which caused progressive toxicity at triple injection (q2d x 3) at almost 20-times lower doses than Doxo-Galactose. For its part, Doxo-Galactose showed 20 to 60 times lower cytotoxicity, using different cell lines, compared to that by doxorubicin.

Another doxorubicin derivative (Doxo-Galactomannan) described in the chapter was its conjugate with a polymeric galactomannan (DAVANAT®). This drug, which carried about 10 percent of doxorubicin residues by weight, had.cytotoxicity some 20 to 90 times lower than that of doxorubicin.

Two other chapters in the same Cancer section of this book describe preclinical and clinical (Phase I) studies using a polysaccharide, a galactomannan of plant origin employing 5-FU (5-fluorouracil), that enhances chemotherapy of colon cancer. Just by mixing 5-FU and DAVANAT® (which was partially degraded under controlled conditions) in the right proportions and introducing the two combined ingredients i.v., median tumor volume of mice bearing human colon cancer (COLO 205 and HT-29) decreased by 17 percent to 65 percent and median survival time (in days) increased to from 100 percent to 150 and even to 190 percent. A radiolabel study has indicated that 5-FU and DAVANAT® apparently showed synergism in entering a cancer tumor, and the same two compounds interfered with each other when entering the liver. This might explain how the combination might show a better antitumor activity and lower toxicity toward healthy cells and organs.

The two chapters also describe a phase I clinical trial of DAVANAT® co-administered with 5-FU in patients with refractory solid tumors. The study was initiated to evaluate the safety and tolerability of escalating doses of DAVANAT® (30 to 280 mg/m^2) in the presence and absence of 500 mg/m^2 5-FU in patients with advances solid tumors. Twenty-six patients had completed the study, and the combination of DAVANAT® and 5-FU was found to be well tolerated.

3. Carbohydrate-based HIV-1 vaccines

Two other chapters in this book offer an overview of the use of carbohydrate-based vaccines against HIV/AIDS.

Generally, vaccines are a large group of approved products that include some of the world's oldest pharmaceuticals. They are targeted against such conditions as adenovirus, MMR (measles, mumps, rubella), influenza, rabies, hepatitis B, polio, DTP (diphtheria, tetanus, pertussis), typhoid, encephalitis, and varicella. Use of vaccines has practically eradicated some historic diseases and led to a lower need for prophylaxis [1]. While traditionally employed to prevent the spread of infectious diseases, a new market is emerging for therapeutic vaccines as potential treatments for a wide range of chronic illnesses [2] such as cancer and HIV; other viral infections like hepatitis, human papilloma virus, and herpes; and autoimmune disorders such as multiple sclerosis, rheumatoid arthritis, lupus, and maybe even drug addiction.

The therapeutic-vaccine industry, which had estimated worldwide revenues of $80 million in 2004, is still in its infancy, and the market awaits its first largely successful product ("blockbuster").Yet there are at least nine vaccines anticipated to launch in the near future, which means that this market may soon experience enormous growth – possibly reaching, according to some predictions, over $2 billion by 2006 [2]. These nine vaccines are all late-stage candidates, in Phase III and above, and are primarily targeted against cancer. They include autologous vaccines (whereby each patient is treated with drug derived from his or her own tumor cell, allowing for truly personalized medicine) and allogeneic vaccines (which consist of antigens specific to a particular type of cancer).

There are no HIV vaccines among these about-to-be newcomers to the market. Although it has been 20 years since the HIV virus was identified, and numerous therapies have emerged, no HIV vaccine yet exists. One vaccine candidate, AIDSVAX, a recombinant HIV *gp120* protein developed by VaxGen, failed to improve HIV protection in clinical trials performed in North America, Europe and Asia, according to results announce in 2003. At least three other vaccine candidates are currently tested in clinical trials, all in Phase I. One of these trials, initiated in May of 2004 by Walter Reed Army Institute of Research and AVANT Immunotherapeutics, is evaluating the safety and immunogenicity of LFn-p24 vaccine. This drug, which uses bacterial vectors removed of their toxins, delivers target antigens into human cells and induces a cell-mediated immune response. Specifically, the vaccine consists of a detoxified anthrax-derived polypeptide fused to the HIV-1 *gag* p24 protein.

Another HIV vaccine, which is currently being developed by Merck, is also in its Phase I international trial (which began in September 2003). This candidate uses replication-defective adenovirus (which causes the common cold) to express the HIV *gag* gene – *gag* being HIV's core protein. Yet another vaccine candidate that began U.S. trials in January 2004 was created by Chiron. It is used in the form of microparticle-delivered *gag* DNA plasmids in combination with a protein-based formulation [3].

A wide variety of other HIV vaccine candidates are being considered on a research level, with investigations so numerous that they cover practically every conceivable target site in the virus and the immune system. Many vaccine candidates contain DNA for the *gag* and *pol* genes; pol includes three enzymes crucial for HIV replication. Gag and pol are considered good candidates for developing AIDS vaccines because they are relatively constant across different virus strains and account for a large percentage of total virus protein [4].

AIDS prophylactics present a new challenge for the vaccine approach. Traditionally, only a few types of vaccines – made primarily from attenuated or killed pathogens or from attenuated or deactivated toxins – have been used as antigens to trigger an adaptive immune response. However, because AIDS is so deadly and frightening, few researchers seriously consider the use of attenuated live or killed virus as a vaccine; and in any case this would not likely be acceptable to the public. Hence the best alternative is to use the new genetic engineering technologies for vaccine development [4]. For instance, fragments of the virus – specifically, viral envelope proteins – and not purified from viruses but instead are synthesized separately in bacteria through genetic engineering techniques. These approaches are surveyed in the abovementioned two chapters.

It makes eminent sense that these two articles are included in this volume. As one of them says: "Pathogen glycosylation is often perceived as a barrier to immune recognition and, by extension, to vaccine design. This notion is perhaps best illustrated in the case of HIV-1, where the glycosylation of the viral surface glycoproteins (gp120 and gp41) appear to profoundly affect the antibody response to the virus and likely contributes to viral immune evasion. The carbohydrate chains, which cover a substantial portion of the antigenic surface of HIV-1, are poorly immunogenic and act as a shield to prevent antibody recognition of the viral particle... Therefore, in principle, the glycosylation of HIV-1 should be considered a target for vaccine design" (D. Calarese et al.).

And according to the other paper: "In order to achieve persistent infection, HIV has evolved strong defense mechanisms to evade immune recognition, including ... heavy glycosylation of the envelope, switch of conformations, and formation of oligomeric envelope spikes. Therefore, the design of an immunogen capable of inducing broadly neutralizing antibody responses remains a major goal in HIV vaccine development... Accumulating data have suggested that the carbohydrate portion of HIV envelope can also serve as attractive targets for HIV-1 vaccine development. Carbohydrates account for about half of the molecular weight of the outer envelope glycoprotein gp120, which cover a large area of the surface of the envelope and play a major protective role for viral immune evasion... The present review intends to provide an overview on our understanding of the structure and biological functions of the HIV-1 carbohydrates, and on how the information might be explored for developing carbohydrate-based vaccines against HIV/AIDS" (Lai-Xi Wang).

4. Polysaccharides and infections

Disseminated fungal infections that generally result from the inhalation of airborne spores of pathogenic molds are called systemic mycoses. These diseases typically start with a primary pulmonary infection and can advance into a secondary, life-threatening conditions. Some patients may develop progressive disease that spreads to other parts of the body.

Systemic mycoses, unlike superficial cutaneous and subcutaneous mycoses (which affect the skin, hair, and nails), involve the blood and internal organs. They are subdivided into primary systemic mycoses, which occur regardless of the host's health; and opportunistic systemic mycoses, that occur in hosts with weakened immune systems. Opportunistic mycoses, typically caused by pathogens such as *Aspergillus*, *Cryptococcus* and *Candida,* are being observed with increasing frequency in patients compromised by disease or drug treatment. Normally, fungi live in a healthy balance with other microorganisms in the human body. However, factors such as antibiotics, prescribed medications, or poor diet can negatively affect this balance, resulting in the accelerated growth of opportunistic microorganisms such as *Candida* yeasts that in turn leads to a systemic fungal infection. Yeast overgrowth, which produces chemicals toxic to the body, is often referred to as candidosis.

Candida infections are thought to affect over 40 million people in the United States, both men and women. *Candida* are single-cell fungal yeasts, and they are considered to be one of the most prolific organisms. There are many *Candida* species, such as *Candida albicans*, *C. krusei*, *C. tropicalis, C. glabrata, C. parapsilosis, C. guilliermondii, C. lusitaniae, C. dubliniensis*. *Candida albicans* is the most frequently isolated causative agent of candidal infection in humans (accounting for more than 50 percent of cases) and is generally accepted as the most pathogenic species of the genus *Candida*. However, in recent years non-*C. albicans* species – such as *C. glabrata* (10 to 30 percent), *C. parapsilopsis* (10 to 20 percent), *C. tropicalis* (10 to 20 percent), *C. krusei*, and *C. lusitaniae* – have been recovered with increasing frequency from cases of candidiasis [5]. All of them produce different enzymes, including proteases, lipases, phospholipases, esterases, phosphatases, that most probably are able to cause damage to host cells *in vivo* [6].

Some *Candida* species are known to rapidly acquire decreased susceptibility to antifungal antibiotics and other agents, such as fluconazole and amphotericin B [5]. Other species are considered to be inherently resistant to the agents. And some *Candida* species can survive in the hospital environment, thus increasing the chance of nosocomial transmission. Hence there is a necessity for targeted and effective antifungal therapy and hospital infection control measures.

This is a subject considered in the chapter titled "Cationic Polysaccharides in the Treatment of Pathogenic *Candida* Infections" (A.M. Ben-Josef at al.) in this book. The paper shows that glucosamine-based cationic polysaccharides isolated and purified from the cell wall of the fungus *Mucor rouxii* and from crab's chitin possess a significant *in vitro* and *in vivo* fungicidal activity against azole-resistant *Candida* species, particularly *Candida albicans.* The authors offer experimental evidence that the polysaccharides bind very rapidly, tightly, and practically irreversibly to the *Candida* cell wall, suggesting that the binding occurs at the wall's carbohydrate-recognition domains. The data show that these cationic polysaccharides possess superior antifungal properties (compared to known antifungal agents such as amphotericin B and azoles) and that they hold great promise of becoming clinically advanced and useful drugs.

5. Aminosugars

Two chapters in this book are dedicated to syntheses of aminosugars, a group of carbohydrates widely represented in prescription drugs. These include the anthracycline antibiotics Doxorubicin, Daunorubicin, Epirubicin, and Idarubicin; the antifungal antibiotic Amphotericin B; antibacterial antibiotics Lincomycin and Clindamycin; the tricyclic glycopeptide antibiotic Vancomycin; the antibacterial aminoglycoside antibiotic Tobramycin; glycosaminoglycans Heparin, Enoxaparin, Tinzaparin and Dalteparin; complex oligosaccharides Streptomycin, Neomycin, and Acarbose; and the macrolide antibiotics Erythromycin, Dirithromycin, Clarithromycin, and Azithromycin.

In their chapter titled "Systematic Synthesis of Aminosugars and Their Stereoselective Glycosylation" J. Wang and C.-W. T. Chang introduce glycodiversification, a concept aimed – in this particular case - at replacing original aminosugars in naturally occurring antibiotics with synthetic aminosugars. To show one way of accomplishing this, the authors discuss systematic synthesis of aminosugars through the regio- and stereoselective incorporation of amino groups on pyranoses.

A related chapter is titled "Synthetic Methods to Incorporate α-Linked 2-Amino-2-Deoxy-D-Glucopyranoside and 2-Amino-2-Deoxy-D-Galactopyranoside Residues into Glycoconjugate Structures" (R. Kerns and P. W. Chang). Here the authors present a similar approach to introducing variably substituted and differentially modified α-D-glucosamine and α-D-galactosamine residues into glycoconjugate structures in order to synthesize many important bioactive glycoconjugates and their structural analogs. The emphasis in the chapter is on 2-amino-2-deoxy-D-hexopyranoside residues and their alpha-linked isomers.

6. Computational studies

Proteins have long been known as a challenge for computer modeling. But the complexity of proteins pales in comparison to that of carbohydrates. Only lately has the enormous task of modeling carbohydrates been made possible, thanks in part to giant strides in computer power and speed.

Computational studies of carbohydrates are described in a chapter titled "Practical Applications of Computational Studies of the Glycosylation Reaction", by D.M. Whitfield and T. Nukada. The authors describe their development of a new algebra – to quantitatively describe the conformations of six-membered rings that dominate the chemistry of carbohydrates – and its use in calculations aimed at understanding the mechanism of glycosylation reactions, both enzymatic and "common chemical", and at optimizing it. More specifically, the authors studied the mechanism of the acyl transfer reaction to the acceptor alcohol with neighboring group participation via a proposed transition state.

The chapter improves our current understanding of the conformations of pyranose sugars with protecting groups, and it is expected to give organic chemists a better tool for designing stereoselective glycosylation reactions.

References

1. Ian Sellick. Streamlining conjugate vaccine production. Genetic Engineering News, 24, No. 16, 48-52, 2004.
2. Rochelle Ellis. Therapeutic vaccines gain growing interest. Genetic Engineering News, 24, No. 21, 25-28, 2004.
3. David Filmore. HIV vaccine testing. Modern Drug Discovery, August 2004, p. 11.
4. Mark S. Lesney. Vaccine futures. Modern Drug Discovery, March 2002, p. 35-40.
5. Koji Yokoyama, Swarajit K. Biswas, Makoto Miyaji, and Kazuko Nishimura. Identification and phylogenetic relationship of the most common pathogenic *Candida* species inferred from mitochondrial cytochrome *b* gene sequences. J. Clin. Microbiology, 38, 4503-4510, 2000.
6. Acacio G. Rodrigues, Cidalia Pina-Vaz, Sofia Costa-de-Oliveira, and Christina Tavares. Expression of plasma coagulase among pathogenic *Candida* species. J. Clin. Microbiology, 41, 5792-5793, 2003.

Chapter 2

Carbohydrate Therapeutics: New Developments and Strategies

Zbigniew J. Witczak

Department of Pharmaceutical Sciences, Nesbitt School of Pharmacy, Wilkes University, 137 S. Franklin Street, Wilkes-Barre, PA 18766

Carbohydrate-based drug development has emerged as a highly promising and exciting area in contemporary medicine and chemistry. Interest in this area is based on a number of factors, including our growing understanding of the role of endogenous carbohydrates in cellular function, our ability to synthesize carbohydrate analogs, and the success of some of these analogs as novel therapeutic agents in treating cancer, diabetes, and infectious diseases. As a result there has been a significant increase in the number of reviews that address the general field of carbohydrate medicinal chemistry. This account is devoted to examining these new developments in carbohydrate-drug design, with a focus on novel synthetic pathways and application of modern synthetic carbohydrate compounds as therapeutic agents.

Introduction

In the early 1980s the first studies of carbohydrate-binding proteins prompted a new era of research and development into the roles of carbohydrates and carbohydrate-binding proteins in biological systems. Since then, many scientific studies have uncovered a number of important structure-activity relationships for carbohydrate-based compounds. These studies demonstrate that the diversity and complexity of carbohydrates permit them to carry out a wide range of biological functions. Studies continue unabated, leading to a number of new developments with significant consequences for the field of carbohydrate chemistry. Research on carbohydrates is now undergoing considerable growth and promises to be a major source of drug-discovery leads *(1-14)*.

New strategies

Carbohydrates, which are dissolved in the aqueous media surrounding all of the body's cells, can be linked to other polymers to form glycoconjugates *(6)*. Glycoconjugates, usually classified as either glycolipids, glycoprotein (O- and N-linked), or proteoglycans, are isolated from mammalian systems and are derived from seven free monosaccharides (D-glucose, D-galactose, D-mannose, D-xylose, L-fucose, D-glucuronic acid, L-iduronic acid) and three functionalized aminosugars (N-acetylgalactosamine, N-acetylglucosanine, and sialic acid [5-N-acetylaminoneuraminic acid]). The near-equal hydrophilicity of the above monosaccharides constitutes a powerful physicochemical characteristic that is important for molecular recognition and interactions with many complex proteins.

Because of their hydrophilic nature, and location on the outside of cell membranes, carbohydrates are primary candidates for intercellular signaling. These characteristics have served as the basis of two very important hypotheses. **First, the location must match the function of carbohydrates in a very specific way. Second, any disruptions in the carbohydrate-carbohydrate or carbohydrate-protein interactions involved in the initiation or development of specific diseases are unique to each particular disease state.**

Diseases for which carbohydrate therapeutics can be particularly effective range from bacterial infection, rheumatoid arthritis, diabetes, and viral infection (influenza) to cancer and immune dysfunction (e.g., AIDS). The proposed usefulness of carbohydrate therapeutics in these disease states is based largely on the principle of "glycotargeting." First demonstrated in 1971 by Rogers and Kornfield *(15)*, glycotargeting utilizes the unique physicochemical properties of carbohydrate-based compounds to target specific and localized cell-surface receptors. The potential of using carbohydrate ligands to create a truly targeted (or actively targeted) drug-delivery system is thus patently clear. However, one particular caveat of this targeting approach is that small-molecule drugs, no matter how heavily glycosylated, will always have the potential to pass into the kidneys, through glomerular filtration, and be rapidly excreted. Consequently, macromolecular assembly—for the construction of compounds that allow for a longer circulation period, additional chemical functionality, and more precise delivery—is an attractive alternative option.

One such example is the design of dendritic molecules used as multivalent scaffolds with specifically targeted biological applications. Indeed, the development of glycodendritic structures as promising new antiviral drugs has recently been published *(16)*.

The mechanisms of glycotargeting

Glycotargeting exploits the highly specific interaction of endogenous lectins with carbohydrates (often, with *multiple* carbohydrates). Because receptor-mediated endocytosis (RME) was the primary biological mechanism targeted, much interest has been focused on the proteins involved in RME, such as the mannose binding protein (MBP). Because of its very high density on hepatocyte surfaces (50,000 to 500,000 per cell), the asialoglycoprotein receptor (ASGPR) in the liver is a particularly attractive target *(9)*. In addition to lectin receptors regularly involved in endocytosis, others that are not so involved may also be targeted. For example, lectin-like "homing" receptors on lymphocytes recognize so-called cell-adhesion molecules (CAM) that contain carbohydrates such as sialyl Lewis-x. *(17)*

Another class of lectins—called galectins—share amino-acid sequences and affinity for β-galactose-containing oligosaccharides. One of the first proteins discovered in the lectin family, galectin-1 *(18)*, has been shown to participate in a variety of events associated with cancer biology. These events include tumor transformation, cell-cycle regulation, apoptosis, cell adhesion, migration, and inflammation. Additionally, new evidence indicates that galectin-1 contributes to tumor evasion of immune responses. Structurally, galectin-1 is a non-covalent dimer composed of subunits with one carbohydrate-recognition domain (CRD). This particular protein binds preferentially to glycoconjugates containing a characteristic disaccharide *N*-acetyllactosamine (Galβ1-3/4GlcNAc). Binding of galectin-1 to individual lactosamine units occurs with relatively low affinity, though arrangement of lactosamine disaccharides in repeating unit chains (polylactosamine) will increase the binding avidity *(19-20)*.

Chronic inflammation is considered to be one of the critical factors contributing to tumor progression. This complex biological process is regulated by a variety of molecules—including the galectins, which are expressed by many different inflammatory cells and are important in regulating their activity. In addition, galectins are also released by tumors and can positively or negatively influence a variety of inflammatory responses.

Expression of galectin-1 has been well documented in many different tumor types, including astrocytoma, melanoma, and prostate, thyroid, colon, bladder, and ovary carcinomas *(21)*. Galectin-1 expression correlates with the unusual aggressiveness of theses tumors and, most importantly, the acquisition of a metastatic phenotype.

Tumor metastasis is a multistep process that includes changes in cell adhesion, increased invasiveness, angiogenesis, and evasion of the immune response. Galectin-1 clearly contributes to all of these processes. For example, it has been shown that galectin-1 increases the adhesion of prostate and ovarian cancer cell lines to the extracellular matrix (ECM) *(22)*. This evidence clearly suggests that galectin-1 might modulate the adhesion between adjacent cancer cells or between cancer cells and ECM.

Based on the established contribution of galectin-1 to tumor growth and metastasis, it is anticipated that inhibitors of galectin-1 will find their way into cancer clinical trials. It is hypothesized that these compounds will cause delays in tumor progression and improvements in the overall survival of cancer patients. The rational design of specific and potent galectin-1 inhibitors that can treat cancer, and that also lack significant side effects, will be a difficult but worthwhile challenge.

Bioavailability

The bioavailabilty of carbohydrate therapeutics is a unique issue that has not been adequately addressed by current molecular-drug-design processes. The inability of a drug molecule to reach the action site—whether because of lack of absorption from the gut, first-pass metabolism in the liver, or lack of penetration of the blood-brain barrier (BBB)—can limit its therapeutic potential unless alternative dosage forms can be formulated that improve bioavailability. For example, carbohydrate S-nitrosothiols (sugar-SNAPs) as nitric oxide donors are poorly absorbed via oral delivery but are quite well absorbed via transdermal (topical) delivery. The factors affecting bioavailability are steadily becoming better understood, but to date we have no hard principles that can be applied across a chemical series of potential analogs with any reasonable degree of predictability.

One of the limiting factors for many new experimental carbohydrate-based therapeutics is oral bioavailability. The anti-inflammatory agent Amiprilose and its hydrochloride salt, Therafectin (developed by Greenwich Pharmaceuticals, Inc., and Boston Life Sciences, Inc.), are classic examples of this constraint. Both compounds are known to exhibit antiproliferative and anti-inflammatory activity. Amiprilose specifically acts as an immunomodulator and therefore has a therapeutic effect on autoimmune disorders such as arthrithis, psoriasis, and, most importantly, systemic lupus. Therafectin has low toxicity and no serious side effects but is required in large doses for effective therapy. *(23)*. This particular factor creates a problem, especially for oral administration, because treatment of inflammatory or autoimmune disorders is often chronic.

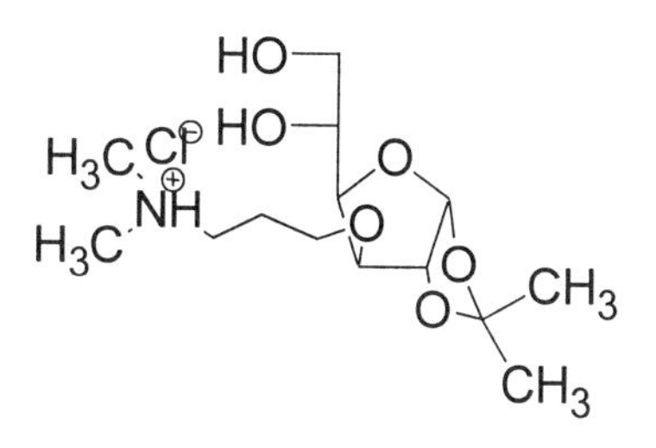

Figure 1. Amiprilose (Therafectin)

Because the bioavailabilities of Amiprilose and Therafectin were not improved despite many efforts, their development was discontinued.

Lipophilicity

Lipophilicity (hydrophobicity) is a molecular property related to the ability of a derivative to partition between water and a nonpolar solvent such as n-octanol. The lipophilicity of a drug candidate can have a dramatic impact on a variety of pharmacokinetic characteristics, including its ability to be absorbed after oral administration and passively diffused across biological membranes so that it may reach its site of action. Typically, the more lipophilic a drug, the better is its absorption following oral administration. This characteristic is particularly relevant to carbohydrate-based compounds, as they tend to be more hydrophilic in nature. Moreover, through functionalization we can increase the lipophilic nature of carbohydrate-based compounds, subsequently improving their bioavailability.

Lipophilicity is also a key factor in drug delivery. This is particularly evident in drugs that are administered transdermally—only highly lipophilic compounds can cross the hydrophobic barrier of the skin. Lipophilicity of free and specifically functionalized carbohydrates varies dramatically, however, depending on the chemical characteristics and π-values of associated functional groups.

Drug delivery

Drug delivery is an important consideration both for new therapeutic entities as well as approved drugs. The delivery system used for a specific drug depends on its chemical composition—whether it consists of whole virus, surface protein, peptide, lipopeptide, glycopeptide, or liposomal formulation. Additionally, critical factors such as drug solubility, membrane permeability, metabolism, and, most importantly, stability need to be considered at the earliest stages of carbohydrate-drug candidate selection.

A large number of drugs are relatively easy to deliver because of their stability and solubility in biological fluids at physiological pH. As such, they can be transported satisfactorily across cellular barriers into the blood and to their site of action. These types of drugs, including carbohydrate therapeutics such as selected aminoglycoside antibiotics (tobramycin), do indeed "deliver themselves" and do not require a sophisticated delivery system. However, for an increasing number of new experimental drugs (including some carbohydrate therapeutics) the situation is more complex and requires consideration of many pharmaceutical, pharmacological, and pathological factors (Table 1).

Table 1. Factors in drug delivery

- Drug polarity
- Drug size
- Drug stability
- First-pass metabolism
- Therapeutic index
- Delivery route
- Pharmacokinetics/Pharmacodynamics
- Toxicity

Drug size (small molecule vs. oligosaccharide/polysaccharide) is one of the few critical factors for the successful development of a delivery system. Various approaches for conjugation of the drug and linking through specifically designated spacers have been developed *(10-12)*. Drug stability, along with first-pass metabolism, are the next most-important factors. Enzymatically stable and nonhydrolyzable carbohydrate mimics, such as C-glycosides, N-iminosugars or S-thiodisaccharides, are logical choices based on their first-pass metabolic conversion, which vary dramatically from their O-glycoside counterparts.

Poor stability is another factor that may preclude the development of certain types of formulations. An unstable drug may be poorly absorbed in some regions of the gastrointestinal tract. Specifically, poor colonic absorption is a special problem for orally administered carbohydrate drugs. This is due primarily to the polar nature of these compounds. As a result it is often difficult, if not impossible, to develop a sensible once-a-day formulation—as was demonstrated in the case of Therafectin (mentioned above in the "Bioavailability" section).

Therapeutic index, a measure that is essentially the ratio of desirable to undesirable drug effects, is another important factor in drug delivery. Carbohydrate drugs with widely varying therapeutic indices can be given by many different delivery routes. The choice of route will depend on the needs of the precise delivery pattern rather than on patient preference. (Therapeutic index of various carbohydrate antibiotics will also be discussed in the "Functional group modification" section.)

The delivery route plays an important role in the required performance specifications of a carbohydrate drug. With compounds for which absorption is not optimal, delivering the drug in a pulsatile fashion can help avoid toxicities, overaccumulation, and nonspecific binding. Experimental technologies for targeting drug delivery to the site of action, such as adenoviral delivery and transdermal (topical) administration, have also been developed to avoid this problem.

Topical administration can deliver a drug directly to its site of action, not only at an effective dose but usually without clinically significant systemic absorption. Indeed, many drugs applied to the skin in clinical practice, including corticosteriods, 5-Fluorouracil, and retinoids, have significant side effects when used systemically but minimally so as topical agents.

As noted above, S-nitrosated thiosugars (sugar-SNAPs) are poorly absorbed via oral delivery but are quite well absorbed via topical delivery. Once in the blood, S-nitrosated thiosugars relase nitric oxide spontaneously. This, in theory, should ensure more consistent NO release than from donors such as glyceryl trinitrate, which need to be metabolized. By modifying the structure of S-nitrosated thiosugars via altering the number of acetyl functional groups, the lipophilicity—and, thus, NO release—could be modified.

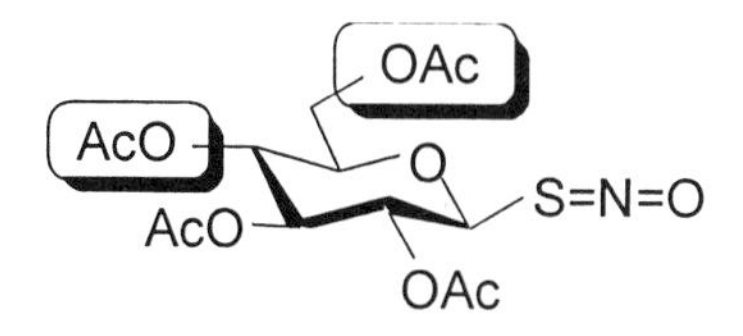

Figure 2. Functional -O-acetyl groups at C-6 and C-4 as potential targets for alteration of polarity and lipophilicity

This class of topical carbohydrate compounds shows therapeutic potential, as the molecules are designed to act quickly and without any serious side effects.

Carbohydrate drugs can currently be delivered to and across mucosal surfaces of the nose (antiviral carbohydrate drug Relenza *[2,11-12]*), but methods of delivery through other mucosal surfaces, such as those of the lungs and intestines, will be required in order for these drugs to reach their therapeutic potential. These drugs will also need to be able to target specific organs, tissues, and cells. For the case of DNA delivery in the emerging field of gene theraphy, there is the additional need to deliver the specifically formulated "therapeutic agent" to a specific site within the cell—its nucleus.

Sugar conjugates

As mentioned earlier, the polyhydroxylated nature of free or functionalized carbohydrate derivatives is an important factor for their potential conjugation with existing drugs. Moreover, the many different classes of sugars currently available, such as deoxy-, thio-, nitro-, and amino-, substantially complicate the conjugation process—much more so than in analogous lipid conjugation.

In general, sugar conjugation can facilitate the active transport of the modified drug across biological membranes and also can modify the physicochemical properties—including polarity, solubility, and stability at physiological pH—of the conjugated drug. Furthermore, pharmacokinetic and pharmacodynamic characteristics of the conjugates can be modified.

Lipid-sugar conjugates

Conjugation of drugs with lipids has been shown to increase their passive transport across intestinal mucosal membranes, whereas sugar-drug conjugates show greater absorption through targeting the sugar transporters in the intestinal epithelium. Consequently, coupling both a sugar and a lipid to a drug molecule can potentially take advantage of both of these important membrane-transport mechanisms.

Sugar transporters

There are several well-known types of monosaccharide transporters, located in different organs within the body *(24-25)*. In particular, transporters important in oral drug delivery are located within the brush-border and basolateral membranes of intestinal epithelial cells. These transport proteins are of the utmost importance and have been very well characterized. *(26)*. Examples include the sodium ion-dependent active transporter SGLT1 and the sodium ion-independent facilitative transporter GLUT5. Moreover, in the basolateral membrane another sodium ion-independent facilitative transporter, GLUT2, has been characterized.

SGLT1 has a high level of specificity for free sugars such as D-glucose, D-galactose, and their deoxy analogues (such as 1-deoxy-D-glucose, 6-deoxy-D-glucose, or methylated 3-O-methyl-D-glucose). Interestingly, L-sugars such as L-glucose and L-galactose do not bind readily to this transporter and therefore are not actively transported. Other free sugars, including D-fructose and sugar alcohols such as D-mannitol and 2-deoxy-D-glucose, also are not actively transported.

The GLUT5 facilitative transporter is primarily responsible for the intestinal absorption of D-fructose, whereas GLUT2 has specificity for simple sugars such as D-glucose, D-galactose, and D-fructose. SGLT1 and GLUT5 transport D-glucose across the brush-border membrane directly into the cell, whereas the GLUT2 receptor transports the above sugars out of the enterocyte, a cross t he specific basolateral membrane, and into the blood.

Through specific functionalization/conjugation reactions with sugars transported by GLUT5 or SGLT1, carbohydrate-based compounds can be modified so that the drug will be transported across the epithelium and then into the bloodstream via GLUT2. For example, the intestinal absorption of β-linked sugar-drug conjugates is greater than that that of their α-anomers. NMR and molecular-modeling studies have been carried out to examine the reasons for this β-anomeric preference *(27)*.

This particular hypothesis has been verified and proven to be effective in increasing absorption and peptide stability (*in vivo*) of the glycopeptides series of Gly-Gly-Tyr-Arg by glycosylation with p-(succunylamido)phenyl α- or β-D-glucopyranoside *(27)*.

Moreover, multiple functionalization, through installation of 3-O-methyl- or 6-deoxy- functional groups on the designated simple sugar molecule, can potentiate active transport of these functionalized derivatives and additionally establish valuable pharmacophores for the desired class of derivatives.

Functional group modification

The ability of a functional group to change the pharmacological profile of a pharmacophore has been clearly documented. This same principle also applies to carbohydrate compounds.

Structure-activity relationships (SAR) studies have uncovered structural modifications that are key to activity and potency manipulations. Standard molecular modification approaches have been developed for the systematic improvement of the therapeutic index. Therapeutic indices among carbohydrate therapeutics can vary significantly, but generally speaking the larger the therapeutic index, the greater the margin of safety of the drug.

Classic examples of low-therapeutic-index drugs are aminoglycoside antibiotics, which are highly ionized molecules. Despite the presence of many functional groups as valuable pharmacophores, the selective side effects and variable toxicity of this class is well known. In the field of amino sugars, the basicity of the amino functional groups is one of the important factors for protein interactions. Consequently, these interactions can promote the formation of allergenic proteins, which are responsible for some of the toxicities associated with functionalized carbohydrates.

Toxicity

Generally speaking, carbohydrate compounds tend to have low toxicity and immunogenicity in comparison to their peptide counterparts. For example, free sugars or minimally functionalized amino-sugars derivatives with protected amino functionality are generally nontoxic or possess low toxicity. A peptide bridge inserted at various positions of amino sugars can specifically change the level of the toxicity of the modified molecule.

In contrast, carbohydrate peptides comprised of two sugars units connected via peptide group possess a variable level of toxicity. Well known examples of natural carbohydrate peptides are marine products, Mycalamides A and B (natural carbohydrates with antitumor and antiviral activity, originally isolated from the New Zealand marine sponge *Mycales*), and Onnamide (isolated from *Theonellia swinhoe*).

New literature on the toxicity *(28)* on Mycalamides clearly demonstrates that they have a very different toxic effect compared to Onnamide (*29*). Mycalamides are structurally similar to the insect toxin Pederin. Remarkably, these structurally similar substances have been isolated from two entirely unrelated groups. Pederin itself is known exclusively as originated from terrestrial *Paederus* and *Paederidus* species of the beetle genera and has been isolated from blister beetle *Paederus fuscipes*. These notorious insects use pederin as a chemical weapon against predators, and the toxin causes severe dermatitis when accidentally exposed to human skin.

In contrast, Mycalamides A and B exhibit IC_{50} values in the subnanomolar range for some tumor model systems, and these peptides prolong the life span of mice bearing a variety of ascitic and solid tumors. Mycalamide A has also been shown to induce apoptosis.

These structurally similar molecules, which nevertheless show completely different toxicity levels, are illustrated below.

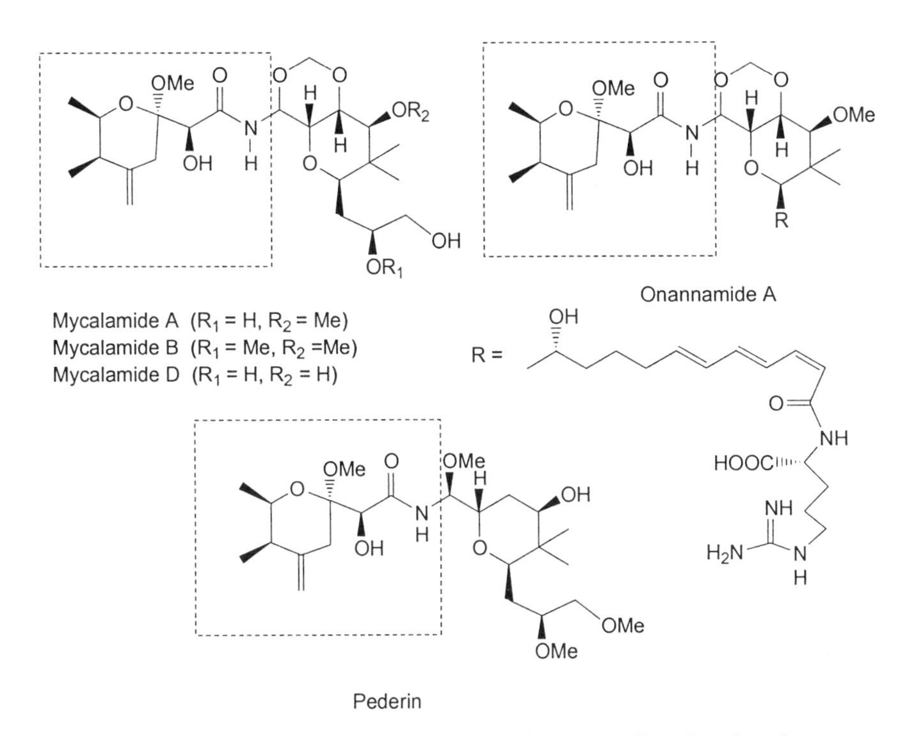

Figure 3. Mycalamides, Onannamide and structurally related Pederin

Pederin is a very weak antibacterial agent but is highly toxic to eukaryotic cells. Ingestion can cause severe internal damage, and intravenous injection causes death at levels comparable to that of cobra venom. This molecule is a rare example, so far, of the extreme toxicity of natural carbohydrates with minimal functionalization and with ordinary functional and protecting groups.

Classification of new targets for carbohydrate therapeutics

α-Methylene-γ-butyrolactones

This particular class of derivatives is represented by a new stable inhibitor, called C75, of mammalian fatty acid synthase (FAS—the enzyme primarily responsible for the synthesis of fatty acids. Compared to normal human tissues, many common human cancers—including breast, prostate, ovary, endometruim, and colon—express high levels of fatty acid synthase. This differential expression of FAS between normal tissue and cancer has led to the hypothesis that FAS may be a possible target for anticancer drug development.

C75 is one such candidate. It is 3-carboxy-4-octyl-2-methylene-butyrolactone *(30)*, conveniently produced from p-methoxy-benzyl itaconate via deprotonation with lithiumhexamethyldisilyl amide, together with aldol condensation with octyl aldehyde, to form diasteromeric mixture of γ-lactones.

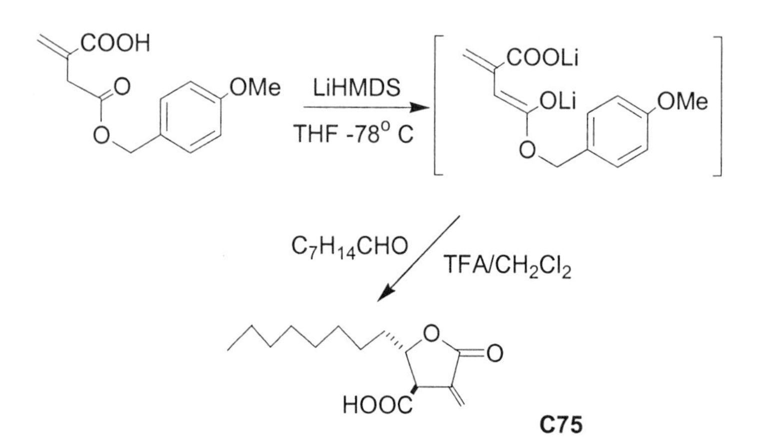

Figure 4. Synthesis of 3-carboxy-4-octyl-2-methylene butyrolactone ***C75*** *from p-methoxybenzyl ithaconate.*

This compound binds to and inhibits mammalian FAS, thus inhibiting fatty-acid synthesis in human cancer cells, as reported by Townsend group *(30)*.

Recent studies also have shown C75 to have significant *in vivo* antitumor activity against human breast-cancer xenografts *(30)*. Therefore the development of C75 should enable extensive in-vivo study of FAS inhibition in human cancer as well as in other diseases associated with dysfunctional fatty acid synthesis activity.

Thiosugars

Sulfur as either a sugar heteroatom or a component of a functional group attached to a strategically important position of sugar ring (C-1, C-6, or C-4) has the ability to drastically change the chemical character of a sugar and, consequently, its biological activity. This change is attributable to the electronegativity of the sulfur atom in comparison to that of the oxygen. Replacement of an oxygen atom by a sulfur atom is tolerated by most biological systems, and this particular replacement increases the stability of the sugar-aglycon linkage against enzymatic cleavage, and against any targeted chemical degradation.

A classic example of the application of this chemical modification is thio-sugars that contain a sulfur bridge between two sugar units. These compounds are extremely good enzyme inhibitors, and they show enormous potential as specific inhibitors of glucosidaases, fucosidases, and sialidases. As a result, there has been enormous interest in the design of a number of new sulfur analogs that potentially could be used as inhibitors of glycosidases. Recently, Withers et.al. *(31)* reported the discovery of a new class of enzymes, thioglycoligases. These enzymes are mutant glycosidases and catalyze thioglycoside synthesis, specifically α-S-thiodisaccharides.

Thiolactomycin and analogs

Thiolactomycin, one of the many natural thiosugars isolated during the last 20 years, was derived from *Nocardia sp. (32)* and later synthesized by Chambers et.al as a racemic mixture *(33)* and by Kremer et.al *(34)* and McFadden and Townsend *(35)* as enantiomerically pure derivatives.

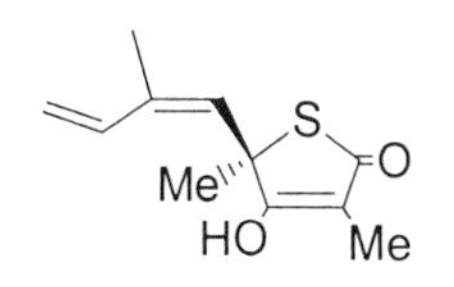

Figure 5. Thiolactomycin

Thiolactomycin and its analogs represent a new class of thiosugars. These compounds contain sulfur as a heteroatom and an unsaturated lactone moiety. The increased level of lipophilicity of thiolactomycin over other thiosugars is due largely to the unsaturated, aliphatic five-carbon chain at the C-5 position and the lack of multiple ionizable functional groups. The importance of the thiolactone moiety and the C-2,3 unsaturation functionality for the biological activity of the thiophene chemical system is well known; their preparative chemistry has been reported by Varela *(36)*.

Thiolactomycin is known to inhibit a variety of enzymes in parasitic organisms, plant cells, and bacteria. Thiolactomycin inhibits bacterial myristate synthesis and selectively inhibits *in vitro* trypanosonal fatty acid synthesis displaying an Ic_{50} value of 150nM. As such, thiolactomycin is a very useful tool for studying and understanding the mechanism underlying parasitic infections and infectious diseases. Specifically, this compound has been shown to inhibit fatty acid synthase (FAS II), an enzyme essential for fatty acid synthesis in bacteria, plants, and some protozoa *(37)* Thiolactomycin is also a selective inhibitor of the bacterial condensation enzymes FabB, FabF, and, to a lesser extent, FabH *(38)*. Additionally, thiolactomycin is capable of blocking long-chain mycolate synthesis, in a dose-dependent manner, in purified-cell wall-containing extracts of *Mycobacterium smegmatis (39)*.

In vitro studies have confirmed that thiolactomycin is active against a wide range of *Mycobacterium tuberculosis* strains, including those resistant to the well-known antimycobacterial drug Isoniazid. Pharmacological studies have shown that thiolactomycin selectively inhibits the mycobacterial acyl carrier protein-dependent type II fatty acid synthase (FAS-II) but not the multifunctional type I fatty acid synthase (FAS-I) present in mammalian cells. *(34, 37)*

Thiolactomycin is a reversible inhibitor of the β-ketoacylsynthase (KAS) of bacterial FAS systems, including KAS I-III in *E. coli*. Specifically, a crystal structure of KAS I (FabB from *E.coli*) with bound thiolactomycin clearly reveals the essential enzyme-ligand binding interactions and establishes the existence of hydrophobic and pantheine binding pockets *(38)*. Indeed, structure-activity relationships clearly indicate that C-5 functionalized analogs of thiolactomycin exhibit enhanced antimicrobial activity *(34)*. Specifically, two new analogs of thiolactomycin synthesized by the Douglas groups *(40)* showed greater activity than the parent compound in inhibiting *Mycobacterium tuberculosis* H37Rv *in vitro*.

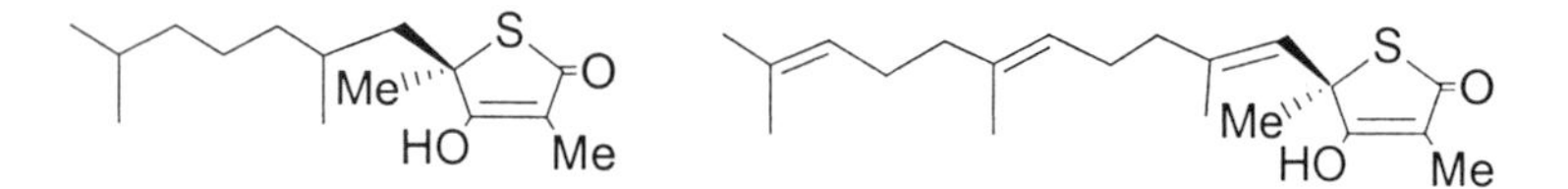

Figure 6. New analogues of thiolactomycin with enhanced in vitro *activity against* M. tuberculosis*H37Rv.*

Other analogs of thiolactomycin, bearing hydrophobic alkyl groups at the C-3 and C-5 positions of the thiolactone ring, are effective against the malaria parasite *Plasmodium falciparum*. The best of these analogs shows a fourteen-fold increase in activity over thiolactomycin. Furthermore, when assayed, other analogs synthesized by Jones group *(41)* showed activity against the parasitic protozoa *Trypanosoma cruzi* and *Trypanosoma brucei*.

Paradoxically, while thiolactomycin exhibits relatively poor *in vitro* antibacterial activity it has good efficacy in animal infection models. It also has activity in mice, both by oral and subcutaneous drug delivery, against various models of bacterial infection *(42)*.

Tagetitoxin

Tagetitoxin was purified from culture filtrates of the plant-pathogenic bacteria *Pseudomonas syringae* pv. *Tagetis*, and its structure was later revised *(43)* as 4-*O*-acetyl-3-amino-1,6-anhydro-3-deoxy-D-gulose 2-phosphate.

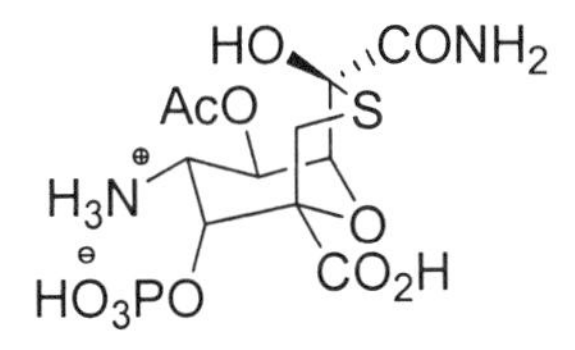

Figure 7. Tagetitoxin

Tagetitoxin is a bacterial phytotoxin *(44)* produced by *P. syringae* pv *Tagetis*. It induces chlorosis in the apex of the host plant by specifically inhibiting chloroplast RNA polymerase, making it potentially useful as an herbicide *(45)*. Interestingly, apical chlorosis was developed on a diverse range of plants following wound inoculation, and the pathogen was identified as *Pseudomonas tagetis*. *(46)*

Evaluations of tagetitoxin precursors and analogs as herbicides and plant-growth regulators have been reported by Furneaux *(47)*. These studies utilize several approaches, starting from convenient templates of specifically functionalized 1,6-anhydrosugars. A recent report by Porter *(48)* on new synthetic approaches to this class of derivatives explored a new ring-expansion reaction of 1,3-oxathiolanes. Tagetitoxin is currently being explored as a potential specific and selective inhibitor of eukaryotic nuclear RNA polymerase III. Moreover, the action of tagetitoxin against RNA polymerase III promoter-directed transcription extends across a broad phylogenetic range, including vertebrates, insects, and yeast *(49)*. Thus tagetitoxin is the first example of a specific RNA polymerase inhibitor that acts against bacterial RNA polymerases and against only one (RNA polymerase III) of the eukaryotic nuclear RNA polymerases.

S-Nitrosothiols as nitric oxide donors

Nitric oxide (NO) is a unique signaling molecule with broad physiologic activity that includes control of vasodilation; cell-cycle regulation, with specific effects on apoptosis, proliferation, and differentiation; and even antimicrobial activity *(50)*. NO-containing compounds have found use in treating angina pectoris and myocardial infarction. Recently a number of new nitric-oxide-donating compounds have been developed, including S-nitrosated thiosugars (sugar-SNAPs) *(51-52)*. S-Nitrosothiols (SNAP) including S-nitrosated thiosugars (sugar-SNAPs) are vasodilators because they release NO. Compared to SNAP, sugar-SNAPs have higher stability and slower NO-releasing properties in aqueous solutions.

The spontaneous release of nitric oxide from an S-nitrosated thiosugar is caused by the formation of a disulphide, as was determined by Moynihan *(53)*. Based on the kinetic data, NO release from S-nitrosated thiosugars occurs predominantly from thermal decomposition, and this chemical reaction is responsible for the vasodilator effect. A series of sugar S-nitrosothiols (sugar-SNAPs) analogs of novel nitric oxide donors have been developed by Wang *(54-57)*. Their chemical activities and biological applications have been reviewed by Wang and coworkers *(58)*.

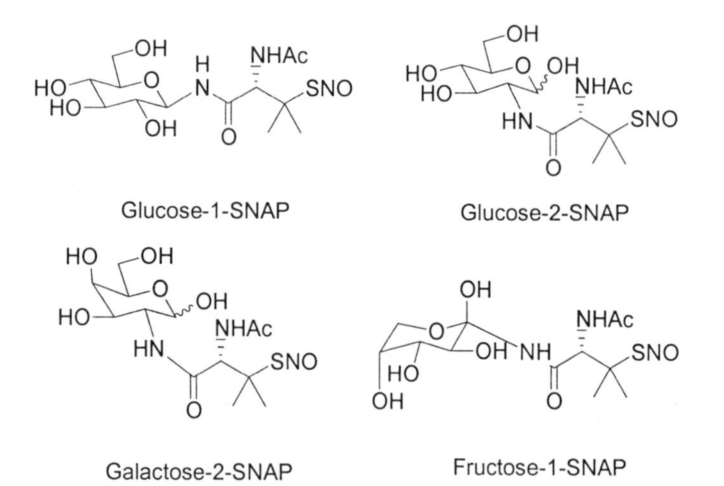

Figure 8. Sugar-SNAPs developed by Wang

The designs of these compounds were based on the observation that facilitated transport of monosaccharides in mammalian cells is accomplished by the glucose-transporter family of transmembrane proteins. *(59-60)*. SNAPs are constructed by an aglycone unit conjugated with a mono- or oligosaccharide.

The aglycone moiety provides the pharmacological activity, whereas the carbohydrate unit enhances water solubility, cell penetration, and a drug-receptor interaction; the carbohydrate unit also influences the dose-response relationship. Preliminary cytotoxic studies of glucose SNAPs against different cancer cells have indicated that glucose SNAPs are more potent than SNAP itself.

An interesting novel analog of sugar-SNAP—N-(S-nitroso-acetylpenicillamine)-2-amino-2-deoxy-1,3,4,6-tetra-O-acetyl-β-D-glucopyranose (RIG200)—was developed by Butler and coworkers *(61-62)*. RIG 200 has been found to prolong vasodilation in endothelium-deduced, isolated, rat-femoral arteries.

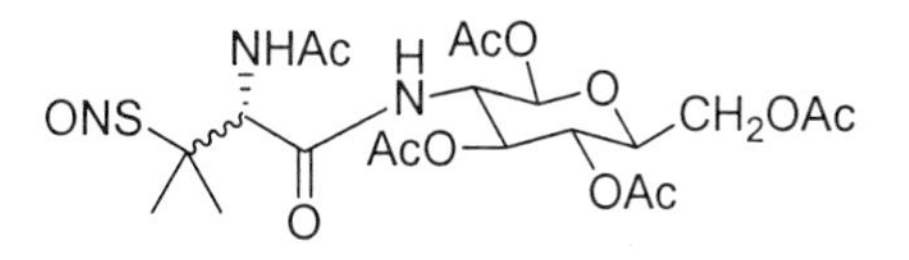

Figure 9. N-*(*S*-nitroso-acetylpenicillamine)-2-amino-2-deoxy-1,3,4,6-tetra-O-acetyl-β-D-glucopyranose* ***(RIG 200)***

Independently, Williams and coworkers *(63)* also synthesized and characterized a series of novel SNAP analogs derived from thiosugars. The 2-thioglycerol sugar-SNAP analog exhibited high stability, whereas the 2-glycerol-2-nitrosothiol decomposed rapidly at room temperature.

Other developments in this area (as mentioned earlier in the Bioavailability section) include improvements in systems designed to deliver NO-bearing compounds to their site of action. Specific advances have been made with experimental technologies such as adenoviral delivery and the much simpler mode of topical administration *(61)*.

Given such advances, Butler and coworkers developed a series of NO-donor analogs derived from 1-thiosugars, such as glucose, galactose, xylose, maltose, and lactose, based on the similarity of glycerol and sugar.

Critically important lipophilicity and the ability to release NO could be controlled by specific alteration (deacetylation) of acetyl functional groups. All novel NO-donor analogs have both hydrophobic and hydrophilic functional groups, which allow them to be delivered transdermally *(64)* This physical-chemical profile lends itself to their effective transdermal absorption. For example, S-nitroso-1-thio-2,3,4,6,-tetra-*O*-acetyl-glucopyranose was shown to be more effective in relaxing human cutaneous vascular smooth muscle than regular SNAP when delivered transdermally. The finding suggests that this type

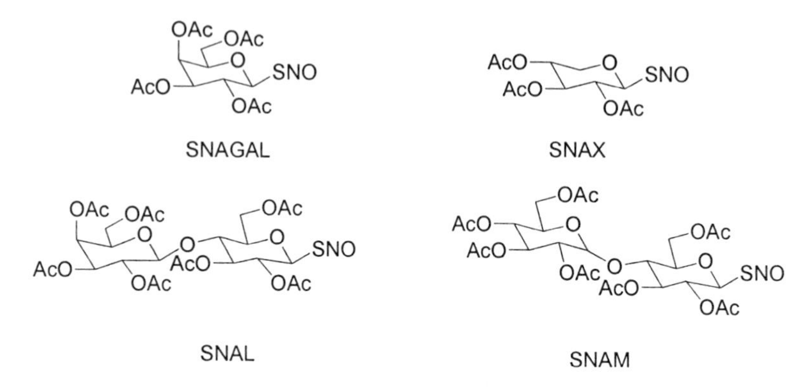

Figure 10. Nitrosated thiosugars as novel NO donors

of sugar-SNAPs is a more effective NO-donor-drug candidate for human smooth-muscle relaxation.

Additionally, it was discovered that the effect of copper ions on the release of NO from glucose SNAP was less than that of regular SNAP.

Delivery of the released NO to the microvessels of the skin most probably occurs by diffusion across the epidermis. This type of novel delivery of NO might provide an attractive source of localized vasodilatation without *(65)* inducing systemic side effects.

Polysaccharides. Tubercin

Tubercin, extracted from *Mycobacterium tuberculosis*, is a complex mixture of low-molecular-weight polysaccharides, free of lipids and proteins, of an unusually branching arabinomannan structure. The average molecular weight of each polysaccharide is about 7,000 or less. The polysaccharides are comprised of mannose, arabinose, glucose, and galactose.

In 1974 Chung *(66-67)* first reported on the therapeutic effect of tubercin on leptomatrous leprosy. The patent literature by Chung *(68)* claimed it as a highly effective compound in treating various cancer patients without incurring any adverse side effects. The specific immunotherapeutic properties of tubercin were shown in many previous studies. When administered as an adjuvant treatment in 500 patients with advanced cancers, including melanoma, they had a significantly longer disease-free survival *(69-71)*.

Conclusions

Recent insights into the biological functions of cellular carbohydrates have stimulated research into the use of glycochemicals as therapeutic agents, and many new and fascinating discoveries have resulted.

All of the new strategies currently available or under development—such as combinatorial synthesis and library screenings; identification of natural products, antibiotics, and mimetics; and the utilization of endogenous proteins as lectins—will constitute a significant milestone in the area of carbohydrate therapeutics. Simultaneously occurring is an increased understanding of structural requirements for the binding and catalytic mechanisms of enzyme inhibitors (thioligases).

The development of carbohydrate-based drugs represents a novel approach to treating life-threatening disorders such as viral infection, immune dysfunction, and various forms of cancer. This activity, which undoubtedly will be closely followed as many new candidates are commercialized and approved for use, has very promising prospects for the future.

References

1. *Complex Carbohydrates in Drug Research Structural and Functional Aspects*, Bock, K.; Clausen, H. Eds. Alfred Benzon Symposium 36, Munksgaard, Copenhagen, **1994**.
2. *Carbohydrates in Drug Design,* Witczak Z.J. Nieforth K. Eds. Marcel Dekker, New York, **1997**.
3. *Carbohydrate Antigens,* Garegg, P. J.; Lindberg, A. A. Ed. ACS Symposium Series No. 519, ACS. Washington, D.C. **1993**.
4. β-Glucosidases: Biochemistry and Molecular Biology, Esen, A.; Ed. ACS Symposium Series No. 533, ACS Washington, D. C. **1993**.
5. *Carbohydrates - Synthetic Methods and Application in Medicinal Chemistry,* Ogura, H.; Hasegawa, A.; Suami, T. Eds., Kodansha, Tokyo, & VCH Weinheim, **1992**.
6. *Glyconjugates Composition, Structure and Functions,* Allen H. J. Kisailus E., Eds. Marcel Dekker, New York, **1992.**
7. *Carbohydrate-based drug discovery,* Wong C-H. Ed, Wiley-VCH, Weinheim, **2003**.
8. Witczak, Z. J. *Curr. Med. Chem.* **1995,** *1*, 392.
9. McAuliffe, J.; Hindsgaul, O. *Chem. & Ind.* **1997**, 170.

10. Yarema, K. J.; Bertozzi C. R. *Current Opinion in Chemical Biology*, **1998**, *2*, 49.
11. Barchi, J. J. *Curr. Pharm. Design,* **2000,** *6*, 485.
12. Osborn, H. M. I.; Evans, P.G.; Gemmell, N.; Osborne S.D. *J. Pharm & Pharmacology,* **2004,** *56,* 691.
13. Varki, A. *Glycobiology*, **1993**, *3*, 97.
14. Varki, A.; Varki, N.M. *Brazillian J. of Med. and Biol.Res.,***2001,** *34,* 711.
15. Rogers, J.C.; Kornfield, S. *Biochem Biophys Res Commun.* **1971,** *45,* 622.
16. Rojo, J.; Delgado, R. *J. Antimicrob Chemotheraphy*, **2004**, *54*, 579.
17. Ohrlein, R. *Mini Review in Medicinal Chem.*, **2001**, *1*, 349: Wang X.; Zhang L-H.; Ye X-S. *Med Res Reviews* **2003**, *23*, 32; Dey, P. M.; Witczak, Z. J. *Mini Review in Medicinal Chem.*, **2003,** *3,* 271.
18. Rabinovich, G.A. *British Journal of Cancer,* **2005**, *92*, 1188.
19. Schwarz, F.P.; Ahmed, H.; Bianchet, M.A.; Amzel, L.M.; Vasta, G.R. *Biochemistry*, **1998**, *37*, 5867.
20. Ahmad, N.; Gabius, H.J.; Sabesan, S.; Oscarson, S.; Brewer, C.F. *Glycobiology*, **2004**, *14*, 817.
21. Danguy, A.; Camby, I.; Kiss, R. *Biochim Biophys Acta* **2002**, *1572* 285.
22. Ellerhorst, J.; Nguyen, T.; Cooper, D.N.; Lotan, D.; Lotan, R. *Int J Oncol* **1999**, *14,* 217.
23. Linhardt, R.J.; Baezinger, N.C.; Ronsen, B. *J Pharm Sci.* **1990,** *79,* 158.
24. Thorens, B. *Ann Rev Physiol* **1993**, *55,* 591.
25. Wright, E.M. *Ann Rev Physiol* **1993**, *55* 575.
26. Tamai, I.; Tsuji, A. *Adv Drug Del Rev,* **1996**, *20,* 5.
27. Mizuma, T.; Nagamine, Y.; Dobashi, A.; Awazu, S. *Biochim Biophys. Acta* **1998,** *1381,* 340: Nomoto,M.; Yamada, K.; Haga, M.; Hayashi, M. *J Pharm Sci.* **1998**, *87*, 326.
28. Perry, N. B.; Blunt, J. W.; Munro, M. H. G.; Pannell, L. K. *J. Am. Chem. Soc.* **1988**, *110*, 4850.
29. Burres, N. S.; Clement, J. J. *Cancer Res.* **1989,** *49*, 2935; Perry, N. B.; Blunt, J. W.; Munro, M. H. G.; Thompson, A. M. *J. Org. Chem.* **1990,** *55*, 223; Nakata, T.; Nagao, S.; Oishi, T. *Tetrahedron Lett.* **1985**, *26*, 6465.
30. Kuhajda, F.P.; Pizer, E.S.; Li, J.N.; Mani, S. N.; Frehywot, G.L.; Townsend C.A. *Proc. Natl. Acad. Sci. USA* **2000**, *97*, 3450; Pizer, E.S.; Thupari, J.; Han, W.F.; Pinn, M.L.; Chrest, F.J.; Freywot, G.L; Townsend, C.A.; Kuhajda, F.K. *Cancer Res.* **2000,** *60,* 213.
31. Jahn, M.; Marles, J.; Warren, R.A.J.; Withers, S.G. *Angew. Chem.,* **2003,** *115*, 366; Jahn, M.; Withers, S.G. *Biocatalysis and Biotransformation* **2003,** *21;* 159.
32. Oishi, H.; Noto, T.; Suzuki, K.; Hayashi, T.; Okazaki, H.; Ando, K.; Sawada, M. *J. Antibiot (Tokyo)* **1982,** *35*, 391.
33. Chambers, M.S.; Thomas, E.J.; Williams, D.J. *J Chem Soc Chem Commun,* **1987**, 1228.

34. Kremer, L.; Douglas, J.D.; Baulard, A.R.; Morehouse, C.; Guy, M.R.; Alland, D.; Dover, L.G.; Lakey, J.H.; Jacobs, W.R. Jr.; Brennan, P.J.; Minnikin, D.E.; Besra, G.S. *J Biol Chem*, **2000,** *275;* 16857.
35. McFadden, J.M.; Frehywot, G.L.; Townsend, C.A. *Org Lett,* **2002,** 4; 3859.
36. Varela, O. *Pure Appl Chem.* **1997,** *69;* 621.
37. Slayden, R.A.; Lee, R.E.; Armour, J.W.; Cooper, A.M.; Orme, I.M.; Brennan, P.J.; Besra, G.S. *Antimicrob Ag Chemother* **1996,** *40;* 2813.
38. Jones, A.L.; Herbert, D.; Rutter, A.J.; Dancer, J.E.; Harwood, J.L.; *Biochem J,* **2000,** *347;* 205.
39. Price, A.C.; Choi, K-H.; Heath, R.J; Li, Z.; White, S.W.; Rock. C.O. *J Biol Chem,* **2001,** *276;* 6551.
40. Douglas, J.D.; Senior, S.J.; Morehouse, C.; Phetsukiri, B.; Campbell, I.B.; Besra, G.S.; Minnikin, D.E. *Microbiology,* **2002,** *148;* 3101.
41. Jones, S.M.; Urch. J.E.; Brun, R.; Harwood. J.L.; Berry, C.; Gilbert, I.H. *Bioorg Med Chem,* **2004,** *12*; 683.
42. Miyakawa, S.; Suzuki, K.; Noto, T.; Haranda, Y.; Okazaki, H. *J Antibiot (Tokyo)* **1982,** *35;* 411.
43. Mitchell, R.E.; Coddington, J.M.; Young, H. *Tetrahedron Lett.,* **1989,** *30;* 501.
44. Mitchell, R.E.; Durbin, R.D. *Physiol Plant Pathol,* **1981,** *18;* 157.
45. Mathews, D.E.; Durbin, R.D. *J Biol Chem*, **1990,** *265*; 493.
46. Trimboli, D.; Fahy, P.C.; Baker, K.F. *Australian J. Agric Res,* **1978,** *29;* 831.
47. Dent, B.R.; Furneaux, R.H.; Gainsford, G.J.; Lynch, G.P. *Tetrahedron,* **1999,** *55;* 6977.
48. Ioannou, M.; Porter, M.J.; Saez, F. *Chem Commun.,* **2002,** 346.
49. Steinberg, T.H.; Mathews, D.E.; Durbin, R.D.; Burgess, R.R. *J Biol Chem,* **1990,** *265;* 499.
50. Fang, F.C, *J Clin. Invest.* **1997,** *99;* 2818.
51. Butler, A.R.; Field, R.A.; Greig, I.R. *Nitric Oxide Biol. Chem.* **1997,** *1,* 211.
52. Khan, F.; Pearson, R.J.; Newton, D.J.; Belch, J.J.; Butler, A.R. *Clinical Science,* **2003,** *105*; 577.
53. Moynihan, H.A.; Roberts, S.M. *J Chem Soc Perkin Trans I.* **1994,** 797.
54. Hou, Y.-C.; Wang, J.-Q.; Andreana, P. R.; Cantauria, G.; Tarasia, S.; Sharp, L.; Braunschweiger, P. G.; Wang, P. G. *Bioorg. Med. Chem. Lett.* **1999,** *9,* 2255.
55. Hou, Y.-C.; Wang, J.-Q.; Ramirez, J.; Wang, P. G. *Methods Enzymol.* **1998,** *301*, 242.
56. Ramirez, J.; Yu, L.-B.; Li, J.; Braunschweiger, P. G.; Wang, P.G. *Bioorg. Med. Chem. Lett.* **1996,** *6,* 2575.
57. Hou, Y. C.; Wu, X. J.; Xie, W. H.; Braunschweiger, P. G.; Wang, P. G. *Tetrahedron Lett.* **2001,** *42,* 825.
58. Wang, P.G.; Xian, M.; Tang, X.; Wu, X.; Wen, Z.; Cai, T.; Janczuk, A. J. *Chem Rev,* **2002,** *102;* 1091.

59. Bell, G. I.; Burant, C. F.; Takeka, J.; Gould, G. W. *J. Biol. Chem.* **1993**, *268*, 19161.;
60. Mellanen, P.; Minn, H.; Greman, R.; Harkonen, P. *Int. J. Cancer* **1994**, *56*, 622.
61. Megson, I. L.; Greig, I. R.; Gray, G. A.; Webb, D. J.; Butler, A. R. *Br. J. Pharmacol.* **1997**, *122*, 1617.
62. Megson, I. L.; Greig, I. R.; Butler, A. R.; Gray, G. A.; Webb, D. J. *Scott. Med. J.* **1997**, *42*, 88.
63. Munro, A. P.; Williams, D. L. H. *Can. J. Chem.* **1999**, *77*, 550.
64. Butler, A. R.; Field, R. A.; Greig, I. R.; Flitney, F. W.; Bisland, S. K.; Khan, F.; Belch, J. J. F. *Nitric Oxide* **1997**, *1*, 211.
65. Al-Sa'doni, H.H.; Ferro, A. *Curr. Med. Chem.* **2004,** *11*, 2679.; Richardson, G.; Benjamin, N. *Clinical Science* **2002,** *102*, 99.; Sogo, N.; Campanella, C.; Webb, D.J.; Megson, I.L *British J. Pharmacology.* **2000,** *131,* 1236.; Megson, I.L; Sogo, N.; Mazzei, F. A.; Butler, A.R. Walton, J.C.; Webb, D.J. British J. Pharmacology, **2000,** *131,* 1391.
66. Chung, T-H. *J Korean Med Assoc.* **1974,** *6,* 427.
67. Chung, T-H.; Song, J. Y.; Hong, S. S. *Yonsei Medical Journal.***1976,** *17*, 131.
68. Chung T-H. U.S. Patent 6,274,356, **2001**.
69. Chung, T-H. *Korean J Biochem,* **1983**, *15,* 65.
70. Lee, B.H.; Gu, B.M.; Eun, H.C. *Kor J Dermatol.* **1979,** *5,* 373.
71. Chung,T-H.; Kim, S.M.; Shin, T. S.; Kim, D.Y. *J.Kor Surg Soc.* **1978,** *20,* 651.

Cancer

Chapter 3

Development of a Polysaccharide as a Vehicle to Improve the Efficacy of Chemotherapeutics

David Platt, Anatole A. Klyosov, and Eliezer Zomer

Pro-Pharmaceuticals, 189 Well Avenue, Newton, MA 02459

The GMP process has been developed for manufacturing DAVANAT®, a modified galactomannan from *Cyamopsis tetragonoloba*, or guar gum. The material was structurally identified and characterized by standard analytical techniques as well as by the following methods: ^{13}C Nuclear Magnetic Resonance (^{13}C-NMR); Size Exclusion Chromatography with Multi-Angle Laser Light Scattering (SEC-MALLS) for absolute molecular weight determination; and Anion Exchange Liquid Chromatography with Pulsed Amperometric Detection (AELC-PAD) for carbohydrate composition and to verify the uniformity and purity of the final products. Both in the manufacturing of the bulk material and the final drug product, precautions were taken to control degradation by thermal or microbial/enzymatic activity. As a parental/i.v., the slow-dissolving polymer was formulated as a sterile solution to make it easy to handle in a hospital. The shelf life of the polymer solution has been estimated to be at least three years at room temperature. Preclinical studies with 5-fluorouracil, doxorubicin, irinotecan, and cisplatin have shown significant degrees of efficacy enhancement both in colon and breast-cancer models in nude mice.

Introduction

A background for galactose-containing compounds' anti-cancer activity has been established in a few studies showing that galactoside-containing carbohydrates and galactoside-specific lectins (carbohydrate-binding proteins) disturbed cell association and may affect tumor development and metastasis (1,2,4).

In vitro studies have shown that galactoside-binding proteins on the tumor cell surface interact with terminal D-galactoside residues located on adjacent cells, thus mediating cellular attachment and the potential for metastasis (1,2).

We report here that a linear galactose-containing polymer —galactomannan — interacts with cancer drugs and positively affects their anti-tumor activity. Although structures like cyclodextrins have been shown to entrap a variety of hydrophobic drugs and improve their solubility and delivery, this was the first time that co-administration of polysaccharides along with chemotherapeutics has been proven effective in animal models.

In particular, a modified galactomannan coined DAVANAT® has consistently produced positive enhancement of the activity of 5-fluorouracil (5-FU) —a widely used chemotherapeutic drug —against human colon tumor models. From the developmental point of view it was achieved by a careful consideration of the raw material (guar gum) and its GMP processing. As a parental/i.v., the polysaccharide was formulated as a sterile solution to make it easy to handle in a hospital. The shelf life of DAVANAT® solution has been confirmed to exceed three years at room temperature. In addition to 5-FU, pre-clinical studies with doxorubicin, irinotecan, and cisplatin have shown reduction in toxicity along with significantly enhanced anticancer activity in human colon, breast, and melanoma xenographic models in nude mice.

Polymeric carbohydrates have been long used in separation techniques to trap specific chemicals in a three-dimension matrix. Structures like cyclodextrins have been shown to entrap a variety of hydrophobic drugs and improve their solubility and delivery. We studied the use of carbohydrate polymers for chemotherapeutics, targeting tumor-lectin binding moieties also known as galectins. A variety of galactomannans exist naturally, and a few were selected because of their high homogeneity and uniform structure. These molecules have been modified (while maintaining the linear structure of the mannose backbone with galactose residues at approximately every second mannose residue) to improve systemic solubility.

Studies have shown that specific lectins on the tumor cell surface interact with terminal D-galactoside adducts of glycoproteins located on adjacent cells, thus mediating various types of cellular interactions (1,2). These lectins also modulate cell–cell and cell–matrix interactions (3). Lectins can act as receptors that are involved in selective intercellular adhesion, cell migration, and recognition of circulating glycoproteins.

Galectins are members of a family of beta-galactoside-binding lectins (1). Many experimental agents—including monoclonal antibodies, simple sugars, and some polysaccharides like pectin fragments—are known to interact *in vitro* with the lectins on the cancer cell surface. Some of these agents inhibit the development of tumor-cell colony formation (4,5). The galactomannan with its exposed galactose residues has previously been shown to interact with hepatoma cell lectins and reversed by lactose, and thus potentially modulate their physiological actions.

It is believed that lectins mediate various types of cellular interactions, and some of them have been studied for the last 15 years. These studies have resulted in identification of more than a dozen lectins, some of them turned out to be tumor-associated galectins. A few glycoproteins that allegedly interact with these lectins have also been identified. Some of them reportedly result in inhibition of tumor cell colony development, as described in (1–5). Even simple carbohydrates such as methyl-α-D-lactoside and lacto-N-tetrose have been shown to slightly inhibit metastasis of B16 melanoma cells, while D-galactose and arabinogalactose inhibit liver metastasis of L-1 sarcoma cells (6).

Despite advances in chemotherapeutic regimens for metastatic colorectal cancer, as well as other solid tumors, patients refractory to these therapies represent a large unmet medical need. For example, 5-FU is used alone or in combination with many other chemotherapeutic agents for treatment of solid tumors, particularly colorectal, breast, hepatic, and gastric cancers (7,8). These combinations, while marginally improving efficacy, may increase toxicity. It is reasonable to hypothesize, however, that more specific targeting of efficacious chemotherapeutic agents such as 5-FU to tumor cells could result in higher response rates along with reduced toxicity.

Compared to many other known polysaccharides, galactomannans have multiple single-sidechain galactose units. These units might readily interact with galactose-specific receptors such as galectins on the tumor cell surface, modulate the tumor surface physiology, and potentially affect delivery of 5-FU to the tumor. This hypothesis has been proven in our studies, and results are described in this chapter and the next.

DAVANAT® is a chemically modified galactomannan from the seeds of *Cyamopsis tetragonoloba*. The natural ("initial") galactomannan has too large a molecular weight to be water-soluble, but after a controlled molecular-weight reduction it acquires good solubility in water, up to 100 mg/mL.

In animal models of colorectal cancer, DAVANAT® enhances the antitumor activity of 5-FU. Pro-Pharmaceuticals, Inc., is developing DAVANAT® for use in humans as a potential enhancer of the antitumor activity of 5-FU against solid tumors that include (but are not limited to) those of colorectal, breast, lung (non-small cell), gastric, pancreatic, hepatic, ovarian, prostate, and head and neck cancers.

With its exposed galactose residues, DAVANAT® might readily interact with carbohydrate-binding proteins on the cell surface and thus block their physiological actions.

In the Phase I study, DAVANAT® was evaluated alone and in combination with 5-FU in patients with solid tumors who have failed proven surgical, radiation, and chemotherapeutic regimens. This study is described in more detail in the next chapter.

Description and structural identification of DAVANAT®

Physical Description

DAVANAT® is a white water-soluble (up to 100 mg/mL) polysaccharide obtained by partial and controlled chemical cleavage of the storage polysaccharide from seeds of *Cyamopsis tetragonoloba*, commercially available as guar gum flour. Its chemical formula can be expressed as [(1→4)-linked β-D-mannopyranosyl]$_{17}$-[(1→6)-linked-α-D-galactopyranose]$_{10}$}$_{12}$ with average molecular weight of 48,000 Dalton, and mannose to galactose ratio of 1.7. In other words, the polymer is a linear (1→4)-β-D-Mannopyranosyl backbone to which a single α-D-Galactopyranosyl residue is attached by (1→6) linkage randomly to every other mannopyranose (Klyosov and Platt, U.S. Patent No. 6,645,946).

Chemical names of the galactomannan are 1,4-β-D-Galactomannan, or [(1→6)-α-D-galacto-(1→4)-β-D-mannan].

Figure 1. The building block of DAVANAT®

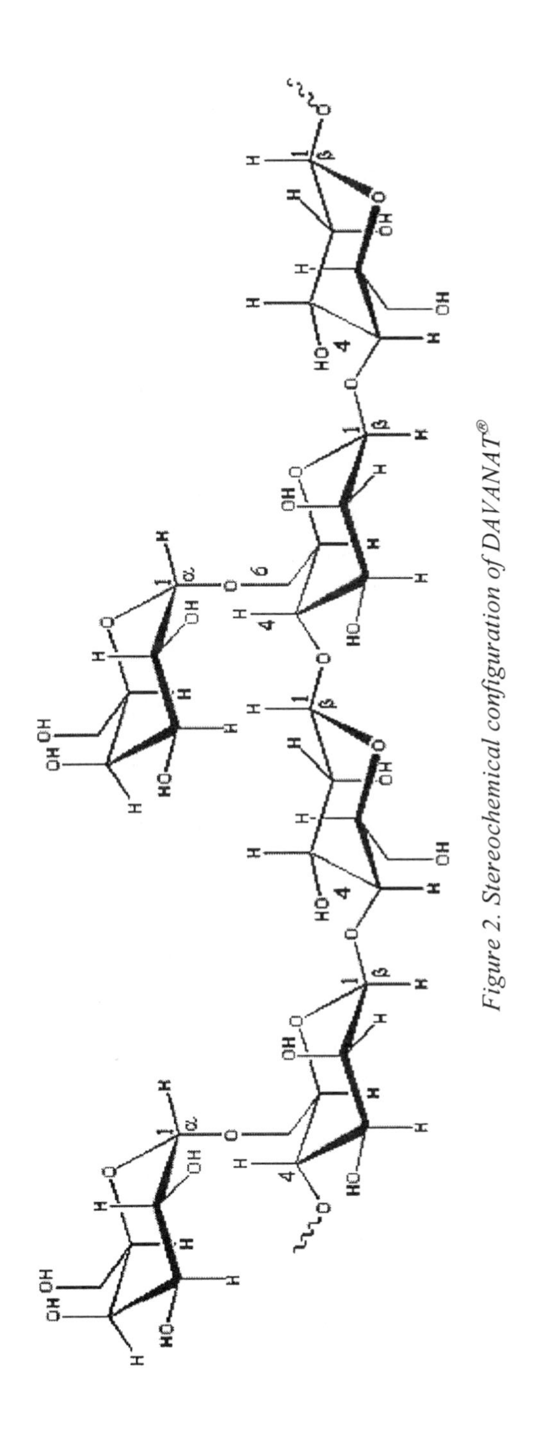

Figure 2. Stereochemical configuration of DAVANAT®

Composition of the final formulation

DAVANAT® is packaged sterile as a 60 mg/mL solution in 10 mL normal saline, and is dispensed in single-use glass vials with rubber stoppers. Each vial has a permanently affixed label on it with the lot number, volume in vial, manufacturing date, concentration, and storage conditions (2–8°C), along with the Pro-Pharmaceutical name and address plus the statement "Caution: New Drug. Limited by Federal Laws to Investigational Use."

This clinical solution should be stored in a locked refrigerated space with access limited to study personnel.

Manufacturing

DAVANAT® was manufactured from a commercial flour of *Cyamopsis tetragonoloba* (guar gum, galactomannan), using controlled thermal degradation and purification through copper salt complex and multiple ethanol washes. The typical yield of DAVANAT® was 50 percent (w/w from the initial guar gum). To assure a pharmaceutical-grade product, extra specifications were put in place. Thus the prime raw material has minimum microbial contamination and consistent chemical structure, and purity suitable for pharmaceutical manufacturing. The purification procedure results in DAVANAT® that is more than 98-percent pure (formalized calculations show purity of 100±2 percent)

Analytical Identification

While the purity and composition of mannose and galactose were established by AELC-PAD method, the polysaccharide structure was established with ^{1}H-NMR and ^{13}C NMR spectra. In Figure 3 the NMR spectrum is given, along with that of a purified galactomannan from carob (locust-bean) gum for comparative purposes. An easy identification of the two principal sugar residues—that is, mannose and galactose—comes from two peaks in ^{1}H-NMR, at 4.8 p.p.m. and 5.0 p.p.m. (doublet), respectively, given that the ratio of mannose to galactose (Man/Gal) in the galactomannan from guar gum is 1.7 and from carob gum is 4.3.

The ^{13}C-NMR spectrum of DAVANAT® (Figure 4) shows detailed positions of the chemical shifts and their intensities, and it confirms the above chemical structure of DAVANAT®: All three sugar structures—that is, α-D-galactopyranosyl, 4-O-β-D-mannopyranosyl (unsubstituted), and 4,6-di-O-β-D-manno-pyranosyl (substituted)— are seen in the NMR spectrum. The positions of signals from C1 to C6 for all three sugars—G-1 to G-6, M-1 to M-6, and GM-1 to GM-6, respectively—are shown in the above figure as well. (The positions of G-2 and G-4, M-1 and GM-1, M-2 and GM-2, M-3 and GM-3, and M-4 and GM-4 are coincided within these pairs.)

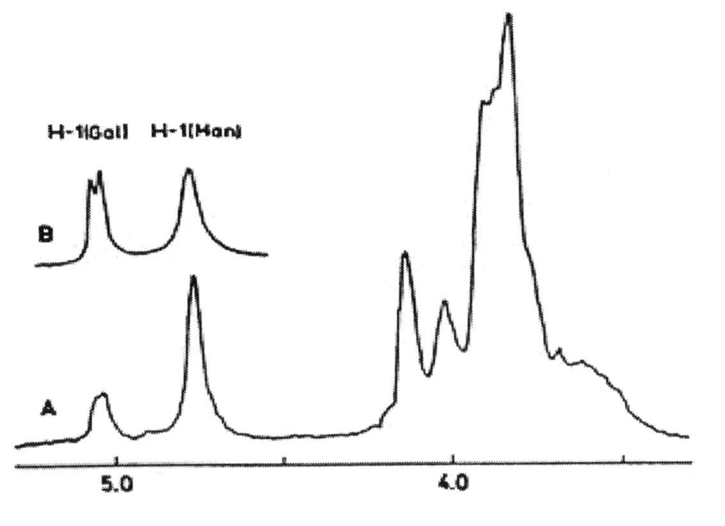

Figure 3. ^{1}H NMR Spectra (99.6 MHz) of galactomannan from carob gum (A) and DAVANAT® (B) in D_2O solutions (10 mg/mL) at pD 7 and 90°C

The positions of signals for galactose in the NMR spectrum completely correspond to those of carbon atoms of free galactopyranose, providing evidence for the absence of substituents at positions of Gal units in the galactomannan. Substitution of certain mannose residues is at C-6 (because of the shift of the signal from 63.6 ppm for "normal" unsubstituted methylene carbon to a "substituted" one at 69.6 ppm, along with a shift of the adjacent C-5 signal from a "normal" 78 ppm to 76.4 ppm). The 63.6 ppm shift for the methylene carbons is well documented.

C-1 of galactose residue is involved in formation of a galactoside bond (compared to C-1 free of galactose that in the NMR spectrum had a lowfield shift of +6.5 ppm). Mannose residues are attached to each other "head-to-tail," forming the β(1→4) backbone chain. There are unsubstituted Man-Man pairs along with substituted Man(Gal)-Man, Man-Man(Gal), and Man(Gal)-Man(Gal).

The NMR of galactomannans from guar gum, carob gum, and clover seeds in the region of C-4 (Man) resonance (split into three peaks) is used to identify each with highest peak at high field (corresponding to unsubstituted Man-Man pairs) and is observed for a galactomannan from carob gum (Man/Gal = 3.8). The lowest peak corresponds to a galactomannan from clover seeds (Man/Gal = 1.4) and the intermediate peak corresponds to DAVANAT® from *Cyamopsis tetragonoloba* (with a Man/Gal = 1.7). That also can be calculated from frequencies for unsubstituted and substituted Man residues, measured from C-4 (Man) resonance.

The relative frequency of unsubstituted Man-Man pairs in DAVANAT® was 22 percent, that of Man(Gal)-Man and Man-Man(Gal) total was 48 percent, and that of totally substituted pairs Man(Gal)-Man(Gal) was 30 percent (from intensities of the split C-4 signals into their respective three lines/peaks).

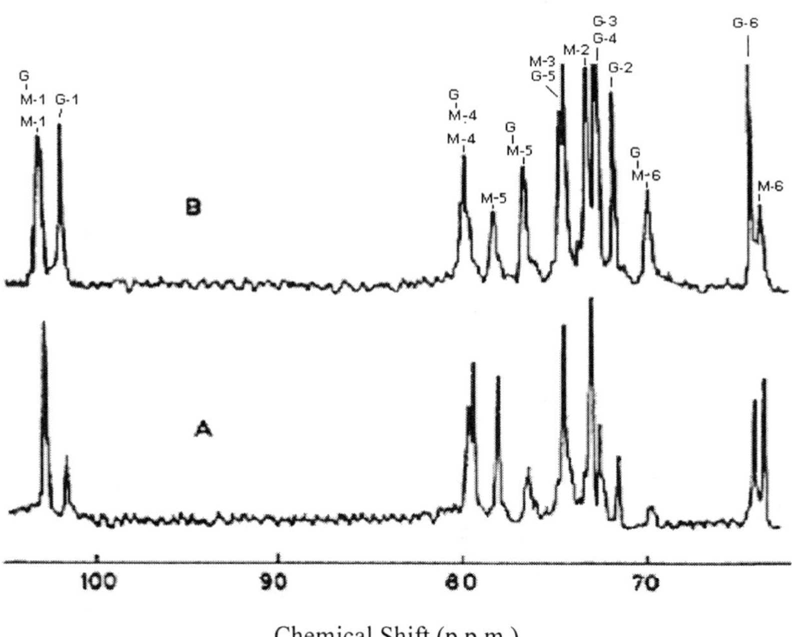

Chemical Shift (p.p.m.)

Figure 4. The ^{13}C-NMR spectrum of DAVANAT® compared with that of purified galactomannan from carob gum (Chemical shifts: G - galactose carbons, M - mannose carbons)

The ^{1}H-NMR signals for the DAVANAT® galactose anomeric protons appear at approximately 4.9 ppm (doublet). The signals for the DAVANAT® mannose anomeric protons appear at approximately 4.6 ppm (broad signal). These signals are completely separated from those of the free monosaccharides: the galactose alpha proton at 5.1 and beta at 4.5 ppm, the mannose alpha at 4.8 ppm and beta at 5.0 ppm. The ratio of mannose to galactose units was calculated at 1.67 for this working reference standard produced at 1 Kg batch size (Figure 5)

Certainly, more details need to be given to fully characterize the alpha linkage between galactose and mannose. However, this type of linkage is a well-established fact in galactomannans of plant origin.

The spectra were recorded at 3-percent solution of D_2O at 30^0C on the Bruker AM-300 NMR spectrometer, with an operating frequency on carbon nuclei of 75.4 MHz. Acetone was used as an internal standard (31.45 ppm).

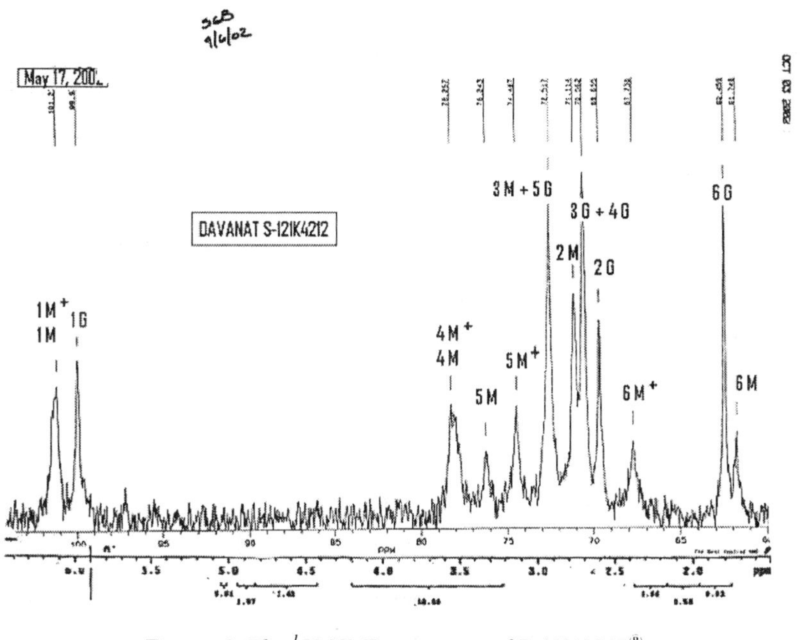

Figure 5. The ^{1}H-NMR spectrum of DAVANAT®.
Lot # S-12K4212

This NMR spectrum clearly shows all three sugar units: D-galactopyranosyl (G on the spectrum); 4-O-β-D-mannopyranosyl- (M); and 4,6-di-O-β-D-mannopyranosyl- (M+).

The positions of signals for galactose (except for C-1) in the NMR spectrum completely correspond to those of carbon atoms free of galactopyranose, providing evidence for the absence of substituents at positions C-2 to C-6. C-1 of galactose residue is bound to the mannan backbone, with the NMR spectrum revealing a low field shift of +6.4 ppm.

Substitution of substantial mannose residues is at C-6 (because of the shift of the signal from 62.3 ppm for a "normal" unsubstituted methylene carbon to a "substituted" one at 67.7 ppm, along with a shift of the adjacent C-5 signal from a "normal" 77.4 ppm to 74.5 ppm). The 62.3 ppm shift for the methylene carbons is well documented (see footnote to Table 1).

A lowfield shift of C-1 and C-4 indicates that mannosyl residues are involved in the formation of a 1-4 glycoside bond. The ratio of mannose to galactose in the galactomannan is 1.67, which is confirmed by the NMR spectrum obtained at 60^{0}C.

Table 1. Position and interpretation of signals of the ^{13}C-NMR spectrum of DAVANAT®

Monosaccharide residue	Chemical shift, *p.p.m.*					
	C-1	C-2	C-3	C-4	C-5	C-6
α-D-Galactopyranosyl-	99.9	69.6	70.6	70.6	72.5	62.5
α-D-Galactopyranose*	93.5	69.6	70.4	70.6	71.7	62.4
4-O-β-D-Mannopyranosyl-	101.2	71.1	72.5	78.3	76.2	61.7
4,6-di-O-β-D-Mannopyranosyl-	101.2	71.1	72.5	78.3	74.5	67.7
β-D-Mannopyranose*	94.9	72.5	74.3	67.9	77.4	62.3

*Data presented for comparison, see Lipkind G.M., Shashkov A.S., Knirel Yu.A., Vinogradov E.V., Kochetkov N.K. (1988) *Carbohydrate Research* 175, 59-75.

Quantitation of DAVANAT® and determination of molecular weight by HPLC/RI-MALLS

In order to characterize the molecular-weight average and distribution throughout the research-and-development and scale-up phases, Pro-Pharmaceuticals has adapted the GPC/RI-MALLS technique by using a ZIMM plot analysis (Figure 6). This analytical method was incorporated as the basic procedure into the Drug Certificate of Analysis specification.

The use of dual monitoring of the HPLC elution profile of DAVANAT® provides us with two important chemical specifications of DAVANAT®—that is, the quantitative measurement by the Refractive Index signal and the absolute molecular weight by the Multi-Angle Laser Light Scattering (MALLS) detector. Furthermore, these methods provide data on molecular stability and breakdown derivatives of DAVANAT®.

Two chromatograms (see Figure 7 below) are given for comparison purposes: 1) Blank (control); and 2) DAVANAT® (Standard B3a), 30µl, injected at 8 mg/ml. The system is as follows: column packed with Phenomenex, Polysep GFC and a mobile phase of 50 mM Acetate, 0.1M NaCl, pH 5.0 at flow rate of 1 ml/min.

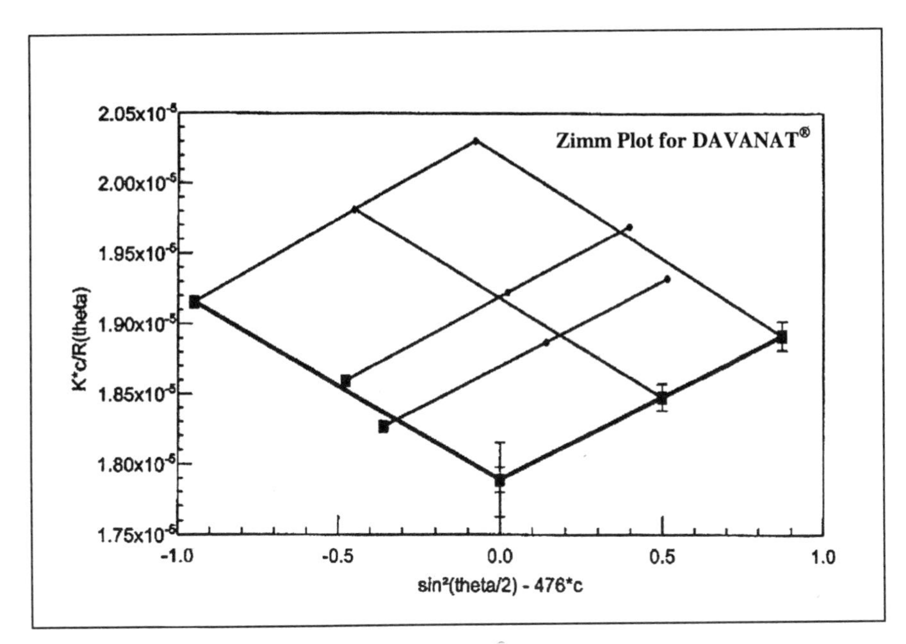

Figure 6. ZIMM plot of DAVANAT® generated with descending concentrations.

High Performance Liquid Chromatography (HPLC) using Gel Permeation Chromatography (GPC) separation technology is a well-established technique for the characterization of polymers. GPC in combination with Refractive Index (RI) detection and Multi-Angle Laser Light Scattering (MALLS) create a powerful tool for the determination of molecular weights of polymeric carbohydrates.

During the chromatographic run, the MALLS detector (placed in turn at many different angles) measures the degree of light scattering of a laser beam. The output of the light-scattering detector is proportional to the multiplication product of the concentration and the molecular weight of macromolecules, whereas the RI detector signal is proportional to the concentration only. Therefore, at any elution time, the molecular weight of the polymer eluting from the column can be calculated from the MALLS and RI signals.

The specification range for the molecular weight of DAVANAT® is 42,000 to 58,000 Da. As an example of a typical measurement, a recent QC test of bulk

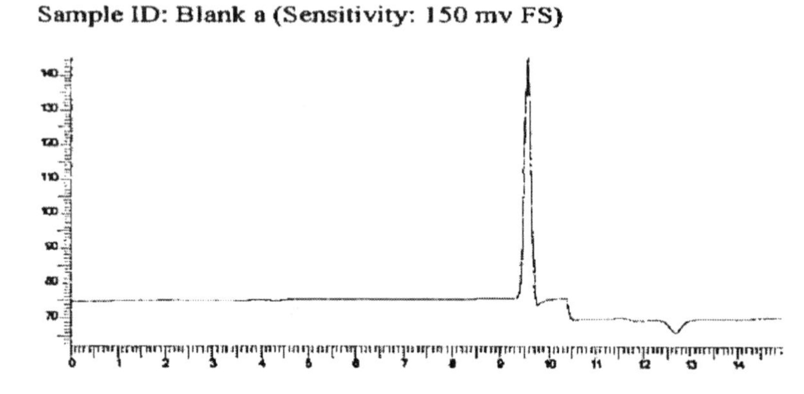

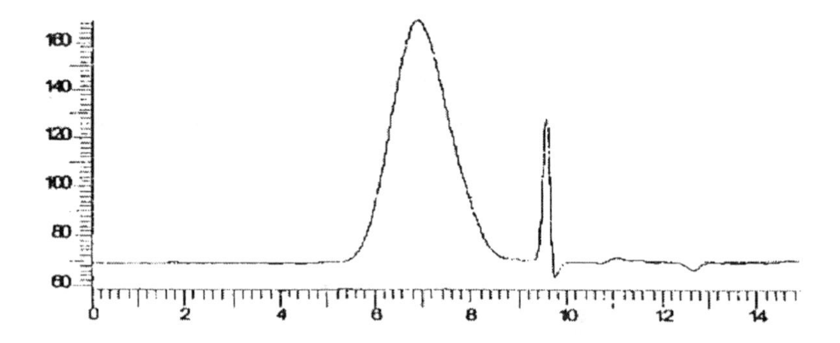

Figure 7. HPLC Profile of DAVANAT® using refractive index monitoring

product produced the following three results: 54,780 Da, 54,500 Da, and 54,360 Da. Average molecular weight was 54,550±210 Da.

Stability studies

DAVANAT® is very stable both in the dry-powder form and in water solutions. We have employed three different experimental techniques for testing stability: a shelf-life study for the dry powder, an accelerated study, and a study of stress under various (and harsh) experimental conditions.

The most recent results of the shelf-life study for DAVANAT® packaged dry powder (GMP Drug Substance) are shown in Table 2.

Table 2: Shelf life and accelerated thermal stability of DAVANAT® by molecular weight data

By GPC-MALLS of DAVANAT® packaged dry powder at room temperature and accelerated storage of 40°C and 65 °C at 75% and 85% relative humidity.

Testing	Analysis Date	Molecular Weight (Daltons)				
		Bulk Storage	% of Initial	25°C 60%RH	40°C 75%RH	65°C 85% RH
Initial Sigma C of A	02-Jun-02	54,760[1]	NA			
10 Day	29-Jul-03			50,537	49,320	47,931
30 Day	19-Aug-03			50,480	50,523	44,436
60 Day	22-Sep-03			49,783	49,603	43,438
90 Day	21-Nov-03			47,643	49,160	19,408[2]
180 Day	21-Jan-04			48,347	49,900	NT
270 Day	16-Apr-04			46,887	47,573	NT
360 Day	27-Jul-04			49,733	44,877	NT
PPD Bulk Release Testing	03-Feb-04	54,547	99.6			

[1] Initial value is the mean.
[2] Separate experiments
NT - not tested.
10 Day values obtained at both storage conditions.

The table shows that changes in molecular weight of DAVANAT® after a year's storage at 25°C occur practically only within a reasonable margin of error.

For the accelerated study, DAVANAT® powder in 1000-mg amounts was placed in polyethylene containers, which were tightly closed in the packaging room at the temperature and humidity of the bulk product. The containers were exposed to the respective thermal and humidity conditions of the accelerated test, while the control was run at an average temperature of 25°C. The molecular weight was used to estimate 10-percent loss of the initial polysaccharide. An Arrhenius calculation was then used to estimate the shelf-life:

$$\ln k = A - E/RT$$

Where:
k= kinetic rate constant, or reaction rate
A= constant

E = activation energy
R = gas constant
T= absolute temperature of incubation

The plotting of reaction rates against 1/T allows the calculation of rate at any temperature, and thus it provides a prediction of shelf-life t_{90} (time to 90-percent potency).

The Arrhenius plot using individual-concentration data points, and assuming the reaction is first-order, calculated the activation energy to be E = 98,000 J/mol and predicted a shelf-life of over 3 years at 25°C.

For the stress study, DAVANAT® solutions at 0.9 to 60 mg/mL were first exposed to temperatures of 80^{0}C and 110^{0}C at pH 5.0, 7.0, and 9.2. At 110^{0}C, the first-order kinetic rate constant for depolymerization of DAVANAT® was 0.021 min^{-1} both at pH 5.0 and pH 9.2, and 0.077 min^{-1} at pH 7.0. This corresponds to the respective half-life times of 33 min at 110^{0}C and pH 5.0 and 9.2, and of 90 min at 110^{0}C and pH 7.0.

Extended stability studies of DAVANAT® sterile solutions of 30 and 60 mg/mL at -70^{0}C to $+65^{0}$C were carried out for a range of durations, from 30 days to two years, in the absence and presence of 5-FU (10 to 50 mg/mL) in saline (pH 6.8 and 9.2). The frozen solutions were tested after thawing for 30 days. The molecular weight of DAVANAT® after these freezing-thawing steps was practically the same, within experimental errors, and no precipitation was observed. These studies predict stability at room temperature for at least two years.

The stability study was also performed by using the AELC/PAD method, which can monitor free galactose and mannose as well as low-molecular-weight oligomers, the likely degradation products of DAVANAT®. This method combines Anion Exchange Liquid Chromatography (AELC) and Pulsed Amperometric Detection (PAD). The data showed that practically no free galactose appeared in the course of depolymerization of DAVANAT®, with Arrhenius plot-based predictions of the stability (in terms of free galactose and mannose formation) reaching 5000 years! Therefore the Man-Gal covalent bond in DAVANAT® is extremely stable compared to the Man-Man bond in DAVANAT®.

In summary, the stability studies indicate that DAVANAT® (dry powder and saline solutions) can be effectively sterilized without damaging its structure, and that the stability at room temperature is in excess of five years at pH 7.

Stability and Quantitative Determination of DAVANAT® and 5-FU

The 5-FU was analyzed in 0.9% saline, pH 9.2, using HPLC with UV detection at 254 nm. Stock solutions for 5-FU and DAVANAT® were in the range of 0.64 to 5.5 mg/mL, and 0.8 to 8.0 mg/mL, respectively. The procedure was applicable (after dilution for the analysis) at concentrations of 5-FU from 0.0128 to 0.128 mg/mL, and DAVANAT® from 0.8 to 8.0 mg/mL. At these concentration ranges, both of the analyses were linear, reproducible, and highly accurate.

Both 5-FU and DAVANAT® were calculated, based on data of Table 3, to likely be stable at room temperature for a long time period, and certainly for more than three years.

Compatibility studies

A study was performed to evaluate compatibility of DAVANAT® and 5-FU with a clinical infusion and filtration system used for preparation of chemotherapeutics. Solutions were prepared in accordance with standard pharmaceutical instructions. Stock solutions of 65 mL of 8.9-mg/mL DAVANAT® and 15.9-mg/mL 5-FU were prepared in duplicate and placed in 60-mL syringes.

A 5-mL aliquot was taken from each preparation, and the remainder was filtered through either a 0.22-mm MILLEX GP or GV filter and injected into an emptied Baxter 100-mL 0.9%-saline bag using two sterile 30-mL syringes with 16-gauge

Table 3: A compatibility study with 5-FU at elevated temperatures and pH of 9.2 predicts the stability of the combination of 5-FU and DAVANAT® for over 12 months

Temperature C°	DAVANAT® Clinical Solution Thermal Stability		
	10 days	**30 days**	**60 days**
5 (Control)	100%	100%	100%
25	100%	100%	100%
40	95%	94%	91%
65	93%	92%	87%

needles. Each preparation was maintained in its respective container at room temperature for 20 hours, at which time collections were made at 5 minutes and 25 minutes by pumping the sample through an 84" Primary Solution Tubing Set at a flow rate of 2 mL per minute. A Sabratek i.v. pump was used for the sample collection. The target formulation concentrations of 8.9 mg/mL of DAVANAT® and 15.9 mg/mL of 5-FU used in the study were selected based on maximum dose expected for humans weighing 75 kg.

All analysis sets met the system-suitability criteria described in the individual methods; the results are shown in Tables 3 and 4. Appearance and pH were noted for each of the samples. All solutions were clear and colorless. No interference was introduced by either filter type that would affect the quantification of DAVANAT® and 5-FU. The actual measures of mg/mL found and percent-recovery obtained are reported in Tables 4 and 5.

Table 4. Recovery results for MILLEX GP, 0.22-μm filter

Component	pH	Sample Time (min)	Actual mg/mL	Theoretical mg/mL	% Recovery
Control	9.01	0	9.773	8.9	109.8
DAVANAT®	9.00	5	9.694	8.9	108.9
	9.00	25	9.094	8.9	102.2
Control	9.01	0	15.576	15.9	98.0
5-FU	9.00	5	15.548	15.9	97.8
	9.00	25	15.302	15.9	96.2

Table 5. Recovery results for MILLEX GV, 0.22-μm filter

Component	pH	Sample Time (min)	Actual mg/mL	Theoretical mg/mL	% Recovery
Control	9.02	0	8.988	8.9	101.0
DAVANAT®	8.98	5	9.172	8.9	103.1
	8.99	25	8.889	8.9	99.9
Control	9.02	0	15.589	15.9	98.0
5-FU	8.98	5	15.751	15.9	99.1
	8.99	25	15.719	15.9	98.9

The quantitative results for DAVANAT® and 5-FU, using either a MILLEX GP 0.22-µm filter or MILLEX GV 0.22, show that the filtered-fraction solutions were within 10 percent of the control solution. Hence both the MILLEX GP 0.22-µm and the MILLEX GV 0.22 filters and the infusion system were compatible with the combined solution of DAVANAT®-5-Fluorouracil, at the doses tested of 576 mg and 1035 mg in 60mL 0.9% sodium chloride for injection.

Pre-clinical Data

Results of the preclinical studies of DAVANAT® alone and in combination with 5-FU are described in the next chapter. Here we briefly mention some other chemotherapy drugs tested in combination with DAVANAT®: Camptosar®, or irinotecan (IRI); Adriamycin®, or doxorubicin (DOX); Platinol®, or cisplatin (CIS); and Eloxatin®, or oxaliplatin (OXA).

The preclinical studies employed nude mice at conventional-chemotherapy drugs' tolerated doses, calculated from published literature to give approximately 50-percent inhibition of tumor growth. DAVANAT® alone and a drug alone were given intravenously once every four days for a total of four injections (q4d x 4). DAVANAT® was used at 30 and 120 mg/kg/dose, or co-administered as one injection on the same treatment schedule. Three tumors—colon tumors COLO 205 and HT-29, and mammary tumor ZR-75-1—were used.

The selected doses were well below LD_{50}, and weight loss was insignificant. Implanted tumors grew well in all mice, and when tumors reached an average size of 100–150 mg, the treatment started. The median time required to quadruple of tumor volume was used to estimate growth rate and the effect of DAVANAT® on the chemotherapy. The average experiment lasted 35 to 60 days. The results for 14 studies with three tumor models can be summarized as follows: The average overall improvement of tumor growth inhibition was equal to 26 percent, with average improvement of 27 percent, 28 percent, 25 percent, 22 percent, and 14 percent for 5-FU, irinotecan, cisplatin, oxaliplatin, and doxorubicin, respectively.

References

1. Barondes SH, Castronovo V, Cooper DN, Cummings RD, Drickamer K, Feizi T, Gitt MA, Hirabayashi J, Hughes C, Kasai K: Galectins: a family of beta-galactosidase-binding lectins. Cell, 1994, 76:597-598.

2. Beuth, J., Ko, H.L., Oette, K., Pulverer, G., Roszkowski, K. and Uhlenbruck G. 1987. Inhibition of liver metastasis in mice by blocking hepatocyte lectins with arabinogalactan infusions and D-galactose. J. Cancer Res. Clin. Oncol. 113: 51-55.
3. Bocci G, Danesi R, Di Paolo AD, Innocenti F, Allegrini G, Falcone A, Melosi A, Battistoni M, Barsanti G, Conte PF and Del Tacca M: Comparative pharmacokinetic analysis of 5-fluorouracil and its major metabolite 5-fluoro-5,6-dihydrouracil after conventional and reduced test dose in cancer patients. Clin. Cancer Res. 2000, 6:3032-3037.
4. Fukuda, M.N., Ohyama, C., Lowitz, K., Matsuo, O., Pasqualini, R. and Ruoslahti, E. 2000. A peptide mimic of E-selectin ligand inhibits sialyl Lewis X-dependent lung colonization of tumor cells. Cancer Res. 60: 450-456.
5. Gabius HJ. Detection and functions of mammalian lectins- with emphasis on membrane lectins. 1991, Biochim. Biophys. Acta, 1071:1-18.
6. IARC Monographs on the Evaluation of the Carcinogenic Risk of Chemicals to Humans. Vol. 26. Some Antineoplastic and Immunosuppressive Agents. WHO, International Agency for Research on Cancer, May 1981, p. 217- 235.
7. Kannagi, R. 1997. Carbohydrate-mediated cell adhesion involved in hematogenous metastasis of cancer. Glycoconj. J. 14: 577-584.
8. Lehmann S, Kuchler S, Theveniau M, Vincendon G, Zanetta JP: An endogenous lectin and one of its neuronal glycoprotein ligands are involved in contact guidance of neuron migration. 1990, Proc. Natl. Acad. Sci. USA, 87:6455-6459.
9. Novel Approaches to Selective Treatments of Human Solid Tumors: Laboratory and Clinical Correlation. 1994. Adv. Exp. Med. Biol., v. 339 (Youcef M. Rustum, Ed.), Plenum Press, 319 p.
10. Ohyama, C., Tsuboi, S. and Fukuda, M. 1999. Dual roles of sialyl Lewis X oligosaccharides in tumor metastasis and rejection by natural killer cells. EMBO J. 18: 1516-1525.
11. Platt, D. and Raz, A. 1992. Modulation of the lung colonization of B16-F1 melanoma cells by citrus pectin. J. Natl. Cancer Inst. 84: 438-442.
12. Raz A, and Lotan R., Endogenous galactoside-binding lectins: a new class of functional tumor cell surface molecules related to metastasis, 1987, Cancer Metastasis Review, 6:433-452.
13. Zhang, J., Nakayama, J., Ohyama, C., Suzuki, A., Fukuda, M., Fukuda, M.N. 2002. Sialyl Lewis X-dependent lung colonization on B16 melanoma cells through a selectin-like endothelian receptor distinct from E- or P-selectin. Cancer Res. 62: 4194-4198.

Chapter 4

DAVANAT® and Colon Cancer: Preclinical and Clinical (Phase I) Studies

Anatole A. Klyosov, Eliezer Zomer, and David Platt

Pro-Pharmaceuticals, 189 Well Avenue, Newton, MA 02459

Introduction

DAVANAT® is being developed to enhance the efficacy of anti-neoplastic drugs by targeting specific receptors, possibly, but not necessarily, galectins, present on cancer cells. Since DAVANAT® consists of a polymeric mannose backbone carrying galactose side chains, and therefore allegedly capable to interact with some galectins, that is receptors specific for galactose residues (both β- and α-galactose, as it was recently shown), the initial idea was to use DAVANAT® as a kind of a chaperon along with 5-fluorouracil (5-FU) to facilitate its delivery into the cancer cell. A choice of 5-FU as a chemotherapy drug in a combination with DAVANAT® was based on a fact that for over 40 years 5-FU has been the standard first-line agent in the treatment of metastatic colorectal cancer.

Preliminary animal studies with a variety of soluble galactomannan oligomers from various plant sources have shown promising response to the combination therapy of 5-FU along with 1,4-β-D-galactomannan of a certain molecular weight, with mannose to galactose ratio of 1.7. Fortunately, such a galactomannan can be obtained from a readily available source, namely Guar gum (*Cyamopsis tetragonoloba*), by controlled partial depolymerization. The water-soluble galactomannan of molecular weight 55,000 Da was thoroughly characterized and selected for process optimization and manufacturing of DAVANAT®, and tested *in vivo* in animals for both efficacy and overall reduction of toxicity. In particular, DAVANAT® has been shown to increase the anti-tumor activity of 5-FU in mice. Based on data obtained it was proposed that DAVANAT® may potentiate the efficacy of 5-FU by changing its tissue distribution and targeting its delivery to the tumor cell.

In June of 2003 the FDA has allowed to test DAVANAT® in a formulation with and without 5-FU in Phase I clinical trials. This multi-center, open label study was designed for cancer patients with different type of solid tumors who have failed standard, approved surgical, radiation, and chemotherapeutical regimen.

In November of 2003 the US Patent (No. 6,645,946) was granted to Pro-Pharmaceuticals that protects methods and compositions with respect to galactomannan co-administered with a therapeutic agent (5-FU among them) for reducing toxicity of the agent.

Chemistry of DAVANAT®

DAVANAT® is the galactomannan isolated from seeds of Guar gum, *Cyamopsis tetragonoloba*, and subjected to a controlled partial chemical degradation (Klyosov & Platt, US Patent No. 6,645,946). A backbone of the galactomannan is composed of (1→4)-linked β-D-mannopyranosyl units, to which single α-D-galactopyranosyl is attached by (1→6)-linkage.

Chemical names of the galactomannan are 1,4-β-D-Galactomannan, or [(1→6)-α-D-galacto-(1→4)-β-D-mannan].

Structural formula of the galactomannan:

α-D-Gal*p* **α-D-Gal*p***
1 **1**
↓ **↓**
6 **6**
→4)-β-D-Man*p*-(1→4)-β-D-Man*p*-(1→4)-β-D-Man*p*-(1→4)-β-D-Man*p*-(1→

The average repeating unit of DAVANAT® consists of seventeen β-D-Man residues and ten α-D-Gal residues (Man/Gal ratio is 1.7), and an average polymeric molecule contains approximately 12 of such repeating units (average molecular weight 60,000 Da).

Stereochemical configuration

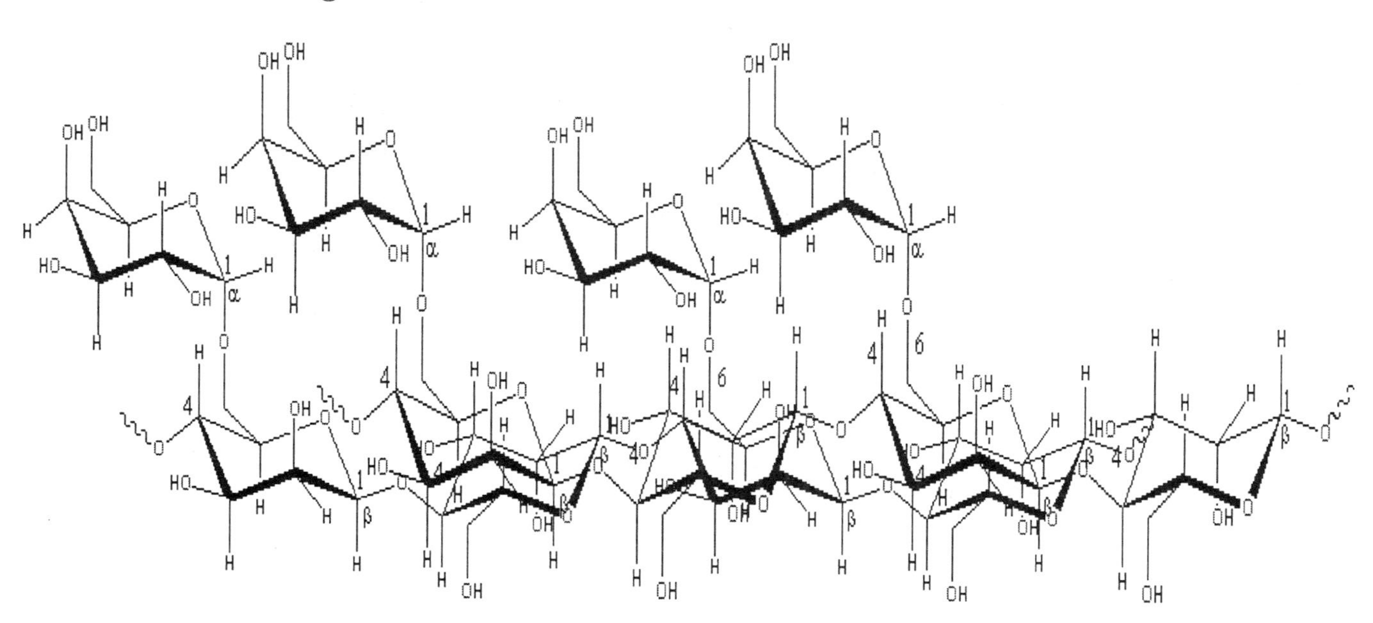

Isolation, Purification, Identification

Isolation and purification procedure of DAVANAT®, when started from a commercial *Cyamopsis tetragonoloba* guar gum flour, contains five principal steps:

- Aqueous extraction of galactomannan
- Controlled partial depolymerization
- Recovering as insoluble copper complex
- Recovering from the copper complex
- Ethanol repeated precipitation

The final yield of DAVANAT® is typically 50% from the weight of guar gum flour

For the manufacturing of DAVANAT®, extra specifications have been established with the supplier of Guar gum to warrant a uniform and consistent product suitable for pharmaceutical manufacturing.

The purification procedure results in a pure DAVANAT® as white powder with solubility in water more than 60 mg/mL, molecular weight of 54,000 Da, and mannose/galactose ratio of 1.7.

^{1}H-NMR and ^{3}H-NMR spectra of DAVANAT® are described in the preceding chapter.

DAVANAT® Mutagenicity Studies

Mutagenicity studies that were conducted included two Ames bacterial reverse mutation assays in which DAVANAT® was evaluated by itself in the first study and combined with 5-FU in the second study. The test articles in both studies were evaluated in bacterial assays using *Salmonella typhimurium* strains TA97a, TA98, TA100, TA1535 and *Escherichia coli* strain WP2 *uvr*A (pKM101), both in the presence and absence of an exogenous metabolic activation system. No evidence of mutagenic activity was detected in either DAVANAT® or DAVANAT® + 5-FU.Evidence of toxicity was not detected at any concentration level, namely from 5 to 2,000 μg per plate.

5-FU: Chemistry and Drug Description

5-FU remains the first line chemotherapeutic agent for the treatment of metastatic colorectal cancer. Colorectal adenocarcinoma is the second leading cause of cancer deaths accounting for 11 percent of the total number. An estimated 146,940 cases occurred in 2004 in the United States (4). Approximately 25% of patients present with distant metastases, and an additional 20% develop metastases during their lifetime (5, 6). The mortality rate is about 22/100,000 compared to 206.5/100,000 total of cancer mortality (4). Although survival from this disease has improved by approximately 8% over the past 2 decades largely due to earlier detection, advances in the treatment of advanced colorectal cancer are largely attributed to modifications of or additions to regimens centered around 5-FU.

5-Fluorouracil (5-FU) is a highly toxic drug with a narrow margin of safety. It is effective against carcinoma of the colon, rectum, breast, stomach and pancreas, slows or stops growth of cancer cells and their spread in the body. 5-FU is a fluorinated pyrimidine. Chemically, it is 5-fluoro-2,4 (1H,3H)-pyrimidinedione, or 2,4-dioxo-5-fluoropyrimidine, a relatively small molecule:

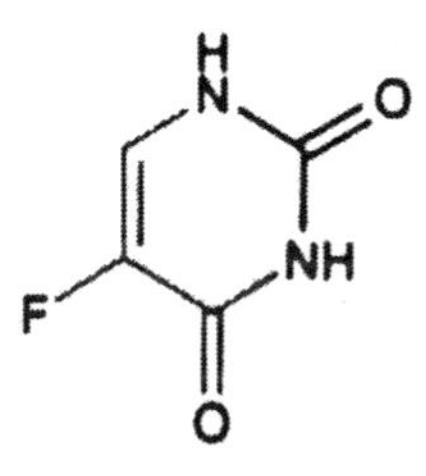

The method of chemical synthesis of 5-FU was described in the US Patent Nos. 2,802,005 and 2,885,396, issued in 1957 and 1959, and the patents' protection have long expired.

The following description of 5-FU is referenced in the Physician's Drug Reference which we consider as a reliable and FDA accepted source. 5-FU blocks the methylation reaction of deoxyuridylic acid to thymidylic acid, and thereby interferes with the synthesis of DNA and to a lesser extent inhibits the formation of RNA. Since DNA and RNA are essential for cell division and grows, the effect of 5-FU creates a thymine deficiency, which provokes unbalanced growth and death of the cell. The effect of DNA and RNA deprivation are most marked on those cells which grow more rapidly and which take up 5-FU at a more rapid rate, that is, cancer cells.

In patients, 7%-12% of 5-FU is excreted unchanged in the urine in six hours; of this 90% is excreted in the first hour. The remaining 88%-93% of the administered dose is metabolized, primarily in the liver. The inactive metabolites are excreted in the urine over the next 3 to 4 hours. Overall, 90% of the dose is accounted for during the first 24 hrs following i.v. administration. The mean half-life of elimination of 5-FU from plasma is approximately 16 minutes, with a range of 8 to 20 minutes, and is dose dependent. No intact 5-FU can be detected in the plasma three hours after i.v. administration.

5-FU is typically manufactured for intravenous administration. Following the injection, 5-FU distributes into tumors, intestinal mucosa, bone marrow, liver and other tissues throughout the body. It diffuses readily across the blood-brain barrier and distributes into cerebrospinal fluid and brain tissue.

The drug shows an extreme toxicity. Severe hematological toxicity, gastrointestinal hemorrhage and even death may result from the use of 5-FU, despite meticulous selection of patients and careful adjustment of dosage. Generally, a therapeutic response is unlikely to occur without some evidence of toxicity. Severe toxicity is more likely in poor risk patients, although fatalities may be encountered occasionally even with patients in relatively good condition.

Acute nausea and vomiting occurs frequently and may be severe. Stomatitis (sores in the mouth) and esophagopharyngitis may occur, and some patients have fever blister-like sores. These effects may be severe leading to gastrointestinal ulceration and bleeding. Anorexia, diarrhea, temporary loss of hair (alopecia), skin rash, increased skin and/or vein darkening, fatigue (2-3 days after injection) are commonly seen during therapy. Almost all patients lose all body hair, though temporarily, since 5-FU kills the hair cells.

Besides, there is a high incidence of bone marrow depression (decreased blood count), primarily of leukocytes. Rapidly falling white blood cell count often occurs during treatment with 5-FU, thereby diminishing the infection fighting activity of the body. 5-FU therapy is counter indicated for patients in a poor nutritional state and those with depressed bone marrow function. It is incompatible with doxorubicin, another powerful anti-cancer chemotherapeutic drug.

The administration of 5-FU has been associated with the occurrence of hand-foot syndrome, resulting in the symmetrical swelling of the palms and soles. This syndrome causes a tingling sensation of hands and feet, which may progress to pain when holding objects or walking in the few days following administration.

Animal studies of 5-FU have shown a number of negative causes, such as teratogenic effects and malformations, including skeletal defects.

Despite the described side effects, 5-FU has been the mainstay for therapy of colorectal cancer for the last 2 decades. Generally, fluoropyrimidines, especially 5-FU, have been used in standard chemotherapy regimens for a number of solid tumors including colorectal, breast, non-small cell carcinoma of the lung (NSCCL), gastric, pancreatic, ovarian and head and neck tumors (1). Schedule modification of 5-FU administration including prolonged intravenous infusion, and pharmacokinetic modulation has produced improved response rates and tolerability; however, this has not always translated into improved survival (2-3).

Bolus intravenous injection (i.v.) therapy, which was the standard of care until approximately a decade ago, produced response rates ranging from 11 to 18% (4). The intensity of exposure to 5-FU as measured by the area under the concentration vs. the time curve, correlates well with anti-tumor activity but also with toxicity. Regimens which include 5 days of treatment every 3 weeks produce mucositis, diarrhea and neutropenia (4). Regimens aimed at enhancing the duration of thymidylate synthase inhibition by continuous i.v. infusion or bolus injection in combination with leucovorin (folinic acid) have improved response rates ranging from 22 to 38% (2-4). Continuous infusion regimens can result in hand-foot syndrome (see above) in approximately 20% of patients, mucositis is similar in continuous infusion compared with bolus therapy, but diarrhea and neutropenia are less frequent by the continuous infusion route (4).

Despite advances in chemotherapeutic regimens for metastatic colorectal cancer as well as other solid tumors, patients refractory to these therapies represent a large unmet medical need. Many clinical trials are currently underway investigating new chemotherapeutic agents and regimens, antisense oligonucleotides, as well as biologics including monoclonal antibodies to various cell surface tumor proteins. It is reasonable to hypothesize that more specific targeting of efficacious chemotherapeutic agents, such as 5-FU, to tumor cells could result in higher response rates along with reduced toxicity. In vitro studies have shown that tumor galactoside-binding proteins on the cell surface interact with terminal D-galactoside residues located on adjacent cells, thus mediating cellular attachment and the potential for metastasis. Hence, the initial idea for developing DAVANAT® which represents a relatively rigid backbone (poly-mannose) carrying galactose residues. Phase I clinical trial described below investigates the safety and tolerability of DAVANAT® , which has been shown in preclinical studies, also described below, to enhance the efficacy of 5-FU in mouse models of human colorectal cancer without increasing toxicity.

Pre-Clinical *in vivo* Toxicity Studies

Non-Clinical Safety Studies

Acute toxicology studies of DAVANAT® were performed in mice, rats and dogs and subchronic toxicology studies were performed in rats and dogs. The following is a summary of the findings of these studies.

Toxicity on mice

LD_{50} for 5-FU in mice is equal to 340 mg/kg, or 1,020 mg/m^2 i.v. (Physician's Desk Reference, 1994, p.1925). In accord with this value, 5-FU in the single i.v. dose of 409 mg/kg (1227 mg/m^2) in our experiments caused death in 3/5 mice after 13-16 days. However, single i.v. doses of 112 mg/kg (336 mg/m^2) of DAVANAT® or 417/222 mg/kg (1251/666 mg/m^2) 5-FU/ DAVANAT® produced no clinical signs of toxicity, death or decreased weight gain in mice (n=5 /group).

Toxicity on rats

Single dose study. LD_{50} for 5-FU in rats is equal to 165 mg/kg, or 990 mg/m^2 i.v. (Physician's Desk Reference, 1994, p.1925). In our experiments, in rats (n=5/sex/group), single i.v. doses of 140 mg/kg (840 mg/m^2) of 5-FU alone, 72 mg/kg (432 mg/m^2) of DAVANAT® alone, or 140/72 mg/kg of the combination produced no deaths in any of the groups. Toxic changes in body weight, feed consumption, and hematology were somewhat less severe in the rats injected with the combination, versus those injected with 5-FU alone.

Subchronic studies. Three subchronic rat studies were performed with i.v. injection for four consecutive days with 28-day or 52-day recovery periods.

In the first, 28-day study, research grade unfiltered DAVANAT® alone and in combination with 5-FU was injected. However, it was found that this low grade DAVANAT® produced mild granulomatous foci in the rat lungs. In this study, low grade DAVANAT® was injected singly daily i.v. dose of 48 mg/kg (288 mg/m^2) for four consecutive days, or 5-FU in the same regime at the same dose of 48 mg/kg, or three following combinations of 5-FU/ DAVANAT® in the same regime at 24/13 mg/kg, 36/19 mg/kg, or 48/25 mg/kg at a rate of 1 mL/min for four consecutive days. The four doses of 5-FU (192 mg/kg total) produced mortality in 7 out of 8 rats. With combinations of 24/13, 36/19 and 48/25 mg/kg 5-FU/ DAVANAT® mortality was 1/8, 2/8 and 7/8, respectively. Alopecia, severe transient decreases in body weight and feed consumption, and transient depression of erythrocyte parameters and platelets were observed in all groups receiving 5-FU. Lesions in unscheduled-death animals were primarily due to the expected action of 5-FU on the hematopoietic and lymphoid systems, with secondary bacterial invasion and disseminated hemorrhage due to effects on coagulation. At the study day 5 necropsy, lesions included pronounced hypocellularity of the bone marrow, lymphoid atrophy of the thymus and atrophy of villi in the various segments of the small intestine. All of it is considered due to the expected action of 5-FU. One can see that the low grade DAVANAT® albeit produced mold complications in lungs, did not increase toxicity in combination with of 5-FU compared with 5-FU alone.

In the second, 56-day study, employing groups of 15 male and female rats, GMP grade DAVANAT® has been used at 48 and 96 mg/kg/day (288 or 576 mg/m^2/day) dose. It was i.v. injected alone for four consecutive days, at a slower rate (0.1 mL/min, rather than 1.0 mL/min in the preceding study). Five animals of each sex were sacrificed at days 5, 28 and 56.

Only one (out of 30 rats examined) isolated and non-consistent histologic finding in the lung was associated with trace-level granulomatous inflammation of the lung of rats that-were i.v. dosed for four consecutive days with 96 mg/kg/day of DAVANAT® and sacrificed at Study Day 5 (but not at Day 28 or 56). Male rats given DAVANAT® at 96 mg/kg (a dose 2-fold greater than the highest dose used in the first study) and sacrificed on Day 28 had an increased incidence of trace-level interstitial inflammation in the lungs but had no histologic evidence of granulomatous inflammation. Male rats given DAVANAT® at 96 mg/kg and sacrificed on Day 56 had an increased incidence of trace-level alveolar macrophage accumulation but again, no histologic evidence of granulomatous inflammation. However, those lesions (pulmonary interstitial inflammation and alveolar macrophage accumulation) were morphologically similar to lesions that are commonly encountered as incidental findings in the lungs of laboratory rats. Except the few rats mentioned, none of others showed any histologic evidence of said lesions.

The third study was extended to a separate 56-day injection and recovery period using GMP grade DAVANAT® at 96 and 192 mg/kg/day (576 and 1152 mg/m^2/day) dose i.v. injected once daily for four consecutive days at a fast rate of 1 mL/min (at a concentration of 9.6 and 19.2 mg/mL, respectively). None of the 30 rats injected with the 96 mg/kg/day dose of DAVANAT® were affected. Rats injected with 192 mg/kg/day of DAVANAT® have shown minimal to mild granulomatous focal inflammation and minimal amounts of birefringent and/or non-birefringent foreign bodies in lung tissues. These findings were most notable in animals injected with the unfiltered DAVANAT® solution at 192 mg/kg and sacrificed on Day 5 (24 hours after the last dose). The inflammation is considered to be reversible in rats, based on the fact that the high incidence of mild interstitial inflammation observed on Study Day 5 was reduced to control levels, and the presence of foreign bodies was also decreased substantially, by Study Day 28. In addition, the cellular constituents of the granulomatous foci changed to a more quiescent pattern with time and became less prominent by Study Day 56. However, complete resolution was not evident by Study Day 56.

Based on these studies it was concluded that GMP DAVANAT® clinical solutions, up to a dose of 15 mg/kg (555 mg/m^2, injection solution 9 mg/mL), is safe for human use and poses no undue risk to health. The starting dose in human subjects was suggested to be 30 mg/m^2, which is approximately 10-fold lower than the minimal 288 mg/m^2 dose of DAVANAT® (48 mg/kg) used in the rat repeat dose safety study described above, that produced the same low level of lung inflammation noted in control animals injected with saline.

Toxicity on dogs

Single dose study. LD_{50} for 5-FU in dogs is equal to 31.5 mg/kg, or 630 mg/m^2 i.v. (Physician's Desk Reference, 1994, p.1925). In our experiments, in beagle dogs (n=2/group), single i.v. injections of 28.5 mg/kg (570 mg/m^2) of 5-FU alone, 15 mg/kg (300 mg/m^2) of DAVANAT® alone, or 28.5/15 mg/kg of the combination resulted in the death of 2/2 dogs injected with 5-FU, 2/4 dogs injected with the combination and 0/2 dogs (no deaths) receiving DAVANAT® alone. The surviving dogs remained clinically normal and had no treatment-related changes in body weight, feed consumption, ECG tracings, clinical pathology or gross tissue changes during the 21-day study period.

Subchronic studies. A 28-day subchronic beagle dog study was performed with i.v. single daily injection for four consecutive days. Dogs in six groups, each group contained 4 males and 4 females, were injected with saline (control group), 5-FU alone (6 mg/kg/day, or 120 mg/m^2/day), DAVANAT® alone (12 mg/kg/day, or 240 mg/m^2/day), or three following combinations of 5-FU/ DAVANAT® in the same regime at 4/2 mg/kg, 6/3.2 mg/kg, or 6/6 mg/kg.

There were no death in dogs injected with DAVANAT® alone. With 5-FU alone (6 mg/kg/day), mortality (dead or moribund sacrificed) occurred in all males (4/4 dead) between Study days 2 and 5. With the 4/2 mg/kg/day combination of 5-FU/ DAVANAT®, mortality occurred in ¾ males. With the 6/3.2 mg/kg/day combination, mortality was in 2/4 males, and with the 6/6 mg/kg/day combination, mortality occurred in ¼ females, while all other females and all males were alive (7/8 were alive).

Adverse effects in 5-FU groups included ataxia, prostration, vocalization, convulsions, tremors, hypersensitivity to touch, aggressive behavior (resulting in only three doses being given to the 6/3.2 of 5-FU/ DAVANAT® group), emesis, salivation, soft stools and decreased red cell parameters and platelet counts. Gross/microscopic changes in unscheduled-death animals included congestion of one or more organs (suggesting cardiovascular dysfunction) and atrophy of the mucosa of the GI tract. At the Study Day 5 necropsy, treatment-related histological changes were largely limited to the 5-FU females. At Study Day 29 necropsy, there were no remarkable changes in the survivors. DAVANAT® at 240 mg/m^2/day produced no observed adverse effects.

Pre-Clinical Efficacy Studies, in mice. Enhancement of Anti-Cancer Efficacy of 5-Fluorouracil When Co-administered with DAVANAT®

In this section, we describe the principal results of three pre-clinical efficacy studies that employed DAVANAT®. We have shown that co-administration of DAVANAT® along with 5-FU by i.v. injection to mice bearing human colon tumors (COLO 205 and HT-29) significantly increased efficacy of the 5-FU. This section describes the principal results of three separate pre-clinical studies, employing (1) COLO 205-bearing mice at one dose of DAVANAT® and 5-FU, (2) COLO 205-bearing mice at escalating DAVANAT® doses in combination with 5-FU, and (3) HT-29-bearing mice at two DAVANAT® doses in combination with 5-FU with and without leucovorin.

The studies have shown a DAVANAT® dose-related effect with a maximum efficacy at 120 mg/kg/dose (360 mg/m^2/dose) of DAVANAT® . Effect of an additional oral administration of leucovorin was minimal. Combination of DAVANAT® with 5-FU, compared to 5-FU alone, resulted in the decrease of median tumor volume in mice to 17%-65% and the increase of mean survival time (days) to 150%-190%, respectively.

Study 1. Effect of DAVANAT® on efficacy of 5-FU in NCr-*nu* mice with subcutaneous (s.c.) implants of COLO 205 human colon tumor at one DAVANAT® dose.

This study employed a relatively high dose of 5-FU (75 mg/kg/dose) that exceeded the maximum tolerated dose in mice. The data are shown in Table 3 and Figure 1. Their statistical analysis is illustrated in Table 4.

Table 3. Tumor volumes, dynamics of tumor volumes, and survival time for COLO 205 human tumor-bearing mice treated with 5-FU (75 mg/kg/dose), DAVANAT® (120 mg/kg/dose) and their combination (75 mg/kg/dose and 120 mg/kg/dose, respectively) on q4d x 3 schedule

	Saline (control)	5-FU 75 mg/kg	DAVANAT® 120 mg/kg	5-FU + DAVANAT® 75 mg/kg + 120 mg/kg
Median time to quadrupling of tumor volume (days)	12.5	23.7	15.5	56.0
Median tumor volume (mm^3) at the endpoint (56 days after treatment initiation)	2058	2254	1813	379
Mean survival time (days)	14.2	23.7	19.2	44.2

Untreated control tumors grew well in all mice, with the median number of days for quadrupling the tumor volume equal to 12.5 days. There was no tumor regression after 56 days of the study, and there was practically no tumor reduction. Median tumor volume increased from 111 mm^3 at treatment initiation (in this case with saline only) to 2058 mm^3 after 5-8 weeks. Mean survival time was equal to 14.2 days.

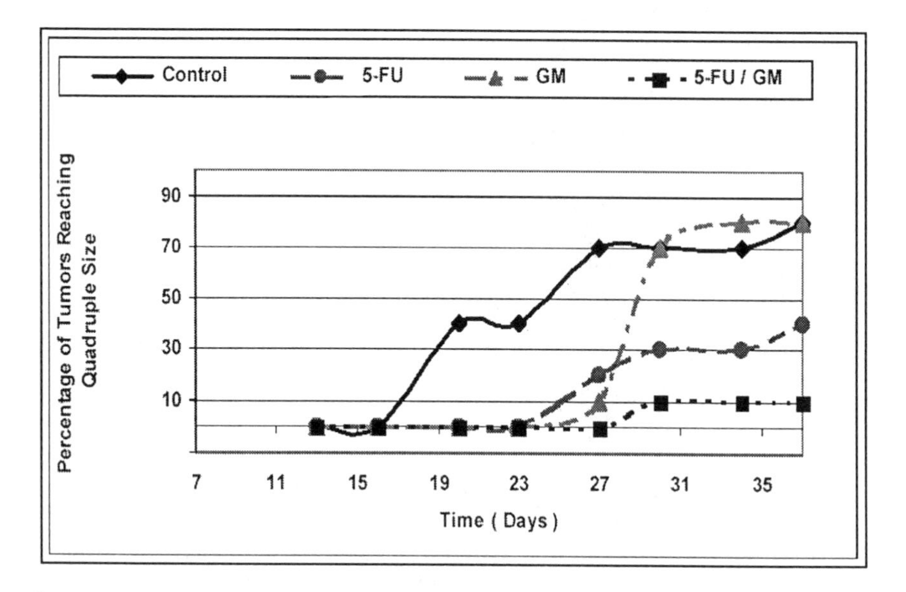

Figure 1. Effect of DAVANAT® (120 mg/kg/dose), 5-FU (75 mg/kg/dose) and their combination in q4dx3 regimen on tumor volume in NCr-nu mice with s.c. implants of COLO 205 human colon tumor (Study 1).

A dosage of 75 mg/kg/dose of 5-FU (that is, 225 mg/kg total dose over 8 days) was in excess of the maximum tolerated dosage and produced treatment-related deaths for three mice out of ten within two weeks. The treatment caused a delay in the median to quadrupling the tumor volume from 12.5 to 23.7 days. Again, there was no tumor regression after 56 days of the study; however, two relatively small tumors were observed that grew from 75 mm^3 each, at initiation of treatment, to 126 mm^3 and 567 mm^3 by the end of the study. Median tumor volume increased from 101 mm^3 at treatment initiation to 2254 mm^3 after 56 days of the study. Mean survival time shifted from 14.2 days (untreated control animals) to 23.7 days.

DAVANAT®, at a dosage of 120 mg/kg/dose administered alone on a q4d (one every 4 days) x 3 schedule, was well-tolerated. No deaths or body weight loss was observed. The median number of days to quadrupling the tumor volume equaled 15.5 days, which is slightly longer than the value for untreated animals (12.5 days). There was no tumor regression after 56 days of study, however, two relatively small tumors (compared to median tumor volume) were observed that grew from 100 mm^3 and 126 mm^3, at initiation of treatment, to 270 mm^3 and 729 m^3, respectively, by the end of the study. Median tumor volume increased from 100 mm^3, at treatment initiation, to 1813 mm^3 after 56 days of the study, which is noticeably less compared to 2058 mm^3 for untreated animals, and 2254 mm^3 for 5-FU (75 mg/kg/dose)-treated animals. Mean survival time was prolonged from 14.2 days (untreated control animals) to 19.2 days.

Table 4. Statistical analysis of tumor volumes for COLO 205 human tumor-bearing mice treated with 5-FU (75 mg/kg/dose), DAVANAT® (120 mg/kg/dose) and their combination on q4d x 3 schedule, for all data points (from the day of treatment initiation onward)

Treatment Group	N (observations)	Mean	p-value*	p-value**
Control	146	1174	--	<0.0001
DAVANAT®	162	962	0.0237	<0.0001
5-FU	118	652	<0.0001	0.0004
5-FU + DAVANAT®	103	358	<0.0001	--

* one-sided t-test p-value, compared to the control (untreated animals)
** one-sided t-test p-value, compared to the 5-FU + DAVANAT® treatment.

Co-administration of DAVANAT® (120 mg/kg/dose) and 5-FU (75 mg/kg/dose) on a q4d x 3 schedule resulted in a remarkable effect, which caused a significant delay in quadrupling of the tumor volume from 12.5 days for untreated animals (control) and 23.7 and 15.5 days for 5-FU alone and GM alone, respectively, to 56.0 days for their combination. There was one tumor that completely disappeared by the end of the study. Two more tumors were relatively small in size, less than 20% that the control value, by the end of the study. Overall, median tumor volume increased from 111 mm^3 at treatment initiation to only 379 mm^3 after 56 days of study, a value significantly less than that for untreated animals or animals treated with 5-FU alone. Mean survival time increased from 14.2 days (untreated control animals) and 23.7 days (5-FU treatment) to 44.2 days for the combination treatment.

Study 2. Effect of DAVANAT® on efficacy of 5-FU in NCr-*nu* mice with s.c. implants of COLO 205 human colon tumor at an escalated DAVANAT® dose.

The principal differences from the first study were: (i) there were four consecutive injections, not three, (ii) there were four doses of DAVANAT® tested, not one, and (iii) 5-FU was administered (i.v.) at a more well-tolerated dose of 5-FU (48 mg/kg), compared to that used in the first study (see above). The data are shown in Table 5 and Figure 2. Results of their statistical analysis are provided below in Table 6.

No mice died in this study. As in the preceding study, untreated control tumors grew well in all mice. The median number of days to quadrupling the tumor volume equaled 7.2 days. No tumor regression or reduction occurred after 13 days of the study. Median tumor volume increased from 162 mm^3, at treatment initiation (in this case with saline only), to 1288 mm^3 after 13 days.

A dosage of 48 mg/kg/dose of 5-FU (192 mg/kg total dose over 12 days) was well tolerated and produced some growth delay in the median number of days to quadrupling the tumor volume, increasing it from 7.2 to 8.7 days. Two tumors in the group of 10 mice were significantly (three times or more) smaller, compared with the median tumor size, after 13 days of treatment, growing from 100 and 163 mm^3 at initiation of treatment to 270 mm^3 and 138 mm^3, respectively, by the end of the study. Median tumor volume increased from 172 mm^3 at treatment initiation to 800 mm^3 after 13 days of the study, which was less than the control value 1288 mm^3.

Co-administration of 5-FU (48 mg/kg/dose) and DAVANAT® (6, 30, 120, and 600 mg/kg/dose) on a q4d x 4 schedule was well-tolerated at all dosages tested and caused a significant delay in quadrupling the tumor volume, from 7.2 days for untreated animals (control) and 8.7 and 6.9 days for 5-FU alone and DAVANAT® alone, to 14.8, 13.5, 16.5, and 16.2 days, respectively (Table 5). The best results were obtained with a combination of 5-FU and 120 mg/kg/dose DAVANAT®, which resulted in a median tumor volume of 540 mm^3 at day 13, the day after the final day of treatment, compared with that of 800 mm^3 for 5-FU treatment alone. Also, the median number of days for quadrupling the tumor volume was almost twice as much for the 5-FU plus DAVANAT® 120 mg/kg/dose than for the 5-FU alone (Table 5).

Table 5. Dynamics of tumor volumes for COLO 205 human tumor-bearing mice treated with 5-FU (48 mg/kg/dose), DAVANAT®, and their combination on q4d x 4 treatment schedule

	Saline, control	5-FU 48 mg/kg	Davanat® 120 mg/kg	5-FU (48 mg/kg) + DAVANAT® in dose (mg/kg):			
				6	30	120	600
Median time to quadrupling of tumor volume (days)	7.2	8.7	6.9	14.8	13.5	16.5	16.2
Median tumor volume (mm^3) at the endpoint (13 days after treatment initiation)	1288	800	1152	715	695	540	588

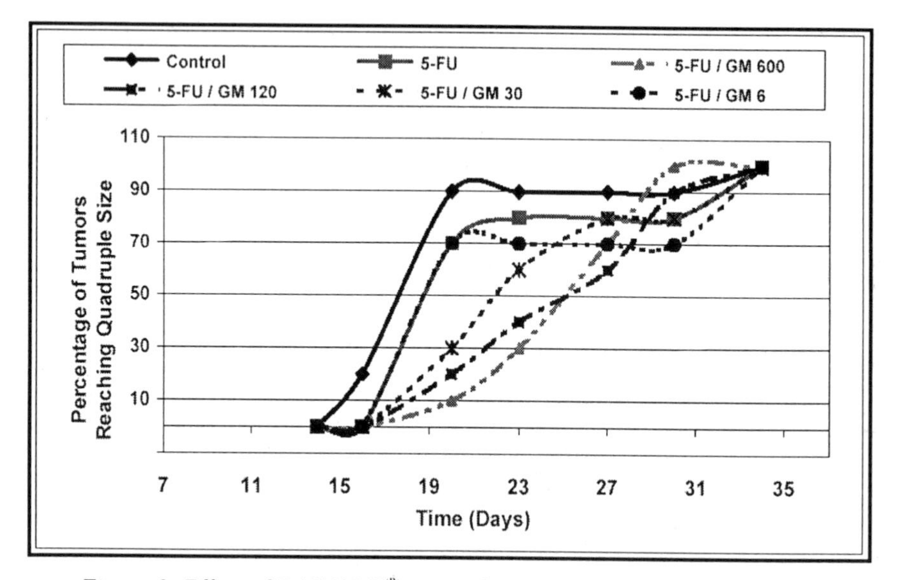

Figure 2. Effect of DAVANAT® at escalating doses (6, 30, 120 and 600 mg/kg/dose), 5-FU (48 mg/kg/dose) and their combination in q4dx4 regimen on tumor volume in NCr-nu mice with s.c. implants of COLO 205 human colon tumor (Study 2).

Table 6. Statistical analysis of tumor volume for COLO 205 human tumor-bearing mice treated with 5-FU (48 mg/kg/dose), DAVANAT® (GM), and their combination on q4dx4 schedule, at Day 13 after treatment initiation, that is the next day after treatment completion, and for all data points (from the day of treatment initiation onward).

Treatment Group	Mean, at Day 13 (N=10)	p-value[1]	Mean, for all data points (N=50)	p-value[2]
Control	1232	0.0055	652	0.00359
5-FU	736	--	419	--
5-FU + DAVANAT® 6 mg/kg/dose	646	0.2496	392	0.3149
5-FU + DAVANAT® 30 mg/kg/dose	594	0.1259	356	0.1201
5-FU + DAVANAT® 120 mg/kg/dose	544	0.0773	342	0.0740
5-FU + DAVANAT® 600 mg/kg/dose	576	0.0904	335	0.0509

[1] one-sided t-test p-value, compared to the 5-FU alone treatment, at the next day after treatment completion (Day 13).

[2] one-sided t-test p-value, compared to the 5-FU alone treatment, for all data points.

Study 3. Effect of DAVANAT® on efficacy of 5-FU in NU/NU-*nu*BR mice with s.c. implants of HT-29 human colon tumor.

The principal differences from the second study were: (i) another tumor (HT-29) was used, and (ii) leucovorin was added to the treatment regimen. The data are shown in Table 7 and Figure 3. Their statistical analysis is provided in Table 8.

As in the two preceding studies, control (untreated) tumors grew well in all mice, with a median of 13.3 days for quadrupling of the tumor volume. Median tumor volume increased from 196 mm^3, at treatment initiation (day 7 after tumor implantation), to 1318 mm^3 after 26 days.

Table 7. Tumor volumes and dynamics of tumor volumes for HT-29 human tumor-bearing mice treated with 5-FU (48 mg/kg/dose), leucovorin (25 mg/kg/dose), DAVANAT® and their combination on q4dx4 treatment schedule

	Saline (control)	DAVANAT® (120 mg/kg) + leucovorin (25 mg/kg)	5-FU (48 mg/kg) + leucovorin (25 mg/kg) + DAVANAT® (mg/kg):		
			0	30	120
Median time to quadrupling of tumor volume (days)	13.3	12.5	15.3	18.1	23.5
Median tumor volume (mm^3) at the endpoint (26 days after treatment initiation)	1318	1595	1120	1521	729

A dosage of 48 mg/kg/dose of 5-FU (192 mg/kg total dose over 12 days of the treatment) along with an oral administration of leucovorin as described above was within the maximum tolerated dosage, producing no treatment-related deaths in the group of 10 mice within three weeks. The treatment caused a delay of two days for the quadrupling of the tumor volume (from 13.3 to 15.3 days). Median tumor volume increased from 179 mm^3, at treatment initiation, to 1120 mm^3 on study day 26. Co-administration of 5-FU with DAVANAT® (30 mg/kg/dose), along with an oral dose of leucovorin as described above, brought further delay in tumor growth, particularly in the first half of the study: quadrupling of the tumor occurred from 15.3 days without DAVANAT® to 18.1 days with DAVANAT® (Table 7).

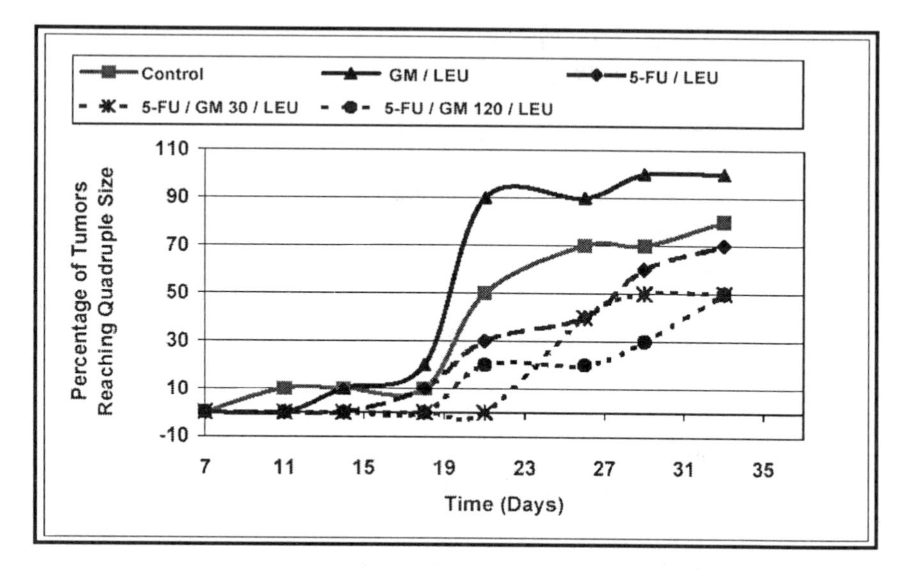

Figure 3. Effect of DAVANAT® (30 and 120 mg/kg/dose), 5-FU (48 mg/kg/dose) and leucovorin (p.o., 25 mg/kg/dose) and their combination in q4dx4 regimen on tumor volume in NU/NU-nuBR mice with s.c. implants of HT-29 human colon tumor carcinoma (Study 3).

Table 8. Statistical analysis of tumor volume for HT-29 human tumor-bearing mice treated with combinations of 5-FU (48 mg/kg/dose), leucovorin (25 mg/kg/dose), and DAVANAT® on q4dx4 schedule, at day 26 after treatment initiation (the final day of the study) and for all data points (from the day of treatment initiation onward).

Treatment Group	Mean, at Day 26 (N=7-10)	Mean, for all data points (N=71-80)	p-value, for all data points
Control	1624	801	0.0778[1] 0.4559[2]
5-FU + leucovorin	1438	633	0.0778[3]
5-FU + leucovorin + DAVANAT® 30 mg/kg/dose	1519	540	0.1796[1]
5-FU + leucovorin + DAVANAT® 120 mg/kg/dose	816	455	0.0259[1] 0.0003[3]

[1] one-sided t-test p-value, compared to the 5-FU + leucovorin treatment.
[2] two-sided t-test p-value, compared to DAVANAT® + leucovorin treatment.
[3] one-sided t-test p-value, compared to control (untreated animals).
[4] two-sided t-test p-value, compared to the 5-FU + leucovorin treatment.

Increasing the DAVANAT® dose to 120 mg/kg/dose in co-administration with 5-FU on a q4d x 4 schedule along with an oral administration of leucovorin, as described above, again produced a significant delay in quadrupling of the tumor volume: from 13.3 days for untreated control animals and 15.3 days for 5-FU/leucovorin-treated animals to 23.5 days for animals treated with all three drugs. Furthermore, when all three drugs were used in combination, one tumor completely disappeared four weeks after treatment initiation, two more tumors were of a relatively small size (269 and 352 mm^3) by the end of the study, and three additional tumors were practically stabilized at a volume of well below 1000 mm^3. Overall, median tumor volume increased from 176 mm^3 at treatment initiation to only 729 mm^3 at study day 26 (significantly less than the 1318 mm^3 for untreated animals and 1120 mm^3 for 5-FU plus leucovorin-treated animals).

Statistical Evaluation

In summary, statistical evaluation of the data indicates the following.

In the first study (COLO 205 tumors treated with a high 5-FU dose), there was a significant advantage for 5-FU plus DAVANAT® versus control ($p<0.0001$) and versus DAVANAT® alone ($p<0.0001$). The difference between 5-FU plus DAVANAT® and 5-FU was significant at a confidence level higher than 99%.

In the second study (COLO 205 tumors treated with a low 5-FU dose), there was a significant advantage for 5-FU plus DAVANAT® 120 mg/kg/dose versus 5-FU alone (with a 93%-95% confidence level), which numerically favored the combination of 5-FU with DAVANAT®.

In the third study (HT-29 tumors), there was a significant advantage for 5-FU/leucovorin plus DAVANAT® 120 mg/kg/dose versus 5-FU/leucovorin with 97% confidence level, which again numerically favored the combination of 5-FU/leucovorin with DAVANAT®.

The described studies form an experimental basis for our long-range goal of discovering mechanism-based chemotherapy agents that reduce the toxicity and increase the efficacy of co-administered antineoplastic agents. Co-administration of 5-FU with DAVANAT® with certain chemical structures (including a specific Man/Gal ratio and molecular size/weight), which might specifically interact with galactose-specific lectins (galectins) on tumor cell surface (on recent data on the specificity of galectins see below), apparently facilitates delivery of 5-FU into cancer cells. While details of this mechanism are not within the scope of the present publication, the current observations were basic for determining the logistics and the experimental design of the preclinical efficacy studies described in this paper.

A few words about ligand specificity of galectins

All galectins exhibit affinity for lactose [4-(β-D-galactosido)-D-glucose, or Galβ1→4Glc] and N-acetyl lactosamine [4-(β-D-galactosido)-D-acetylglucosamine, or Galβ1→ 4GlcNAc]. That is why initially a common opinion was formed in the literature that galectins are specific for beta-anomer of galactose. DAVANAT®, however, contains alpha-anomer of galactose. Nevertheless, the discoverers of DAVANAT® believed that a stereoconfiguration of small saccharides such as lactose and its low-molecular weight derivatives cannot fully characterize the specificity of galectins, particularly related to polysaccharides.

This consideration turned out to be true. In 2002 it was found (7) that alpha-linked galactose rather than beta-linked galactose (as in N-acetyl lactosamine) serves as a specific ligand for some galectins. Furthermore, it was shown that the N-acetyl galactosamine motif is not cancer tumor specific (8).

Recently it was shown indeed, that some galectins are specific to long carbohydrate sequences, such as poly-N-acetyllactosamine chains (9), with the binding being much stronger compared to N-acetyllactosamine.

A physical combination of DAVANAT® with 5-FU was filed with the FDA as a part of a specific IND and was approved for clinical trials. Summarized in Figure 4, DAVANAT® increases the efficacy of 5-FU in tumor-bearing mice, in terms of decreasing median tumor volume and increasing mean survival time. DAVANAT® co-administered along with 5-FU by i.v. once every four days for a total of three injections (study 1) or four injections (studies 2 and 3) results in the decrease of median tumor volume to 18%, 42%, and 55% compared to control (untreated animals), and to 17%, 68% and 65%, respectively, compared to 5-FU alone. Furthermore, the co-administration of DAVANAT® and 5-FU compared to 5-FU alone results in the increase of mean survival time (days) to 190%, 190%, and 150%, for the three studies, respectively. The significance of these findings is increased because the effects were reproduced in kind using two different human colon tumors (COLO 205 and HT-29), as documented by two independent commercial testing facilities (Southern Research Institute and Charles River Laboratories).

Experimental details

DAVANAT®, employed in this work, was thoroughly characterized using high-performance liquid chromatography (HPLC), multi-angle laser light scattering (MALLS), nuclear magnetic resonance (NMR), and quantitative chemical analysis, as described above. All of the solutions for i.v. injections were subjected to a tight quality control, using HPLC analysis in a linear range of

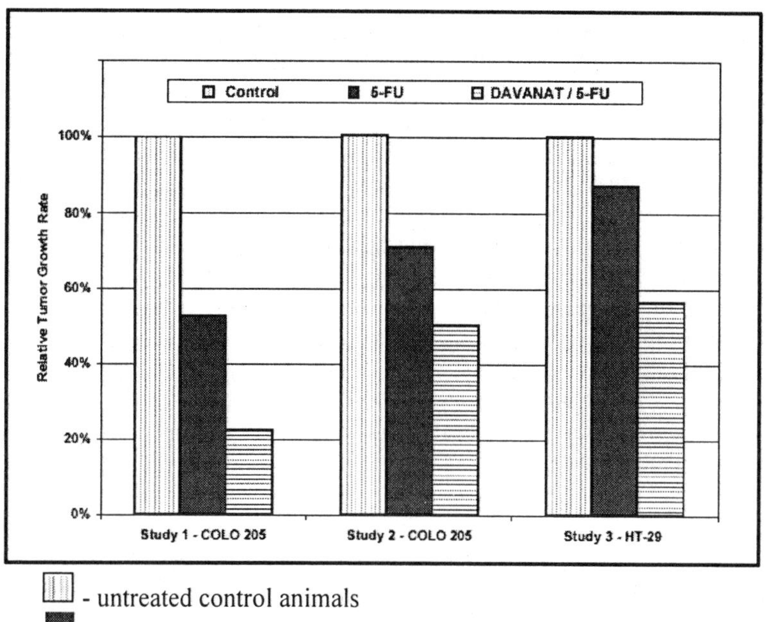

- untreated control animals

- 5-FU alone or in the presence of leucovorin (study 3)

- 5-FU + GM (studies 1 and 2) or 5-FU + GM + leucovorin (study 3)

Figure 4. Enhancement of 5-FU efficacy with DAVANAT® on two human colon cancer tumors in mice in three separate efficacy studies:

Study 1, mice bearing COLO 205 human colon tumor (5-FU 75 mg/kg/dose, DAVANAT® 120 mg/kg/dose);

Study 2, mice bearing the same tumor (5-FU 48 mg/kg/dose, DAVANAT® 120 mg/kg/dose),

Study 3, mice bearing HT-29 human colon carcinoma (5-FU 48 mg/kg/dose, DAVANAT® 120 mg/kg/dose, leucovorin p.o. 25 mg/kg/dose).

DAVANAT® and 5-FU concentrations. In addition, leucovorin did not interfere with DAVANAT® when both compounds were combined with 5-FU (study 3). This non-clinical data has important clinical implications, because DAVANAT® /5-FU may be administered in the future to patients treated with the 5-FU/leucovorin combinations.

Efficacy Experiments on Tumor-bearing Mice: DAVANAT®, 5-FU, Leucovorin (in Study 3), and Their Combinations. In the first two studies, male NCr-*nu* athymic nude mice (Frederick Cancer Research and Development Center, Frederick, MD, and Taconic Farms, Germantown, NY, respectively) were acclimated in the laboratory one week prior to experimentation. The animals were housed in microisolator cages, five per cage, and using a 12-hour light/dark cycle. Weight of the animals was in the range of: (i) 25-34 g on day 13 of the first study (day of treatment initiation), (ii) 21-33 g on day 14 of the second study (day of treatment initiation), and (iii) 24-25 g on day 7 of the third study (day of treatment initiation). In the third study, female NU/NU-*nu*BR athymic nude mice were housed as in the previous two studies. All animals in the three studies received sterilized tap water and sterile rodent food *ad libitum.* The animals were observed daily and clinical signs were noted. The mice were healthy and had not been previously used in other experimental procedures. The mice were randomized and were comparable at the initiation of treatment in all three studies.

The first two efficacy studies were conducted as follows. Thirty- to -forty milligram fragments from an *in vivo* passage of COLO 205 human colon tumor were implanted s.c. in mice near the right axillary area using a 12-gauge trocar needle. Tumors were allowed to reach 75-198 mm^3 in size/volume before the start of treatment. A sufficient number of mice were implanted, so that tumors in a volume range as narrow as possible were selected for the trial on the day of treatment initiation (day 13 or 14 after tumor implantation in the first and second study, respectively). Those animals selected with tumors in the proper size range were divided into the various treatment groups of 10 mice in each. The median tumor volumes in each treatment group ranged from 94 to 117 mm^3 in the first study and from 150 to 172 mm^3 in the second study.

In the third efficacy study, human colon carcinoma (HT-29) cells were injected (with $5x10^6$ cells) into the right lateral thorax of 70 female nude mice, of which 50 tumor-bearing mice were used on the study. Tumors were allowed to reach 100-200 mm^3 in volume before the start of treatment. Those animals selected with tumors in the proper size range were divided into five treatment groups of 10 mice in each. The median tumor volumes in each treatment group ranged from 172 to 196 mm^3.

Study duration was 70 days after tumor implantation or 56 days after treatment initiation in the first study; 27 days after tumor implantation or 13 days after treatment initiation in the second study; and 33 days after tumor implantation or 26 days after treatment initiation in the third study.

The s.c. tumors were measured, and the animals were weighed twice weekly starting with the first day of treatment. Tumor volume was determined by caliper measurements (mm) and using the formula for an ellipsoid sphere: $LxW^2/2=mm^3$, where L and W refer to the dimensions for length and width collected at each measurement.

Drug formulation and administration

Study 1. There were a total of four groups of 10 animals each, s.c.-implanted with COLO 205 human colon tumor xenografts. The groups were treated i.v. on day 13 after tumor implantation on q4d x 3 schedule as follows: (i) saline (NaCl, 0.9%), (ii) 5-FU, 75 mg/kg/dose (225 mg/m^2/dose), (iii) DAVANAT®, 120 mg/kg/dose (360 mg/m^2/dose), and (iv) 5-FU (75 mg/kg/dose) plus DAVANAT® (120 mg/kg/dose).

5-FU was formulated in fresh saline on each day of treatment at a concentration of 3.75 mg/mL, pH 9.2. In the groups where DAVANAT® and 5-FU were co-administered, DAVANAT® powder was dissolved in the 5-FU solution to yield DAVANAT® concentration of 6 mg/mL and 5-FU concentration of 3.75 mg/mL. Both individual compounds and their mixture were administered according to exact body weight, with the injection volume being 0.2 mL/10 g body weight.

Study 2. There were a total of seven groups of 10 animals each, implanted s.c. with COLO 205 human colon tumor xenografts. The groups were treated by i.v. on day 14 after tumor implantation on q4d x 4 schedule as follows: (i) Saline (NaCl, 0.9%), (ii) 5-FU (48 mg/kg/dose), (iii) DAVANAT® (120 mg/kg/dose), (iv) 5-FU (48 mg/kg/dose) plus DAVANAT® (6 mg/kg/dose), (v) 5-FU (48 mg/kg/dose) plus DAVANAT® (30 mg/kg/dose), (vi) 5-FU (48 mg/kg/dose) plus DAVANAT® (120 mg/kg/dose), and (vii) 5-FU (48 mg/kg/dose) plus DAVANAT® (600 mg/kg/dose).

5-FU was formulated in fresh saline on each day of treatment at a concentration of 4.8 mg/mL, at pH 9.2. In the groups where DAVANAT® and 5-FU were co-administered, DAVANAT® powder was dissolved in the 5-FU solution to yield DAVANAT® concentration of 0.6, 3.0, 12, and 60 mg/mL and 5-FU concentration of 4.8 mg/mL.

Study 3. There were a total of five groups of 10 animals each, implanted s.c. with HT-29 human colon carcinoma xenografts. The groups were treated (i.v., except leucovorin) on day 7 after tumor implantation on q4d x 4 schedule as follows: (i) saline (NaCl, 0.9%), (ii) DAVANAT® (120 mg/kg/dose) plus leucovorin (by mouth [p.o.], 25 mg/kg/dose), (iii) 5-FU (48 mg/kg/dose) plus leucovorin (p.o., 25 mg/kg/dose), (iv) 5-FU (48 mg/kg/dose) plus DAVANAT® (30 mg/kg/dose) plus leucovorin (p.o., 25 mg/kg/dose), and (v) 5-FU (48 mg/kg/dose) plus DAVANAT® (120 mg/kg/dose) plus leucovorin (p.o., 25 mg/kg/dose). Leucovorin was administered by oral gavage (p.o.) two hours after the injection (*via* tail vein), at a dose of 25 mg/kg/dose on the same q4d x 4 schedule.

DAVANAT® was formulated in 0.9% sterile fresh saline on each day of treatment at a concentration of 12 mg/mL. Leucovorin powder (clinical formulation, leucovorin calcium for injection) was reconstituted with 0.9% sterile saline to yield a concentration of 2.5 mg/mL. 5-FU and combinations of 5-FU with DAVANAT® were formulated as in study 2. Both individual compounds and their mixture were administered according to exact body weight with injection volume being 0.1 mL/10 g body weight.

Statistical Methods

Study 1. All available tumor volume data points in all animal groups (control and treatment groups), from the day of treatment initiation onward, were evaluated using unpaired one-sided Student's t-test. This was justified, since the research hypothesis (and its experimental proof) that combination of 5-FU and DAVANAT® decreases tumor volumes compared to control, DAVANAT® alone, and 5-FU alone was directional and permitted a one-tail test of significance. In these calculations, only animals that were alive at the time of the post-baseline evaluation were included, as well as in the calculations of the means and medians. The major comparisons of interest were against the 5-FU plus DAVANAT® combination. For these comparisons, the combination treatment served as the reference group, while remaining treatment variables were converted into three indicator variables.

The data were modeled using PROC MIXED with the indicator treatment variables, time, and the resulting interaction terms. The CONTRAST and LSMEANS (using Dunnett's adjustment for multiple comparisons) procedures were used to evaluate the differences between the combination treatment group and the three other treatment groups.

Study 2. The mean tumor weighs were calculated for day 27, which was the day after treatment was complete, and at each post-baseline time. The serial tumor volumes were each analyzed using a longitudinal growth model to evaluate the post-baseline slopes for each treatment group from day 16 through day 27. The model used all available tumor volume data, since no animals died or were sacrificed before the final visit in this study. The major comparisons of interest were against the 5-FU alone group; comparisons against the control (untreated group of animals) and DAVANAT® alone groups are also provided for completeness. Statistical evaluation was performed as described above for the preceding study.

Study 3. The mean tumor volumes were calculated for each post-baseline time. In these calculations, only animals that were alive at the time of the post-baseline evaluation were included. The serial tumor volumes were analyzed using a longitudinal growth model to evaluate the post-baseline slopes for each treatment group from day 11 through day 33. The model used all available tumor volume assessments, however, no data was included if animals died or were sacrificed before the final visit. The major comparisons of interest were against the 5-FU plus leucovorin group. The data were modeled using statistical tools as described above.

These results, observed for treatment of cancer-bearing mice (COLO 205 human colon tumor-bearing male athymic NCr-*nu* mice and HT-29 human colon tumor-bearing female CRL:NU-NU-*nu*BR mice) with a combination of 5-FU and DAVANAT®in the presence and absence of leucovorin are in marked contrast to the results in cancerous mice treated with 5-FU alone or in the presence of leucovorin. Hence, the data, described in this Chapter, show that DAVANAT®combined with 5-fluorouracil (in the presence or absence of leucovorin), increases efficacy of the drug.

Radiolabel Studies

Tissue Distribution of Radiolabeled (Tritiated) DAVANAT® in Tumored (COLO 205-Treated) and Non-Tumored Male Athymic NCr-nu Mice in the Presence and Without 5-FU.

Experimental Details

Male NCr-*nu* athymic nude mice (Charles Rivers Laboratories, Raleigh, NC) were acclimated and housed as described above. The first set of animals (a total of 18 mice) were non-tumored mice. The second set of 18 mice were tumored as follows. Thirty- to -forty mg fragments from an *in vivo* passage of COLO 205 human colon tumor were implanted subcutaneously (s.c.) in mice as described above, and allowed to grow. Tumors were allowed to reach 245-392 mg in weight before the start of treatment. A sufficient number of mice were implanted so that tumors in a weight range as narrow as possible were selected for the trial on the day of treatment initiation. Those animals selected with tumors in the proper size range were divided into the various treatment groups.

Tritiation of DAVANAT® was performed as follows. 12.8 mg of DAVANAT® was dissolved in 2.0 mL of water and exposed to 25 Curies of tritium gas in the presence of $Pd/BaSO_4$ catalyst (120 mg, totally insoluble in water). After one hour the gas supply was removed, the catalyst was filtered away, and the aqueous solution of DAVANAT® was evaporated to dryness repeatedly (four-fold, adding water), until no labile tritium was found. Total yield of the labeled DAVANAT® was 3.8 mC_i, specific radioactivity was 300 μC_i/mg.

All 36 mice, divided into 18 groups, were given a single intravenous injection of cold or tritiated DAVANAT® (either 6 or 60 mg/kg) or of a combination of DAVANAT® (60 mg/kg, cold or tritiated) and 5-FU (114 mg/kg) on the same day. Non-labeled DAVANAT® was formulated in saline, and tritiated DAVANAT® was added to the solution so that each animal received 10 μC_i of radioactivity. 5-FU was dissolved in the solution containing DAVANAT® (at a concentration of 6 mg/mL). All dosing solutions (100 μL each) were counted in duplicate.

Two mice per group were bled at 1, 6, and 24 hrs after injection, and plasma was prepared. Animals were then sacrificed; livers, kidneys, lungs, and tumors (from tumored animals) were collected, weighed and flash-frozen for further analysis.

After weighing, livers were dissolved in 10 mL of Soluene 350 (Packard Instruments, Downers Grove, IL) and incubated first for 4 hrs at 50^0C, and at room temperature, until tissues were solubilized. One mL of the resulting solution was counted in a scintillation counter as a single sample. Based on tissue weight and the sample volume, the number of μC_i of tritiated DAVANAT® per gram of tissue was calculated.

Kidneys were treated in the same manner, but dissolved in two mL of Soluene. After the tissue was solubilized at room temperature, 15 mL of Safety Solve scintillation fluid (Research Products International, Mount Prospect, IL) was added and samples were incubated overnight. Five mL of the resulting solution were diluted in 15 mL of Safety Solve and counted in a scintillation counter as a single sample. Lungs were treated in the same manner but dissolved in one mL of Soluene. Plasma samples (50 μL each) were placed direct into Safety Solve and counted as a single sample.

After weighing, tumors were dissolved in one or two mL of Soluene and incubated for three days at 50^0C to solubilize. 15 mL of Safety Solve were then added and samples were incubated overnight at room temperature. Two mL of water were then added and samples were counted in a scintillation counter as a single sample.

The data are shown in Table 9.

Table 9. Distribution of radiolabeled DAVANAT® alone (at 6 or 60 mg/kg), and in the presence of 5-FU (60 mg/kg of DAVANAT® and 114 mg/kg of 5-FU) in tissues of tumored (by implanting COLO 205 human colon cancer) and non-tumored mice

	1 hr after injection	6 hr after injection	24 hrs after injection
Liver			
μC_i of ^{3}H- DAVANAT® per liver, total	0.262	0.196	0.169
Same, with 5-FU	0.173	0.155	0.132
Same, per 1 g of liver (no 5-FU)	0.219	0.146	0.129
Same, with 5-FU	0.134	0.134	0.106
Tumor			
μC_i of ^{3}H- DAVANAT® per tumor, total	0.203	0.138	0.114
Same, with 5-FU	0.218	0.249	0.146
Same, per 1 g of tumor (no 5-FU)	0.123	0.092	0.066
Same, with 5-FU	0.154	0.104	0.075
Plasma			
μC_i of ^{3}H- DAVANAT® per mL of plasma	0.201	0.095	0.106
Same, with 5-FU	0.192	0.116	0.078
Kidney			
μC_i of ^{3}H- DAVANAT® per kidney, total	0.120	0.060	0.040
Same, with 5-FU	0.125	0.065	0.032
Same, per 1 g of kidney (no 5-FU)	0.302	0.131	0.087
Same, with 5-FU	0.286	0.168	0.080
Lungs			
μC_i of ^{3}H- DAVANAT® per lungs, total	0.019	0.012	0.011
Same, with 5-FU	0.017	0.011	0.008
Same, per 1 g of lungs (no 5-FU)	0.117	0.063	0.060
Same, with 5-FU	0.092	0.075	0.049

Statistical Evaluation.

The statistical evaluation was based on two separate series of 18 mice per series to trace and quantify the labeled DAVANAT® in organs/tissues (tumor, liver, kidneys, lung, and plasma); one series consisted of 18 healthy mice, while the other series consisted of 18 tumor-bearing mice. Each series of 18 mice was treated with 2 mice for each of the following nine possible treatments with DAVANAT® at 60 mg/kg with each of 2 mice sacrificed at 1, 6, and 24 hours; DAVANAT® at 6 mg/kg with each of 2 mice sacrificed at 1, 6, and 24 hours; DAVANAT® at 60 mg/kg + 5-FU 114mg/kg with each of 2 mice sacrificed at 1, 6, and 24 hours, as described above. The percentages of ^{3}H- DAVANAT® recovered per organ and tumor (μC_i /gram) was used for analyses; plasma outcome (μC_i /ml) was counted from the entire sample.

The assessment of radiolabel uptake was challenging with only two mice per treatment-sacrifice time combination and occasional outliers. To address the number of mice per group, pooling tests were performed to determine if non-tumored mice could be pooled with tumored mice in the evaluation of kidneys, liver, lungs, and plasma; successful pooling would increase the number of mice per group to four. Further pooling was assessed for the two DAVANAT® treatment groups; this would potentially increase sample size per group to eight. To address possible outliers, all outcomes for each individual mouse was visually compared to the other mice in the series-treatment-time combination; the four possible tumored and non-tumored mice outcomes were considered in this evaluation since any sigma-based rule would have excluded none or both observations if applied separately to the two different series.

Analysis of variance was used to test for pooling across the two series and across the two DAVANAT® treatments. Each organ (as well as plasma) was assessed separately; tumor could only be assessed for the two DAVANAT® treatments. The sample sizes were large enough to detect a 1.0 effect size difference between groups being compared. Results were then combined to the maximum extent allowed by pooling. Results were run both including and excluding the outliers. The primary analyses were those that excluded the outliers.

SAS (Version 6.12, Cary, NC) was used for all analyses. PROC ANOVA was used for pooling tests, while PROC FREQ was used to estimate means and standard deviations.

Principal Results

It was observed that DAVANAT®freely binds to liver, kidney, lung, tumor, and plasma, and did not reach limits of the binding, e.g. did not reach saturation of the binding between the 6 mg/kg and 60 mg/kg doses. When 6 mg/kg (with a relative radioactivity of 1.0) and 60 mg/kg (with a relative radioactivity of 0.1) doses of DAVANAT®were administered, the amount of bound radioactive DAVANAT® was the same; that is, the amount of bound DAVANAT® increased 10-fold for the 10-times higher dose.

The distribution of radioactivity in whole tissues as well as per weight or volume (in plasma) was practically identical for 6 and 60 mg/kg of DAVANAT®, hence, the respective data were pooled for Table 9. Also, the distribution of radioactivity in whole tissues as well as per weight or volume (in plasma) was practically identical for tumored and non-tumored animals (except in tumors, that obviously were present only in tumored animals), hence, the respective data were pooled for Table 9. Overall, the data in Table 9 are average for eight animals, except the data for tumors, that are average for four animals.

The principal results following from Table 9 are:

(1) In the presence of 5-FU the amount of DAVANAT® in the tumor increases, and stays increased in the course of clearance of DAVANAT®.
(2) In the presence of 5-FU the amount of DAVANAT® in the liver decreases, and stays decreased in the course of clearance of DAVANAT®.

In summary, major findings from this study indicated that tritiated DAVANAT® freely binds to liver, kidney, lung, tumor, and plasma. Saturation of binding did not occur at concentrations of 6 to 60 mg/kg (18 to 180 mg/m^2). Tissue distribution of ^{3}H-DAVANAT® was independent of the injected dose and did not change its pattern when 5-FU (342 mg/m^2) was combined with ^{3}H-DAVANAT®. ^{3}H-DAVANAT® elimination from plasma, kidneys, lungs and tumor in the various groups was rapid. An average of approximately 50% of the one-hour radioactivity was detected at 6 hours except in tumor-bearing mice, where the radioactivity in tumor samples from mice treated with 6 or 60 mg/kg of ^{3}H- DAVANAT® with or without 5-FU averaged approximately 72% remaining after 6 hours. Elimination of ^{3}H-DAVANAT® from the liver was more gradual than in other tissues, and on average, more than 50% of the radioactivity detected at one hour after injection was still present at 24 hours. 5-FU and DAVANAT® work in a synergism when delivered into the tumor. This might explain why DAVANAT® in a combination with 5-FU **increases efficacy** of the drug against COLO 205 human colon tumor, bearing by mice (see above).

At the same time, 5-FU and DAVANAT® work as antagonists (apparently, compete with each other for the same binding sites in the liver) when delivered into the liver.

Statistical Analysis.

*Outlier Assessments***.** There were four outliers identified at the mice level. These outliers were:

- Two tumored mice at 24 hours, DAVANAT® 60 mg/kg; values were in the 0.000-0.002 range for liver, kidneys, lungs, tumor, and plasma.
- One tumored mouse at 1 hour, DAVANAT® 60 mg/kg + 5-FU 114 mg/kg; values consistently elevated relative to other values for liver, kidneys, lungs, tumor, and plasma.
- One non-tumored mouse at 1 hour, DAVANAT® 60 mg/kg; values consistently elevated relative to other values for liver, kidneys, lungs, and plasma.

No other outliers were identified at the organ, plasma, or tumor level. Analyses were run including and excluding these 4 mice.

Pooling Assessments. The goal was to test if:

1. Data from non-tumored and tumored mice could be merged for organs and plasma.
2. Both DAVANAT® 6 and 60 mg/kg doses data could be merged for organs, plasma, and for tumor.

As it was demonstrated by including outliers and excluding outliers, pooling is warranted for combining data for non-tumored and tumored mice for liver and lungs (the unpaired differences between non-tumored and tumored data are $\leq$0.02 for all treatment groups at 1, 6, and 24 hours), while kidneys and plasma remained skewed with higher levels in tumored mice after outlier removal (largest differences seen in the DAVANAT® 60 mg/kg groups at 1 hour). The key point is that pooling is justified for the non-tumored and tumored data for liver, the most critical organ. Pooled data for liver and lungs are provided in Table 9. Is it was shown, ANOVA analyses did not reject the null hypothesis of no difference (implying that the two series could be pooled) for each organ and plasma, despite the small sample sizes and low power.

Then, as it was demonstrated by including outliers and excluding outliers, pooling is warranted for combining DAVANAT® doses of 6 mg/kg vs 60 mg/kg for tumors (<0.01 differences after outlier exclusions), while it is not as evident for lungs (up to 0.04 differences after outlier exclusions), liver (up to 0.06 differences after outlier exclusions), kidneys (up to 0.06 differences after outlier exclusions), and plasma (up to 0.07 differences after outlier exclusions). There was no skewing favoring either dose. The key point is that pooling is justified for the two DAVANAT® doses for tumor, the other critical site for assessing uptake. ANOVA analyses confirmed the ability to pool DAVANAT® 6 mg/kg and 60 mg/kg data for each organ, plasma, and tumor.

Pooled data (across DAVANAT® doses) for tumors were considered including outliers and excluding outliers. The resulting standard deviations of the combined tumor data (after DAVANAT® doses pooling) were 0.0044 at 1 hour, 0.0333 at 6 hours, and 0.0057 at 24 hours after outlier exclusions; corresponding standard deviations for the DAVANAT® + 5-FU group were 0.090 at 1 hour, 0.0141 at 6 hours, and 0.0014 at 24 hours after outlier exclusions. Higher levels were seen for DAVANAT® + 5-FU versus pooled DAVANAT® doses at 1 hour (0.032), at 6 hours (0.012), and at 24 hours (0.009); the 24 hour difference reached statistical significance after outlier exclusions. As it was found with including outliers and excluding outliers, statistical significance also held at 24 hours if data were limited to just the DAVANAT® 60 mg/kg group after outliers were excluded. Thus, **more DAVANAT® reached the tumor in the presence of 5-FU.**

In summary, *in vivo* pharmacology studies indicate that DAVANAT® enhances the antineoplastic effects of 5-FU. DAVANAT® is not mutagenic, either by itself or combined with 5-FU, in bacterial reverse mutation assays. Toxicity studies in mice, rats and dogs have shown DAVANAT® to have a very low potential for toxicity and a capacity for ameliorating some of the toxic side effects of 5-FU.

Compared to many other known polysaccharides, DAVANAT® has multiple side-chain galactose units that should readily interact with galactose-specific receptors such as galectins on the tumor cell surface, modulate the tumor surface physiology and potentially affect delivery of 5-FU to the tumor.

To the best of our knowledge, (a) no simple sugar drugs, aimed at interactions with lectins, such as DAVANAT® , are available in clinical practice, (b) none of these carbohydrates are able to increase *in vivo* efficacy of known chemotherapy drugs, such as 5-fluorouracil, doxorubicin, or others, widely used in cancer chemotherapy, and (c) none of them are able to increase *in vivo* efficacy of any other drug when co-administered.

Despite advances in chemotherapeutic regimens for metastatic colorectal cancer as well as other solid tumors, patients refractory to these therapies represent a large unmet medical need. One widely used chemotherapeutic drug is 5-Fluorouracil (5-FU). It is used alone or in a combination with many other chemotherapeutic agents for treatment of solid tumors, particularly colorectal, breast, hepatic and gastric cancers [10,11]. These combinations, while marginally improving efficacy, may increase toxicity. It is reasonable to hypothesize that more specific targeting of efficacious chemotherapeutic agents such as 5-FU to tumor cells could result in higher response rates along with reduced toxicity.

Clinical Studies, Phase I

In January 2003, a Phase I open-label study was initiated to evaluate the safety and tolerability of escalating doses of DAVANAT® (30 mg/m^2 to 280 mg/m^2) in the presence and absence of 500 mg/m^2 5-FU in subjects with advanced solid tumors.

The study type was interventional, the study design: treatment, non-randomized, open label, uncontrolled, single group assignment, safety/efficacy study. Types of cancer – colorectal, lung, breast, head and neck, and prostate cancer. Excluded were patients with central nervous system (CNS) metastases or primary CNS tumors.

Enrolled in the study were patients with different types of solid tumors who have failed standard, approved treatments. Besides general eligibility criteria (at least 18 years of age, not pregnant or breast feeding, no current alcohol or illicit drugs abuse, patients having other significant medical, psychiatric, or social conditions which, in the investigators' opinion, may compromise the patient's safety in participating in this study, etc), enrolled patients should have had a documented histologic or cytologic recurrent or metastatic solid tumor that is not amenable to curative surgery, radiotherapy, or conventional chemotherapy of proven value. Also, patients must have had completed previous therapy (chemotherapeutic agents or other therapies including radiation) at least four weeks prior to study entry, the same time period must have been elapsed after a major surgery and patients must have been recovered from effects, at least two weeks must have elapsed after minor surgery, and the patient should have had a life expectancy of at least 12 weeks. Excluded from the study were the patients who had congestive heart failure or any other medical condition that could be adversely affected by intravenous infusion of up to approximately 200 mL of fluid over 60 minutes.

The study consisted of two cycles: different doses of DAVANAT® (as described below) were given alone in Cycle 1, and in combination with 5-FU (500 mg/m^2) in Cycle 2. Patients were on study from February 10, 2003. In patients who had cancer that could be measured by MRI (magnetic resonance imaging) and CT (computerized tomography) scan, it was determined whether the tumors changed in size (got bigger, smaller or stayed the same) after Cycle 2.

The study was designed to determine:

- The Maximum Tolerated Dose (MTD) and Dose Limiting Toxicities (DLT) of DAVANAT® alone.
- The safety and side effects of DAVANAT® when given alone and in combination with 5-FU. To define the MTD and DLT of DAVANAT® when administered concurrently with 500 mg/m^2 of 5-FU.

- To determine the pharmacokinetic profile of 5-FU in the presence of DAVANAT®.
- To determine the pharmacokinetic profile of DAVANAT® (the effect of DAVANAT® on the amount of 5-FU at different times in the body) at doses from 30 mg/m^2 to 280 mg/m^2 of DAVANAT®.
- The effect of DAVANAT® plus 5-FU on tumor size in patients with measurable disease.

Patients had a screening period followed by two 28 day cycles. In cycle 1, DAVANAT® alone was administered i.v. on days 1 to 4, and in cycle 2, 5-FU plus DAVANAT® was administered on days 1 to 4, for a total of approximately 90 days on study.

Prior to administration of study drug, a female patient should have a negative pregnancy test. Besides, the patients should have had AST and ALT (liver function) < 2.5 times the upper limit of normal (ULN); total bilirubin < 1.5, hematopoietic parameters WBC > 3000 per mm3; granulocyte count > 1,500 per mm3; platelet count > 100,000 per mm^3, creatinine (renal) less or equal to ULN, and Dlco (pulmonary, carbon dioxide diffusing capacity) higher or equal 60% of predicted.

A letter directed to patients stated: "You have been invited to participate in a clinical research study because you have a solid tumor that has returned or a metastatic (spreading through your body) solid tumor that cannot be cured by surgery, radiotherapy, or standard chemotherapy. (....) The study will involve up to 40 subjects".

DAVANAT® was supplied as a 60 mg/mL solution in normal saline with 10 mL contained in a single use glass vial which is stored at 4-8^{0}C. Dose escalation was proceeded in cohorts of 3 patients with ascending doses of DAVANAT® alone ranging from 30 to 280 mg/m^2 and in combination with 500 mg/m^2 5-FU.

According to the study schedule, three to six patients are to be entered into the study at each of six doses of DAVANAT®, or until the Maximum Tolerated Dose (MTD) is reached. Then, a total of ten patients are to be entered at this determined MTD, or at the highest dosing level if a MTD is not determined. A total time period for the participation for each patient was scheduled to approximately 90 days. If the cancer has not spread further, the patient would have received additional cycles of DAVANAT® plus 5-FU.

An evaluation of tumor size was made by an MRI and/or CT scan (see above).

A more detailed schedule for a first cohort (30 mg/m^2 dose of DAVANAT®) was as follows (after a 30-day patients screening):

Cycle 1. Treatment of the first cohort (three to six patients) with DAVANAT® alone (through an intravenous line into the vein once a day for 30 min for four consecutive days), record of all side effects

- Day 1 – blood laboratory tests (hematology, chemistry, coagulation)
- Day 1 - DAVANAT® i.v. infusion at a starting dose of 30 mg/m^2
- Day 2 - DAVANAT® i.v. infusion (30 mg/m^2)
- Day 3 - DAVANAT® i.v. infusion (30 mg/m^2)
- Day 4 – blood laboratory tests (hematology, chemistry, coagulation)
- Day 4 - DAVANAT® i.v. infusion (30 mg/m^2)
- Day 14 (±2) – blood laboratory tests (hematology and chemistry), record of all side effects
- Day 15-27 – Follow-up.
- Day 28(±2) - blood laboratory tests (hematology, chemistry, coagulation, pregnancy test if female), record of all side effects, limited physical examination, CEA (a special marker for cancer), electrocardiogram, a tumor evaluation, MRI and/or CT scan of the chest, abdomen and pelvis, a lung test, and record of adverse events.

If no side effects were noted that would have prevented the patient from proceeding to Cycle 2, the 5-FU/ DAVANAT®mixture was administered through an intravenous line as follows:

Cycle 2. Treatment with DAVANAT® (30 mg/m^2) in combination with 500 mg/m^2 of 5-FU (once a day for 30 min for four consecutive days), record of all side effects.

- Day 1 – blood laboratory tests (hematology, chemistry, coagulation)
- Day 1 - DAVANAT®/5-FU (30/500 mg/m^2) i.v. infusion, and a blood laboratory test to determine 5-FU concentration at different times following infusion: immediately following infusion (at 0 minutes), 5, 10, 20, 30, 45, 60 minutes, 2, 4, 6, and 8 hours after DAVANAT® and 5-FU administration.
- Day 2 - DAVANAT®/5-FU (30/500 mg/m^2) i.v. infusion, and a blood laboratory test to determine 5-FU concentration immediately following the infusion
- Day 3 - DAVANAT®/5-FU (30/500 mg/m^2) i.v. infusion, and a blood laboratory test to determine 5-FU concentration immediately following the infusion
- Day 4 – blood laboratory tests (hematology, chemistry, coagulation)
- Day 4 - DAVANAT®/5-FU (30/500 mg/m^2) i.v. infusion, and a blood laboratory test to determine 5-FU concentration at different times during 8 hours following infusion, as indicated above.
- Day 14 (±2) – blood laboratory tests (hematology and chemistry), record of all side effects

- Day 15-27 – Follow-up.
- Day 28 - blood laboratory tests (hematology, chemistry, coagulation, pregnancy test if female), record of all side effects, limited physical examination, CEA (a special marker for cancer), electrocardiogram, a tumor evaluation, CT scan of the chest, abdomen and pelvis, a lung test, and record of adverse events.

Interpatient dose escalation was proceeded with the following doses of DAVANAT®: 30, 60, 100, 150, 210, and 280 mg/m^2. There was no intrapatient dose escalation of DAVANAT®. The dose of 5-FU remained at 500 mg/m^2 for all six cohorts throughout the study.

Chemistry included electrolytes (sodium, potassium, chloride, bicarbonate, calcium, phosphate), blood urea nitrogen (BUN), creatinine, total cholesterol, albumin, total protein, uric acid, lactate dihydrogenase (LDH), liver enzymes - aspartate aminotransferase (AST), alanine aminotransferase (ALT), alkaline phosphatase (ALK), total bilirubin, glucose, carcinoembrionic antigen (CEA).

Key pharmacokinetic parameters of 5-FU included an area under the plasma concentration-time curve (AUC), systemic clearance (CL), apparent volume of distribution (Vd), and eliminationhalf-time ($t_{1/2}$). Anti-tumor efficacy in patients with measurable disease was descriptively evaluated for response at the end of Cycle 2.

Dose escalation of DAVANAT® alone for the next cohort was proceeded after the preceding cohort (three to six patients) complete Cycle 1 with no patient experienced DLT. In this case the preceding cohort was advanced to Cycle 2, and the next cohort of three to six patients was escalated to next dose of DAVANAT®.

If, however, one of patients in the preceding cohort experienced DLT by Day 28 in Cycle 1, three more patients were added to the same dose level of DAVANAT® alone, and six patients total were treated at the same dose and the remaining patients with no DLT were advanced to Cycle 2.

If two patients in the preceding cohort experienced DLT by Day 28 in Cycle 1, the dose escalation stopped and this current dose of DAVANAT® was defined as MTD. If the drug was tolerated by five of six patients, the dose was escalated in the next cohort of three patients.

Once MTD is established (or highest dose was administered), total of ten patients were enrolled at MTD or the highest dose to further define the pharmacokinetics at this dose. Any new lung abnormality observed on CT scan was considered as DLT.

The same dose escalation criteria and assessment of DLT and MTD applied to DAVANAT® alone was applied to the combination of each dose of DAVANAT® plus 5-FU. When DLT was observed in one patient per cohort in Cycle 2 with the combination of DAVANAT® plus 5-FU, then the subsequent cohort of patients completing Cycle 1 at the next dose level was de-escalated to the DAVANAT® dose level at which DLT occurred in combination with 5-FU for their second cycle.

The clinical trials were conducted at four clinical sites:

- Florida Oncology Associates(Jacksonville, Florida)
- Ochsner Cancer Institute (New Orleans, Louisiana)
- University of Michigan Comprehensive Cancer Center (Ann Arbor, Michigan)
- Dartmouth-Hitchcock Medical Center (Lebanon, New Hampshire).

Of the first 25 patients with advanced solid neoplasms not amenable to curative surgery, radiotherapy, or chemotherapy of proven value, injected with DAVANAT®, 20 completed Cycle 2 of the study through November 2004.

Intravenous infusion of up to 280 mg/m^2 of DAVANAT® alone and in combination with 5-FU was found to be well tolerated. DLT has not been reached at the final dose level of 280 mg/m^2. Toxicity of DAVANAT® plus 5-FU were similar to those expected for 5-FU alone at this dose and schedule.

Table 10. Patients tolerance with respect to DAVANAT® and DAVANAT®/5-FU doses in Cycles 1 and 2. 5-FU doses in Cycle 2 were of 500 mg/m^2

Dose of DAVANAT® (mg/m^2)	Patients Enrolled	Diagnoses	Completed Cycle 1	Completed Cycle 2	DLT Event
30	4	3 Colorectal 1 Hepatocellular	3	3	None
60	6	5 Colorectal 1 Hepatocellular	6	6	Grade 3
100	5	4 Colorectal 1 Pancreatic	5	3	ALT/ SGPT
150	3	2 Colorectal 1 Prostate	3	3	None
210	4	3 Colorectal 1 Ovarian	4	3	None
280	3	1 Colorectal 1 Pancreatic 1 Spindle Cell Tumor	3	2	None

Response Evaluation Criteria in solid tumors at the end of Cycle 2 showed nine patients with stable disease, nine with progressive disease and two with non-measurable disease. Six of nine stabilized patients received additional cycles of DAVANAT® plus 5-FU. It was also established that DAVANAT® did not alter the pharmacokinetics of 5-FU in patient samples.

Hence, data in progress show that DAVANAT®/5-FU is well tolerated and dose limiting toxicity was not reached in this study. Disease was stabilized in 45% of the initial 20 patients who completed the trial through November 2004. The tolerability of DAVANAT® given with 5-FU warrants further testing for safety and efficacy in a Phase II clinical trial.

Acknowledgements

We thank colleagues from Southern Research Institute (SRI) and Charles River Laboratories for conducting animal studies, Drs. Yulia Maxuitenko (SRI, Birmingham, AL) and Philip T. Lavin (Averion, Framingham, MA) for conducting statistical analysis of the data, Drs. Mildred Christian and Robert Diener (Argus International, Horsham, PA) for discussing the data and the manuscript, and Drs. Marilyn Pike, Jyotsna Fuloria, Yousif Abubaker, Raymond Perez, and Merk Zalupsli for conducting Phase I clinical trials of DAVANAT®.

References

1. Platt, D. and Raz, A. 1992. Modulation of the lung colonization of B16-F1 melanoma cells by citrus pectin. J. Natl. Cancer Inst. 84: 438-442.
2. Kannagi, R. 1997. Carbohydrate-mediated cell adhesion involved in hematogenous metastasis of cancer. Glycoconj. J. 14: 577-584.
3. Ohyama, C., Tsuboi, S. and Fukuda, M. 1999. Dual roles of sialyl Lewis X oligosaccharides in tumor metastasis and rejection by natural killer cells. EMBO J. 18: 1516-1525.
4. The World Almanac and Book of Facts, 2005, p. 87.
5. Fukuda, M.N., Ohyama, C., Lowitz, K., Matsuo, O., Pasqualini, R. and Ruoslahti, E. 2000. A peptide mimic of E-selectin ligand inhibits sialyl Lewis X-dependent lung colonization of tumor cells. Cancer Res. 60: 450-456.
6. Zhang, J., Nakayama, J., Ohyama, C., Suzuki, A., Fukuda, M., Fukuda, M.N. 2002. Sialyl Lewis X-dependent lung colonization on B16 melanoma cells through a selectin-like endothelian receptor distinct from E- or P-selectin. Cancer Res. 62: 4194-4198.
7. Appukuttan, P.S. 2002. Terminal alpha-linked galactose rather than N-acetyl lactosamine is ligand for bovine heart galectin-1 in N-linked

oligosaccharides of glycoproteins. Journal of Molecular Recognition, 15: 180-187.

8. Chacko, B.K. and Appukuttan, P.S. 2001. Peanut (Arachis hypogaea) lectin recognizes alpha-linked galactose, but not N-acetyl lactosamine in N-linked oligosaccharide terminals. Int. J. Biol. Macromol., 28: 365-371.
9. Stowell, S.R., Dias-Baruffi, M., Pentilla, L., Renkonen, O., Kwame Nyame, A., and Cummings, R. 2004. Human galectin-1 recognition of poly-N-acetyllactosamine and chimeric polysaccharides. Glycobiology, 14: 157-167.
10. IARC Monographs on the Evaluation of the Carcinogenic Risk of Chemicals to Humans. Vol. 26. Some Antineoplastic and Immunosuppressive Agents. WHO, International Agency for Research on Cancer, May 1981, p. 217- 235.
11. Novel Approaches to Selective Treatments of Human Solid Tumors: Laboratory and Clinical Correlation. 1994. Adv. Exp. Med. Biol., v. 339 (Youcef M. Rustum, Ed.), Plenum Press, 319 p.

Chapter 5

Synthesis and Biological Activity of Doxo-Galactose and Doxo-Galactomannan, New Conjugates of Doxorubicin with D-Galactose and 1,4,β-D-Galactomannan

Anatole A. Klyosov[1], Anna N. Tevyashova[2], Eugenia N. Olsufyeva[2], Maria N. Preobrazhenskaya[2], Eliezer Zomer[1], and David Platt[1]

[1]Pro-Pharmaceuticals, 189 Well Avenue, Newton, MA 02459
[2]Gause Institute of New Antibiotics, Russian Academy of Medical Sciences, Moscow 119021, Russia

Introduction

Anthracyclines are members of a very important class of antitumor agents that have been used for many years in the treatment of different types of cancer. Among them doxorubicin (**1**, Fig.1) is considered one of the most potent anticancer agents [1-3]. However, therapeutic activity of doxorubicin is dose-limited by serious toxicity, effecting blood and bone marrow functions, and especially by cumulative cardiotoxicity [4-8]. Besides, the efficacy of anthracyclines is particularly low when tumor cells express multi-drug resistance mediated by a decreased intracellular drug accumulation [9]. The development of more selective treatment *via* site-specific drug delivery or targeting employing prodrug therapy is one of the major goals in contemporary drug design.

Doxorubicin (**1**) consists of adriamycinone (an aglycone) and the aminosugar residue daunosamine, chemically attached via $\alpha(1\rightarrow4)$ linkage to 7-hydroxy group of the cyclohexene ring. Its natural analog, daunorubicin (**2**, Fig.1) lacks 14-hydroxy group, and the respective aglycone called daunomycinone. Yet another semisynthetic analog, idarubicin, lacks both 14-hydroxy and 4-methoxy groups (**3**, Fig.1).

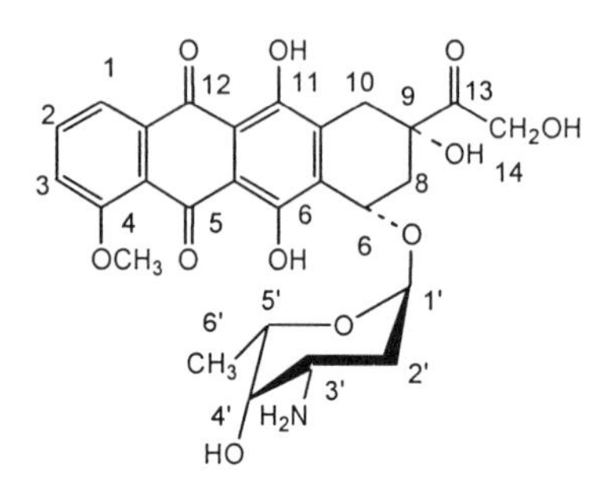

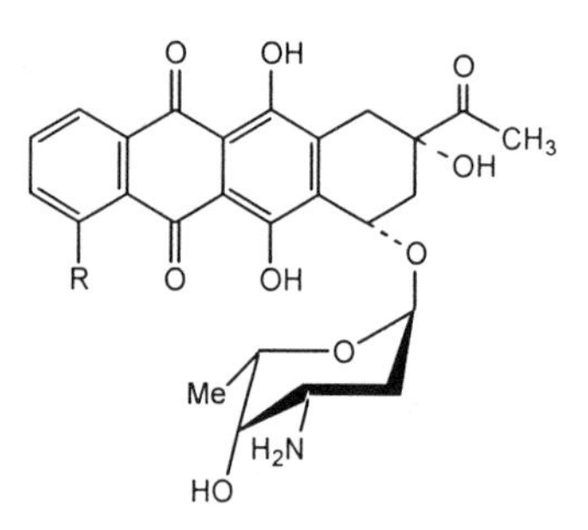

1. Doxorubicin

2. R = OCH_3, Daunorubicin
3. R = H, Idarubicin

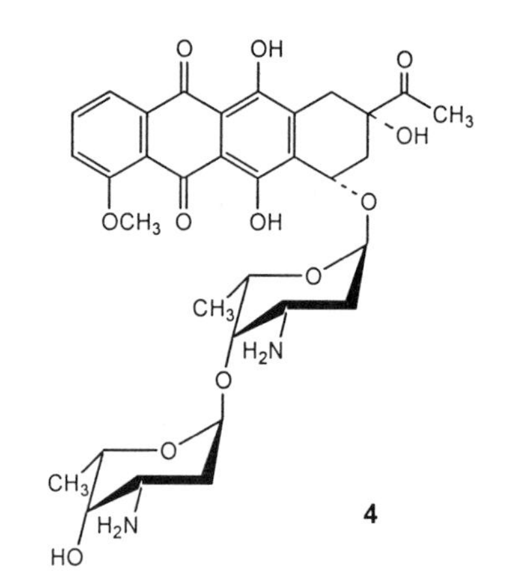

Figure 1. Structure of anthracycline antibiotics (## 1-4).

Anthracyclinone part is responsible for the DNA intercalation, while daunosamine moiety and the cyclohexene ring take part in minor groove binding and interaction with enzymes.

Hence, an essential fragment of anthracycline antibiotics is their carbohydrate moiety. It is recognized to be crucial for the cytotoxic activity of anthracyclines, directed at a nuclear enzyme DNA topoisomerase II, that in turn is involved in the processes of replication, transcription and recombination of DNA. The efficacy of doxorubicin has stimulated several studies aimed at chemical modification of its carbohydrate moiety in hope to enhance the recognition potential of the drug at the target level.

A natural disaccharide compound 4'-O-α-daunosaminyl daunorubicin (**4**, Fig.1) is known, in which a second daunosamine residue is bound by α(1→4) glycosidic linkage to the first one; however, the second sugar in this case does not provide improvement in cytotoxic or antitumor activity [1, pp. 38-46; 10]. On the contrary, it was shown that natural mono-, di- and three-saccharide derivatives of ε-pyrromycinone (Fig.2), namely pyrromycin (**5**), musettamycin (**6**) and marcellomycin (**7**) possess a progressively increased antitumor activity with the increase in the length of the oligosaccharide chain.

Pyrromycin (**5**) contains a 3'-amino sugar (L-rhodosamine, or N,N-dimethyl-L-daunosamine) chemically attached via α(1→4) linkage to 7-hydroxy group of the cyclohexene ring (such as in doxorubicin); the addition of a second sugar residue, 2-deoxy-L-fucose, at the 4'-position of the first aminated sugar makes musettamycin (**6**), and a further addition of a third sugar residue, also 2-deoxy-L-fucose, at the 4'-position of the second sugar makes marcellomycin (7) [10, 11]. The authors note that the increase in the length of the oligosaccharide chain in this series of natural anthracyclines is correlated with an increase in their DNA-binding affinity. However, DNA-binding activity of trisaccharide derivatives of the pyromycinone series does not correlate with its antitumor potency.

In addition to the desired effect of destroying cancer cells, anthracycline antibiotics, and doxorubicin among them, also damage non-cancer cells resulting in side effects for the patient. These side effects limit the dose and duration of treatment with these agents. Generally, side effects are the hallmark of many chemotherapeutic agents which otherwise are effective in reducing tumor size. Attempts have been made to reduce the side effects of this class of

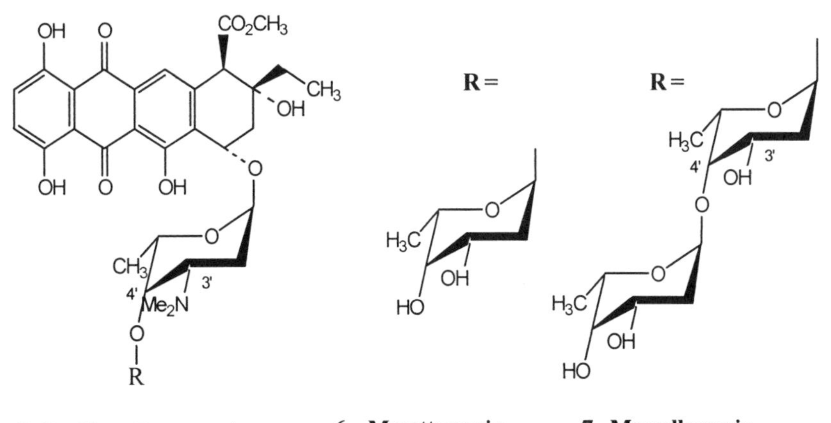

5. R=H, Pyrromycin **6. Musettamycin** **7. Marcellomycin**

Figure 2. Structures of anthracycline antibiotics (## 5-7).

therapeutic agent. However, *in vitro* assays that have been used for evaluation of drug candidates neither measure cytotoxic effects on non-target cells or side effects which arise in a patient, nor indeed provide information on chemotherapeutic benefit of the drug.

One of the approaches aiming at obtaining anthracycline derivatives with better chemotherapeutical index is synthesis of different conjugates of anthracyclines with carbohydrates.

Novel anthracyclines with disaccharides lacking the amino group in the first (aglycone linked) sugar, were designed (Fig.3, **8a,b-10a,b**) [10, 12-18]. The 3'-amino group in the first sugar was replaced by a hydroxyl group, and the second sugar residue was bound to the first one *via* an α(1→4) linkage.

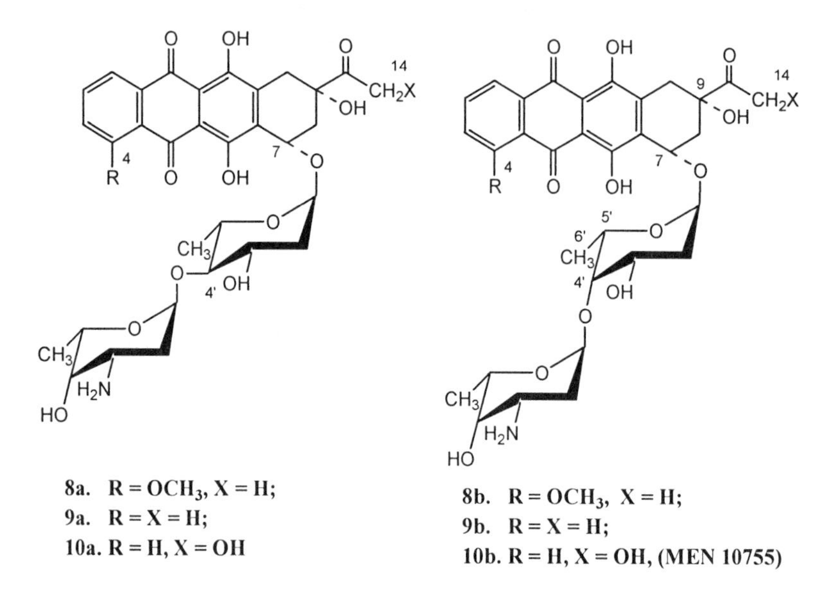

Figure 3. Structures of anthracycline antibiotics (## 8a,b-10a,b).

Analogs of daunorubicin (**8a,b**), idarubicin (**9a,b**) and 4-demethoxydoxorubicin (**10a,b**) in which daunosamine residue was substituted with the disaccharide L-daunosaminyl-α-(1→4)-2-deoxy-L-rhamnosyl (**a**-series) or L-daunosaminyl-α-(1→4)-2-deoxy-L-fucosyl (**b**-series), with the rhamnosyl and fucosyl residues directly linked to the aglycone *via* α(1→4) linkage were obtained [10, 12-16]. *In vitro* studies of the antitumor activity (using large panel of human tumor cell lines of different histotypes) of the obtained compounds revealed that in most cases the substitution of the daunosamine for the disaccharide moiety dramatically reduced cytotoxicity, by 20 to 1000 times. The only idarubicin derivative, 3'-(L-daunosaminyl-α-(1→4)-2-deoxy-L-fucosyl)-idarubicin (**9b**), conferred a cytotoxic potency and antitumor efficacy (on mice bearing cervical carcinoma and ovarian carcinoma) not significantly lower than that of doxorubicin [16].

Further studies showed that doxorubicin analog, 4-demethoxy-3'-(L-daunosaminyl-α-(1→4)-2-deoxy-L-fucosyl)-doxorubicin (MEN 10755) (**10b**) was the best candidate in this series of disaccharide anthracycline analogs. This new compound showed comparable or higher cytotoxic activity and antitumor efficacy compared to doxorubicin, but also a reduced accumulation compared with the latter in all tissues investigated, such as heart, kidney, liver [10, 12, 17]. As to the drug toxicity, MEN 10755 (**10b**) was better tolerated than doxorubicin, but was still cardiotoxic [18]. It was also reported that *i.v.* doses higher than 7 mg/kg as a single injection may cause immediate death in mice [12, 19]. MEN 10755 (**10b**) is currently investigated in phase I clinical trial.

Another approach is based on usage of various conjugates of anthracyclines with carbohydrates as anthracycline prodrugs. Upon activation, the prodrugs are undergoing spontaneous reaction leading to formation of active anti-cancer drugs. The authors' principal premise is that doxorubicin and its analogs modified at the 3'-amino group are inactive, and their cytotoxic activity develops only after the hydrolysis of the ester or glycoside group and "activation" of the prodrug to active drug.

A monosaccharide moiety is generally attached to 3'-amino group of the daunosamine through self-eliminating benzyloxycarbonyl spacer. These prodrugs were designed as a part of a two-step directed enzyme prodrug therapy (DEPT). Anthracycline prodrugs, designed for their activation by human glucosidases, including glucuronidases and esterases were described. [20, 21]

For example, anthracycline prodrugs as possible substrates for human β-glucuronidase, contain *p*-hydroxybenzyloxycarbonyl spacer incorporated between the anthracycline moiety and β-glucuronic acid (for example, daunorubicin prodrug **11**, Fig.4). [20]. The spacer is self-eliminated after hydrolysis by β-glucuronidase *via* 1,6-elimination. These prodrugs showed a 100-fold reduction in toxicity *in vitro* when compared to doxorubicin. *In vivo* data suggested that these prodrugs were much less toxic compared to the parent agent (daunorubicin or doxorubicin respectively) and showed a somewhat better chemotherapeutic activity compared to the parent antibiotic. However, as noted in [21], these prodrugs did not make it to the clinic due to problems of large doses required and increased cost of synthesis.

A similar approach was used for the design of the anthracycline prodrugs, activated by *E. coli* β-galactosidase. Self-eliminating *p*-hydroxybenzyloxycarbonyl spacer (with different substituents in aromatic

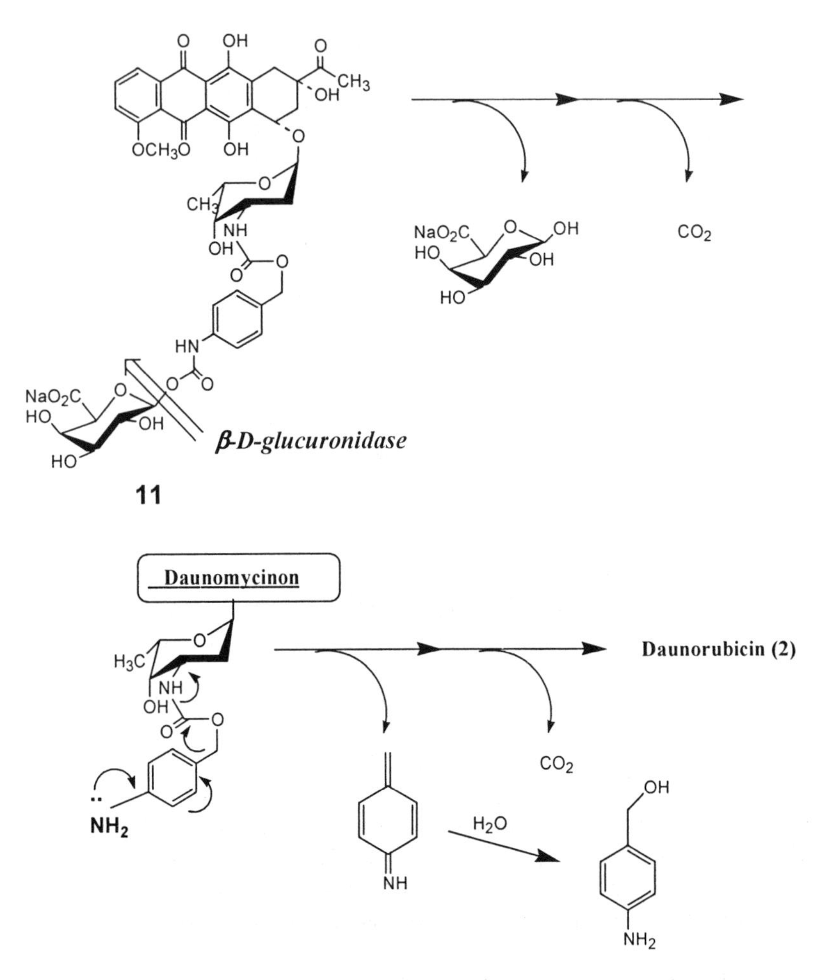

Figure 4. Activation of β-glucuronide daunorubicin prodrug (# 11).

moiety or without them) was incorporated between the anthracycline moiety and β-galactose (for example, daunorubicin prodrug **12**, Fig.5)[22]. These prodrugs were nearly one million-fold more toxic to human A375 melanoma cells in culture in the presence of *E. coli* β-galactosidase than in the absence of the enzyme [23]. However, antitumor evaluation of these analogs in the experiments *in vivo* was not published, probably because of the complexity of localization of the enzyme in tumor.

Serious disadvantages, such as increased cost of the synthesis, large doses required, problems with the localization of the enzyme used for the prodrug activation and others, limit the use of such prodrugs.

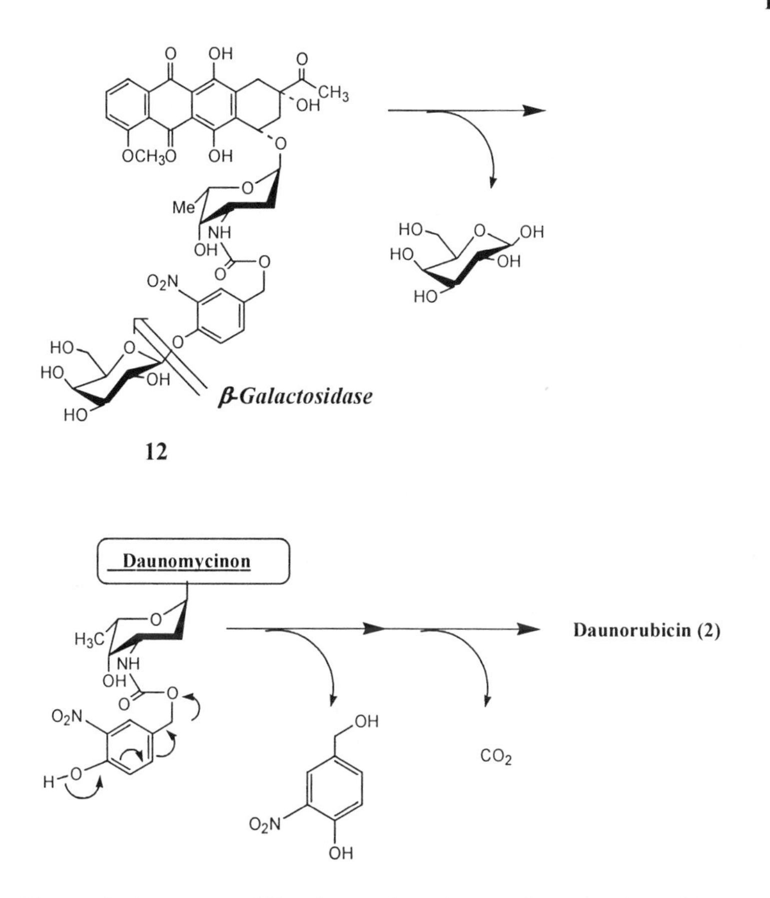

Figure 5. Activation of β-galactosidase daunorubicin prodrug (# 12).

Only in a few cases, polysaccharides were used as drug-carriers for the anthracycline antibiotics [24].

In this study we have considered two different types of doxorubicin derivatives, aiming at reducing the drug toxicity, increasing its efficacy, or both. In one case we modified the initial drug with a stable galactose residue, which was chosen because of the important role that galactose-specific receptors, galectins, play in tumor development [25, 26]. This synthesized and preliminary tested drug candidate was named Doxo-Galactose. In the second case we have synthesized a conjugate of doxorubicin with a water-soluble 1,4-β-D-galactomannan of a certain structure, the latter showed a noticeable antitumor effect when injected along with a chemotherapy drug, 5-fluorouracil (5-FU) [27]. This galactomannan product under trade name of DAVANAT® is currently being studied in a combination with 5-FU in Phase I/ II clinical trial. The conjugates of doxorubicin and DAVANAT® were synthesized in two different chemical forms, and tentatively named Doxo-Galactomannan (Doxo-GM)-1 and -2.

It should be noted that polymer conjugation is of increasing interest in pharmaceutical chemistry as vehicle for target delivery of drugs having biological or physicochemical limitations [28]. These compounds may show extended circulating life *in vivo* and, more importantly, make use of the enhanced permeability and retention effects observed in tumor cells. Currently many polymer-conjugated drugs, including anthracycline polymeric prodrugs, are investigated. Among them are also monoclonal antibodies, polyethylene glycol, polyglutamate. However, the most prospective candidates for polymeric carriers are polysaccharides. They are in general non-toxic, hypoallergenic in nature, and may have a favorable pharmacokinetic.

Polysaccharides have advantage over other polymers as they may be naturally hydrolyzed by enzymes and metabolized in specific organs, and can provide sustained effect with target delivery. Orally administrated drugs frequently use polysaccharide vehicle for sustained delivery and colon specific delivery, while parenteral administration is used to target specific organs like joints (hyaluronic acid) or liver (pullulan) [29].

A new approach that is being examined in our work, aims at a double-functional anticancer effect of drug candidates, that is composed of a "core" chemotherapy drug (doxorubicin in this case) which is covalently attached to a galactose-branched polymer, that can interact with galactose-specific receptors of tumor surfaces, such as galectins, and results in a better drug delivery.

Doxo-Galactose(s)

In recent years, a series of anthracycline derivatives containing sugar moieties connected to the antibiotic through alkyl- or acyl-type spacers were synthesized in connection with gene-directed enzyme prodrug therapy (GDEPT) [22, 30]. In these cases, the sugar moieties serve as enzyme-specific functional groups of the anthracycline substrate. When a spacer is hydrolyzed by specific enzymes in the target tissue, the inactive prodrug is activated into the initial drug. The goal of our research, on the contrary, was to modify the initial drug with a stable galactose residue aiming at an increased chemotherapeutic activity.

We have synthesized a series of new N-substituted derivatives of doxorubicin containing *D*-galactose with hydrophilic (1-deoxyglucit-1-yl) or hydrophobic (benzyl or 3-methoxybenzyl) spacers [31], and test their cytotoxic and antitumor activity compared to the parent doxorubicin. Up to now no methods for the conjugation of the 3'-amino group of daunosamine with aldoses have been published.

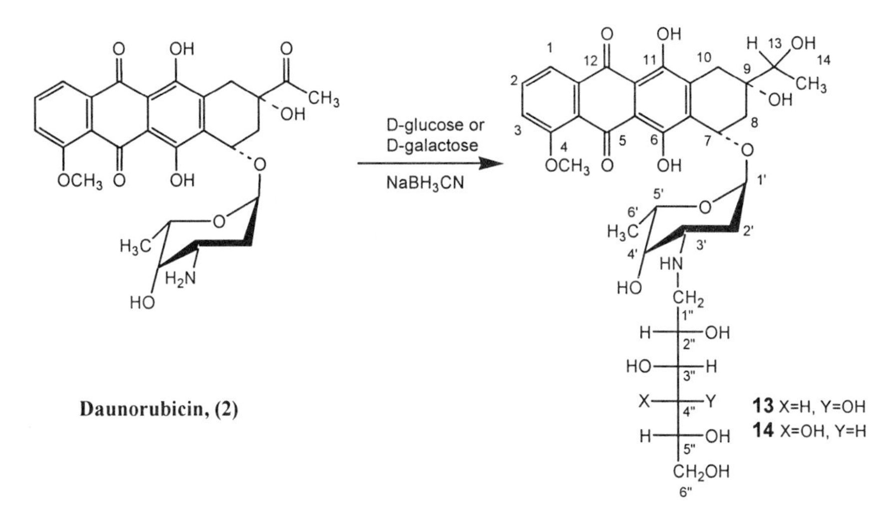

Scheme 1. Synthesis monosaccharide conjugates of 13-(R,S)-dihydrodaunorubicin (## 13,14).

In the first stage of our research, we studied the possibility of 3'-*N*-substitution of daunorubicin **(1)** by reductive alkylation using *D*-glucose or *D*-galactose and $NaBH_3CN$. Although the reductive *N*-alkylation of anthracycline antibiotics has been widely studied neither mono- nor disaccharides have been used in this reaction. By this method 3'-(1-deoxy-*D*-glucit-1-yl)- and 3'-(1-deoxy-*D*-galactit-1-yl) derivatives of 13-(R,S)-hydrodaunorubicin **(13, 14)** were isolated in ~ 20 % yields (Scheme 1). To protect the 13-CO group of the antibiotic from the reduction, the 13-dimethyl ketal of 14-bromodaunorubicin **(15)** was used as the starting compound, obtained from daunorubicin **(2)**. To introduce the *D*-galactose substituent, the disaccharides 6-*O*-α-*D*-galactopyranosyl-*D*-glucose (melibiose) **(16)** and 4-*O*-β-*D*-galactopyranosyl-*D*-glucose (lactose) **(17)** were used. 3'*N*-[α-*D*-galactopyranosyl-(1→6)-*O*-1-deoxy-*D*-glucit-1-yl]doxorubicin **(18)** and 3'*N*-[β-*D*-galactopyranosyl-(1→4)-*O*-1-deoxy-*D*-glucit-1-yl]doxorubicin **(19)** were obtained in 20 and 8% yields, respectively starting from **15** and melibiose **(16)** or lactose **(17)** with the use of $NaBCNH_3$ after hydrolysis of the intermediate bromoketals **18a** and **19a** (Scheme 2). *D*-Galactose has the α-anomeric configuration in compound **18** and the β-configuration in compound **19**; the polyhydroxylated hexit-1-yl spacer in compound **19** is shorter and more branched as compared to that in compound **18.** Compound **18** that showed the best antitumor properties (Table 1) was named **Doxo-Galactose**.

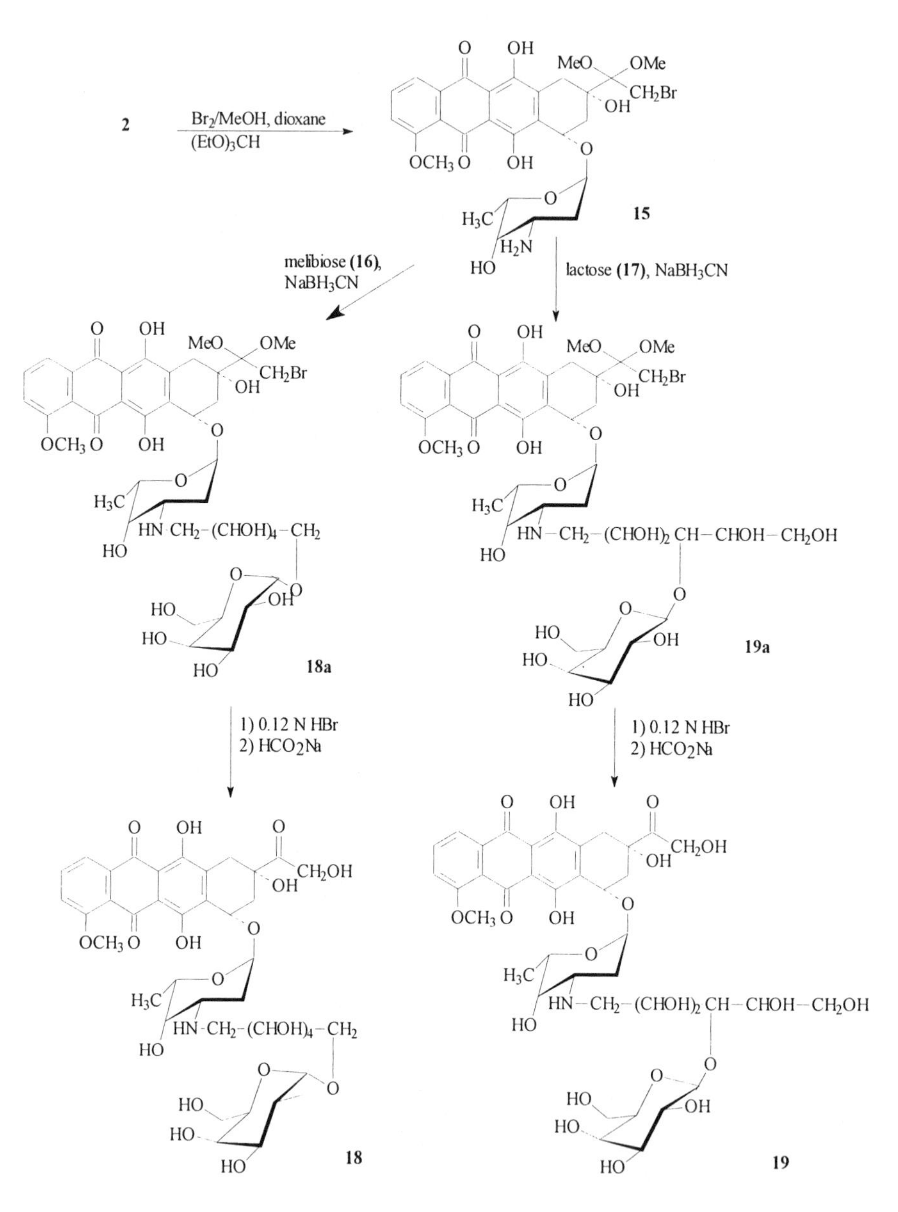

Scheme 2. Synthesis of conjugates of doxorubicin with melibiose (# 18) and lactose (# 19).

Table 1. Antitumor activity of Doxo-Galactose (18) and compounds 19, 27, and 28 [*i.v.* injection to mice (BDF_1 C_{57} B1 x DBA_2, males) with P388, 24 h *i.p.* post implantation of tumor] in comparison with doxorubicin

Compound	Dose (mg/kg)	Injection (*i.v.*) Regimen	Toxic death	ILS, %*
Control	--	--	0/10	0
Doxorubicin	7	Single	0/6	70
Doxorubicin	14	Single	2/6	--
Doxo-Galactose 18	20	Single	0/6	79
Doxo-Galactose 18	40	Single	0/6	118
Doxo-Galactose 18	60	Single	1/6	--
19	40	Single	0/6	65
19	80	Single	1/6	--
27	40	Single	0/6	35
27	60	Single	0/6	44
28	40	Single	0/6	52
28	60	Single	0/6	59
Doxorubicin	2.3	q2 x 3**	4/7	--
Doxo-Galactose 18	20	q2 x 3**	0/10	79
Doxo-Galactose 18	40	q2 x 3**	0/10	133

*Increase of lifespan
**24 h post *i.p.* tumor implanting

In the first stage of the synthesis of the conjugates **27** and **28** of doxorubicin with *D*-galactose linked to the antibiotic through the hydrophobic spacer, 3-methoxy-4-*O*-(2,3,4,6-tetra-*O*-acetyl-β-*D*-galactopyranosyl)- oxybenzaldehyde **(23)** and 4-*O*-(2,3,4,6-tetra-*O*-acetyl-β-*D*-galactopyranosyl)oxybenzaldehyde **(24)** were obtained by the reaction of 2,3,4,6-tetra-*O*-acetyl-α-*D*-galactopyranosyl bromide **(22)** with vanillin **(20)** or 4-hydroxybenzaldehyde **(21)**, respectively (Scheme 3). Reductive alkylation of the 3' amino group of **15** with compounds **23** or **24** by the use of $NaBCNH_3$ gave the corresponding derivatives **25** and **26** of the 13-dimethylketal of 14-bromodaunorubicin. After desacetylation of the galactose moiety in **25** and **26** with $NaOCH_3$ in methanol followed by acid hydrolysis of intermediates **25a** and **26a**, the desired doxorubicin derivatives **27** and **28** were obtained (Scheme 3).

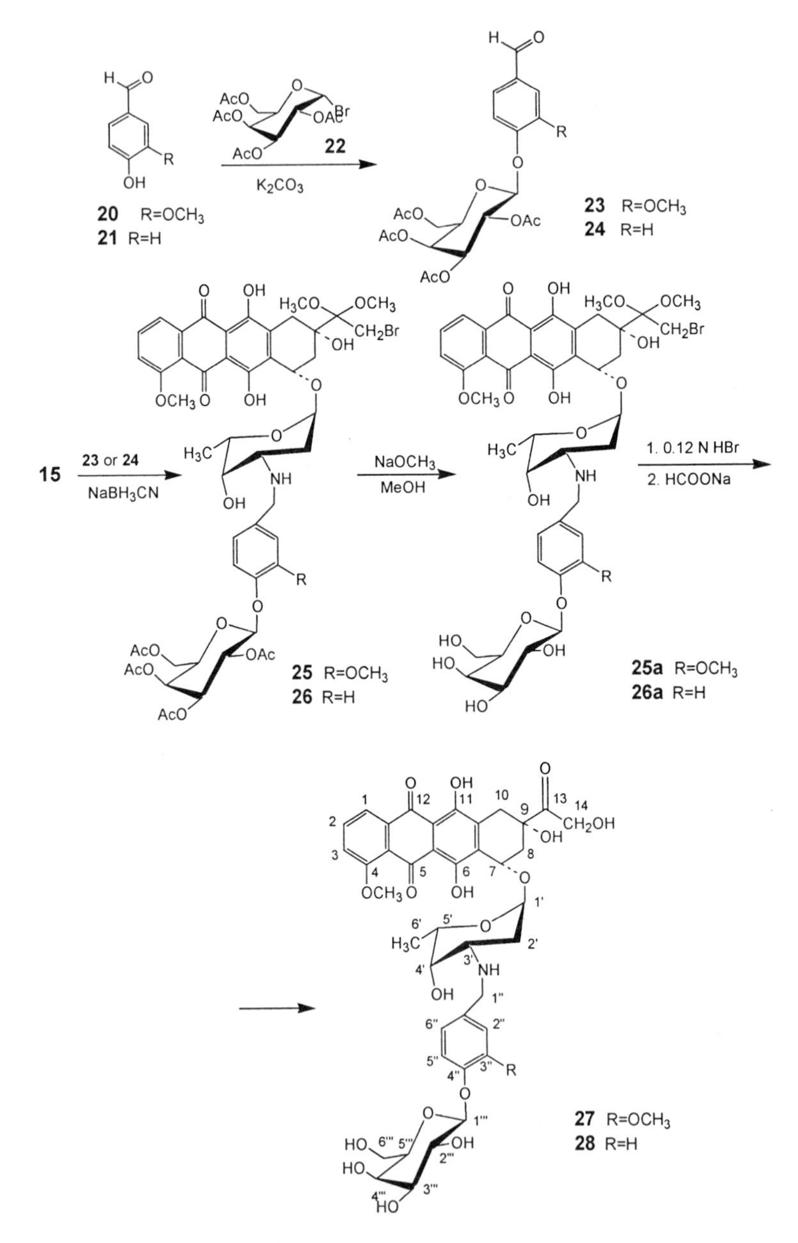

Scheme 3. Synthesis of conjugates of doxorubicin with galactose with hydrophobic spacer (## 27, 28).

3'*N*-(1-Deoxy-*D*-glucit-1-yl)doxorubicin (**29**) and 3'*N*-(1-deoxy-*D*-galact-1-yl)doxorubicin (**30**) were obtained from **15** and *D*-glucose or *D*-galactose by the similar scheme (Scheme 2) each in 5% yields (Fig. 6).

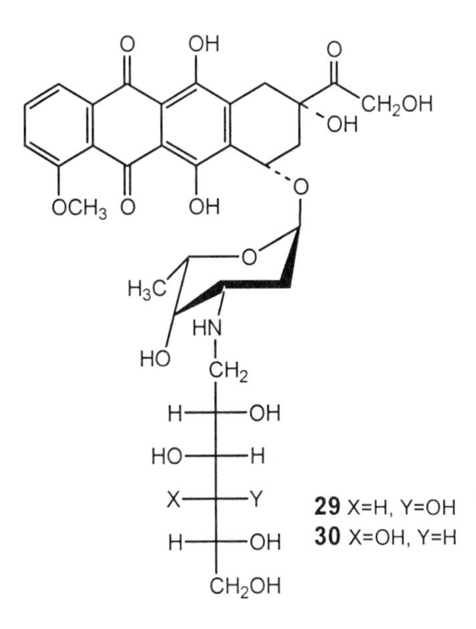

Figure 6. Structure of monosaccharide derivatives of doxorubicin (## 29,30).

Thin-layer chromatography (TLC) and high-performance liquid chromatography (HPLC) analyses showed that Doxo-Galactose (compound **18**), as well as compounds **19**, and **27** through **30** were homogeneous, and contain no admixed daunorubicin or doxorubicin. Under conditions of drastic acid hydrolysis (1N HCl, 105^0C, 1 h.) compounds **18, 19, 27,** and **28** produce the aglycon adriamycinone plus galactose, as demonstrated by TLC and paper chromatography, using authentic compounds as standards. Compounds **13** and **14** under similar conditions produced 13-(*R,S*)-dihydrodaunomycinone, and compounds **29** and **30** gave adriamycinone. NMR investigation permitted identification of all signals in the aglycon, spacers, and carbohydrate moieties, and mass-spectral data showed the correct molecular weights [31].

Antitumor activity

The *in vitro* inhibitory effects of Doxo-Galactose (**18**) and compound **19** on the proliferation of murine leukemia L1210/0 showed that these compounds are two orders less cytotoxic than doxorubicin: for Doxo-Galactose IC_{50} (50 %

inhibitory concentration) = 21 μM (L1210/0) and for **19** IC_{50} = 24 μM, whereas for doxorubicin IC_{50}= 0.213 μM.

In vivo studies revealed differences in the antitumor properties of Doxo-Galactose and compound **19.** The maximum tolerated dosages (MTD) were 40-60 mg/kg for Doxo-Galactose, 60-80 mg/kg for **19**, and >60 mg/kg for both **27** and **28** (single i.v. injection, BDF_1 mice ($C_{57}Bl$ x DBA_2, males), as compared to 7-10 mg/kg for doxorubicin. The antitumor activity for these compounds was studied on mice bearing lymphocyte leukemia P-388 at single and multiple (q2d x 3) i.v. injection regimens (Table 1).

3.1. Single injection regimen.

When **19** was *i.v.* injected to mice with P-388 (BDF_1 mice) 24 h post *i.p.* implantation of the tumor, 65 % ILS at the dose of 40 mg/kg was achieved. Doxo-Galactose was more active than compound **19**: at the dose of 20 mg/kg it induced 79 % ILS and at 40 mg/kg 118 % ILS (without toxic effects). A dose of 60 mg/kg showed some toxicity. The maximal antitumor effect of doxorubicin was 70 % at the MTD dose of 7 mg/kg. Compounds **27** and **28** at doses of 40 and 60 mg/kg induced ILS in the range of 35-44% (**27**) and 52-59% (**28**), respectively.

3.2 Multiple injection regimens.

Multiple injections of doxorubicin revealed an increased toxicity of the drug, apparently due to its cumulative toxic effect. Compared to a single dose regimen, when at the dose of 7 mg/kg there was no toxic deaths (70% ILS), triple injection of doxorubicin with 2-day intervals (q2d x 3) at 2.3 mg/kg each dose (6.9 mg/kg total dose) resulted in four toxic deaths out of seven animals. However, Doxo-Galactose did not show any cumulative toxic effect at the triple dose of 40 mg/kg (120 mg/kg total dose), and induced ILS equal to 133%. Hence, Doxo-Galactose may be of interest for supportive therapy.

This study shows that doxorubicin derivatives containing at the nitrogen atom of the daunosamine moiety a polyhydroxylated spacer connected in turn with the galactose moiety, and particularly Doxo-Galactose, may afford compounds having lower toxicity and better pharmaceutical index compared to the parent doxorubicin. Furthermore, this study shows that the α-anomeric configuration of D-galactose rather than β-, and/or a longer length of the spacer (six carbon atoms rather than four in this particular case of Doxo-Galactose and compound **19** may result in a more efficacious drug candidate in this series.

Doxo-Galactomannan(s)

Earlier it was shown [27], and described in chapter "DAVANAT® and colon cancer: Preclinical and clinical (Phase I) studies" above, that co-administration of a soluble 1,4-β-D-galactomannan (GM, Scheme 4; Man/Gal ratio = 1.7) along with antitumor drug 5-fluorouracil (5-FU) by *i.v.* injection to mice bearing human colon tumors (COLO 205 and HT-29) significantly increased efficacy of the 5-FU. Combination of the galactomannan with 5-FU, compared to 5-FU alone, resulted in the decrease of median tumor volume to 17-65% and the increase of mean survival time (days) to 150-190%.

Compared to many other polysaccharides, galactomannans have multiple side-chain galactose units that may readily interact with galactose-specific receptors such as galectins on the tumor cell surface, modulate the tumor surface physiology and potentially affect delivery of drug to the tumor.

The aim of our research was the synthesis of a conjugate of doxorubicin and galactomannan that would act like a double prodrug, that is liberating both doxorubicin and DAVANAT® in *in vivo* conditions and, importantly, in proximity to galactose/manose specific receptors.

Earlier conjugates of doxorubicin with different polysaccharides were described [24]. In most cases doxorubicin was bound to a polysaccharide through amine bond. Typically, a polysaccharide was activated to a dialdehyde derivative by periodate oxidation. In other words, for the introduction of the anthracycline antibiotic, the reductive alkylation of doxorubicin with the oxidized polysaccharide was used [32]. However, the reduction of the Schiff base led to simultaneous reduction of 13-keto group of doxorubicin resulting in an N-alkyl derivative of 13-(R,S)-dihydrodoxorubicin, as it takes place in the case of daunorubicin (see Scheme 1). 13-(R,S)-dihydrodoxorubicin is less active then doxorubicin. Besides, doxorubicin conjugates obtained by this method cannot be considered as doxorubicin prodrugs, as no proof that the antibiotic can be released from the polymer in physiological conditions was shown. That is why we selected binding doxorubicin to the galactomannan *via* imine bond (Schiff base), in anticipation that the conjugate will be stable in both mild basic and neutral conditions, and will slowly liberate doxorubicin in mild acid conditions.

In this Chapter we describe synthesis and biological activity of two DAVANAT®-doxorubicin conjugates, tentatively named Doxo-Galactomannan (GM)-1 and -2, and resulting from (a) a direct covalent binding of the two compounds, and (b) the linking of doxorubicin to DAVANAT® via an L-lysyl-bridge.

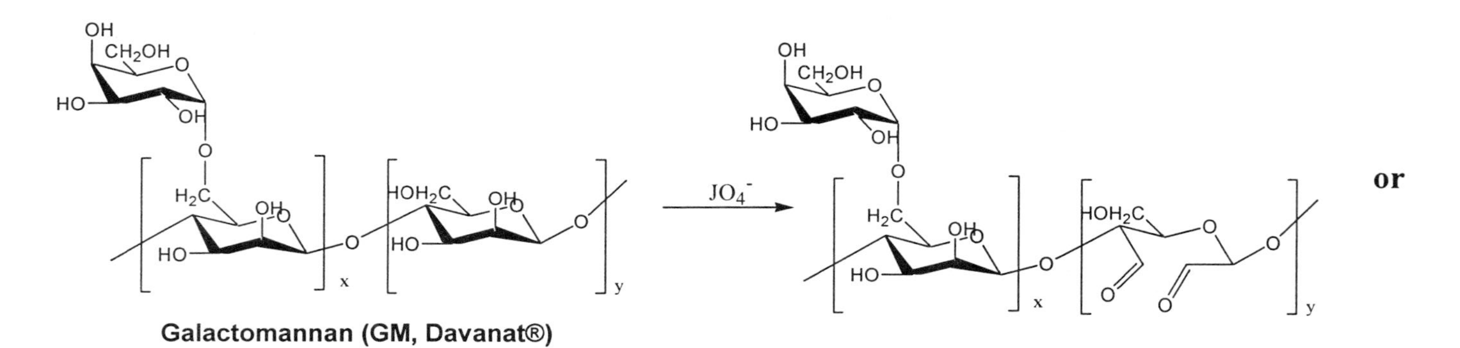

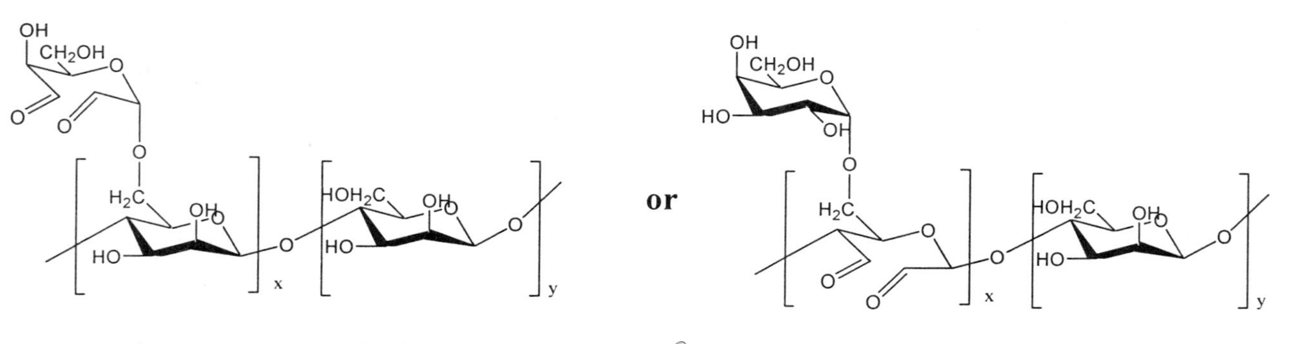

Scheme 4. Activation of galactomannan (DAVANAT®) to a dialdehyde derivative by periodate oxidation.

The initial water-soluble 1,4-β-D-galactomannan (DAVANAT®) was obtained by a controlled partial chemical degradation of a high molecular weight galactomannan from *Cyamopsis tetragonoloba*, or guar gum [27, 33] and further purification under Good Manufacturing Practice (GMP) conditions (Chapter "DAVANAT®: A modified galactomannan that enhances chemotherapeutics. Chemistry, manufacturing, control" in this book). Molecular weight of the obtained DAVANAT® was 54,000 Da (MALLS, or Multi-Angle Laser Light Scattering), 95,000 Da (viscometry), or 92,000 Da (GPC, or gel permeation chromatography). Mannose/galactose ratio (by the NMR and chemical analysis after complete acid hydrolysis) of DAVANAT® was 1.7 [27]. DAVANAT® was activated to a dialdehyde derivative by periodate oxidation (Scheme 4) with a subsequent purification using gel permeation chromatography.

Purified activated DAVANAT® was coupled with doxorubicin to form a Schiff base (Scheme 5 exemplifies one of the variants of dialdehyde-GM). A series of experiments aimed at various degrees of oxidation of DAVANAT® and performed with various amounts of doxorubicin showed that a conjugate was soluble in water only at a relatively low content of doxorubicin in the conjugate, about 5% by weight. The obtained Doxo-GM-1 behaved at gel-filtration and dialyses similarly with the initial DAVANAT®.

Molecular weight of Doxo-GM-1 was 95,000 Da (GPC), while for the starting DAVANAT® it was 92,000 Da when determined on the same column.

The final preparation of Doxo-GM-1 was obtained as a pink powder, and it contained 5% of doxorubicin covalently attached to the initial DAVANAT®, as evaluated by UV absorption at 490 nm. It did not contain free doxorubicin. UV absorbance spectra of Doxo-GM-1 and free doxorubicin revealed that their characteristic absorbance peaks were practically identical.

A plausible cyclic imine structure for the obtained Doxo-GM-1 is shown on Scheme 5. Linear structure, or a combination of the two cannot be excluded.

In order to further increase the amount of doxorubicin in the conjugate, Doxo-GM-2 was synthesized, in which a doxorubicin derivative carried an additional amino group. For this purpose 3'N-L-Lysyl-doxorubicin was chosen, taking into account its higher water-solubility compared to doxorubicin.

Doxo-GM-2 was synthesized from the initial DAVANAT® and 3'N-L-lysyl-doxorubicin (**34**). L-Lysine was converted to N^{α}, N^{ε}-Di-(9*H*-fluorenylmethoxycarbonyl)-L-lysine (**31**), using N-(9*H*-fluorenyl-methoxycarbonyloxy)-succinimide in the presence of dicyclohexylcarbodiimide (Scheme 6). Then, N^{ε}-Di-(9*H*-fluorenylmethoxycarbonyl)-L-lysyl]oxy-

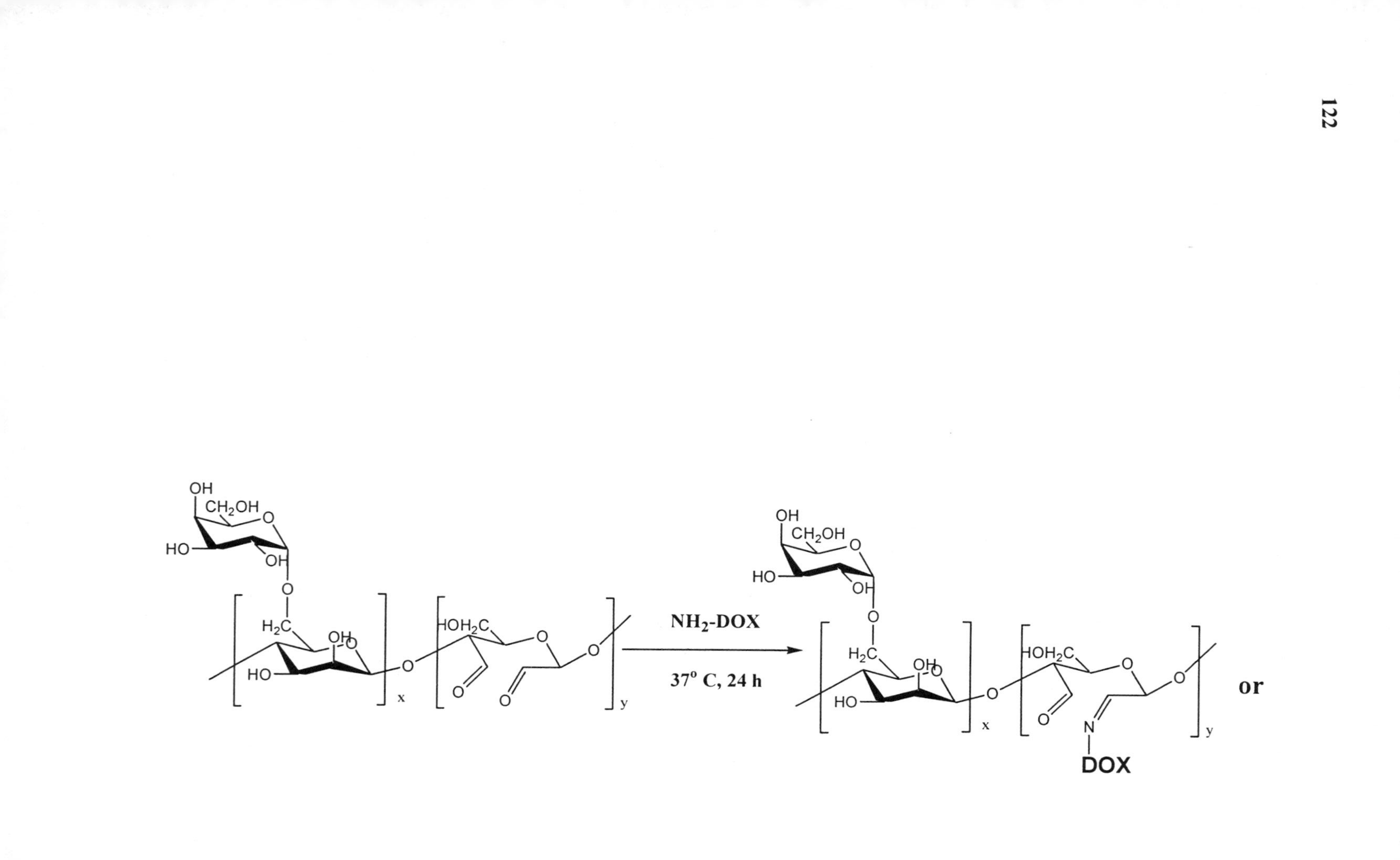

OH
CH2OH
HO
OH
H2C
HOH2C
x
y
NH2-DOX
37° C, 24 h
N
DOX
or

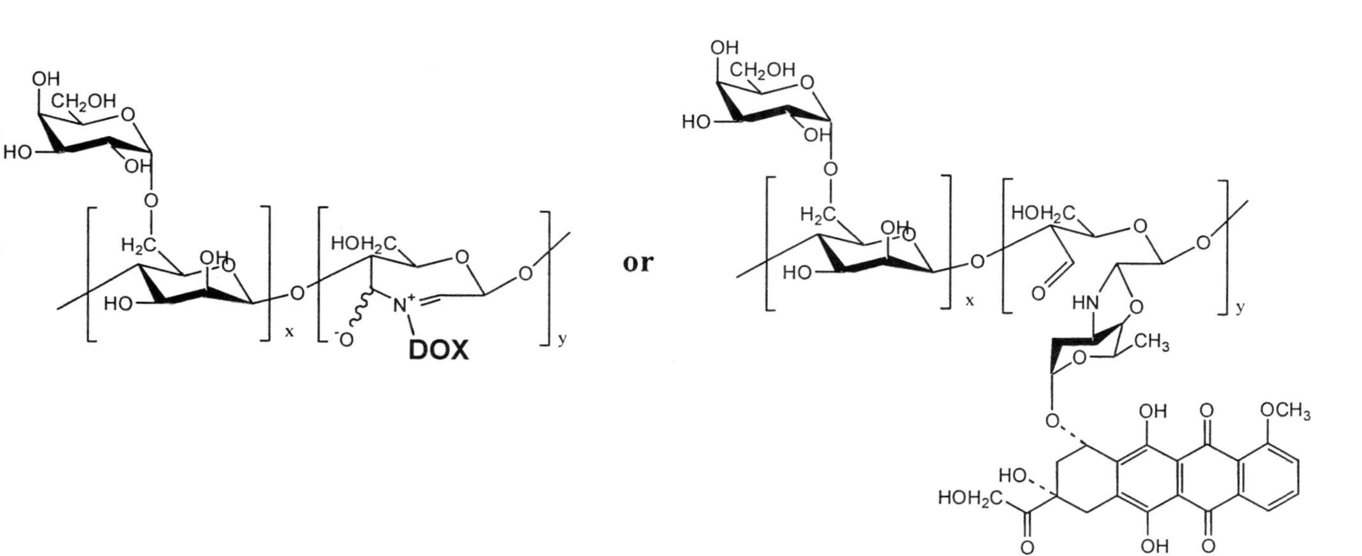

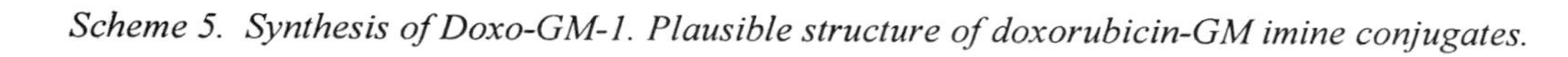

Scheme 5. Synthesis of Doxo-GM-1. Plausible structure of doxorubicin-GM imine conjugates.

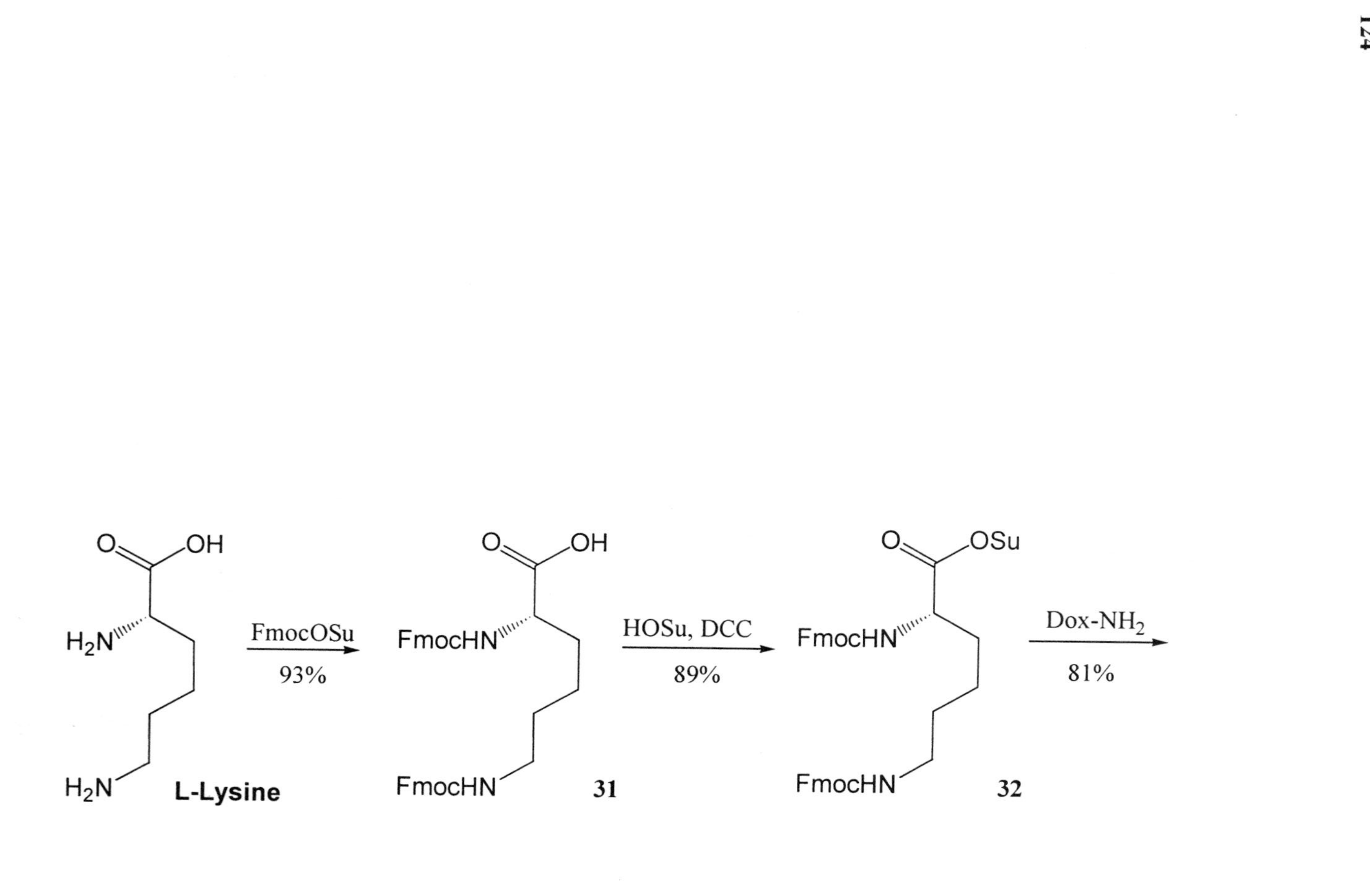

O
OH
H2N
H2N
L-Lysine
FmocOSu
93%
O
OH
FmocHN
FmocHN
31
HOSu, DCC
89%
O
OSu
FmocHN
FmocHN
32
Dox-NH2
81%

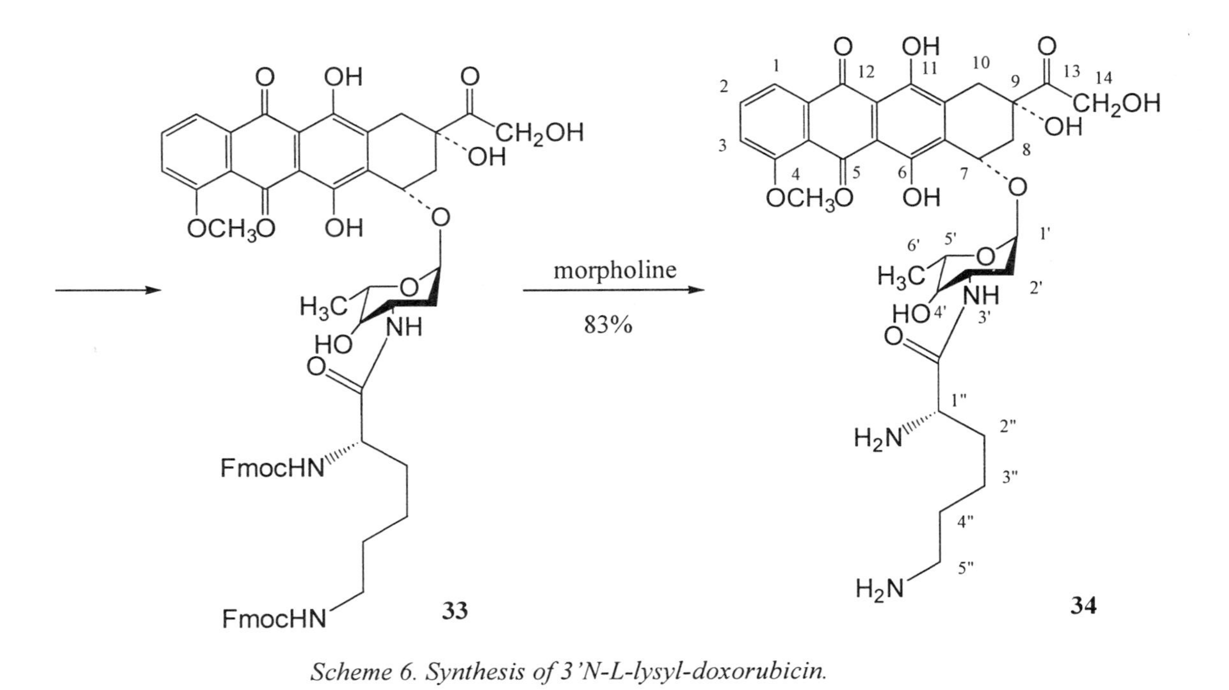

Scheme 6. Synthesis of 3'N-L-lysyl-doxorubicin.

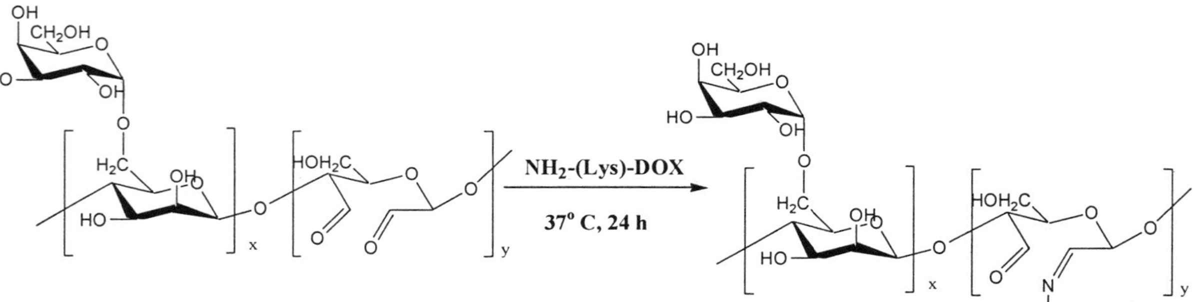

or

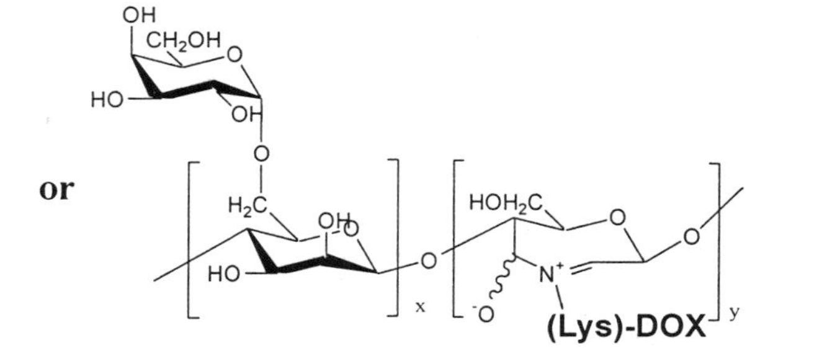

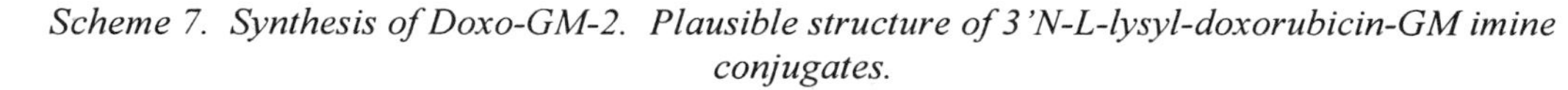

Scheme 7. Synthesis of Doxo-GM-2. Plausible structure of 3'N-L-lysyl-doxorubicin-GM imine conjugates.

succinimide (**32**), *in situ* was obtained. It was coupled with doxorubicin in the next step, into 3N'-[N$^{\alpha}$, N$^{\varepsilon}$-Di-(9*H*-fluorenylmethoxycarbonyl)-L-lysyl]doxorubicin (**33**), which was then, in the next step, deblocked into 3'N-L-lysyl-doxorubicin (**34**) with 67% yield (re. doxorubicin). The structure of 3'N-L-lysyl-doxorubicin (**34**) was confirmed using NMR and ESI-mass spectrometry. DAVANAT® was oxidized (activated) as described above (Scheme 4) and conjugated with hydrochloride of **34** to give Doxo-GM-2 (Scheme 7).

Doxo-GM-2 was purified using dialysis and GPC. The final preparation was obtained as dark red powder containing 10% of covalently bound doxorubicin, as evaluated by UV absorption at 490 nm. Molecular weight of Doxo-GM-2 was 98,000 Da (GPC), compared to 92,000 Da for the initial DAVANAT®.

Cytotoxicity

Cytotoxic activities of Doxorubicin, Doxo-Galactose, Doxo-GM-1 and Doxo-GM-2 have been tested using the B16-F1 mouse melanoma cells, MCF-7 (HTB-22) human breast cancer cells, and HT-29 (HTB-38) human colon cancer cells. Data are shown in Table 2. One can see that Doxorubicin is much more cytotoxic compared to all its derivatives. *In vivo* experiments are in progress.

Table 2. ED_{50} (growth inhibition activity, per weight of the compounds) for doxorubicin, Doxo-Galactose, Doxo-GM-1 and -2 towards three cancer cell lines

Doxorubicin and its derivatives	**ED_{50}, µg/mL**		
	B16-F1 mouse melanoma cells	**MCF-7 (HTB-22) human breast cancer cells**	**HT-29 (HTB-38) human colon cancer cells**
Doxorubicin	0.01-0.02	0.08-0.12	0.2-0.3
Doxo-Galactose	0.4-0.6	1.7-2.6	1.5-2.2
Doxo-GM-1	0.5-0.8	3.0-4.4	13-20
Doxo-GM-2	>50	>50	>50

References

1. Arcamone F, "Doxorubicin: Anti-Cancer Antibiotics", Medicinal Chemistry Series, 1981, Vol. 17, Academic Press
2. Hutchinson C. R., "The Biosynthesis of Tetracycline and Anthracycline Antibiotics," in Antibiotics IV Biosynthesis, 1981, pp. 1-11, Ed.: J. W. Corcoran, Pub.: Springer-Verlag.
3. R. J. White, "Anthracyclines," in Biochemistry and Genetic Regulation of Commercially Important Antibiotics, 1983, pp. 277-291, Ed.: L. C. Vining, Pub.: Addison Wesley
4. Reynolds J. E. F., ed. Martindale: The Extra Pharmacopoeia, 28th ed. London: The Pharmaceutical Press, 1982, pp. 205-208.
5. McEvoy G.K, ed. American Hospital Formulary Service: Drug information 1991. Bethesda: American Society of Hospital Pharmacists, 1991, pp.527-530.
6. Pinedo H.M.,. Chabner B.A , eds. Cancer Chemotherapy/7: The EORTC Cancer Chemotherapy Annual. New York: Elsevier Science Publishing Co Inc, 1985, pp. 57-75.
7. Creasey W.A., McIntosh L.S., Brescia T., Odujinrin O., Aspnes G.T., Murray E., and Marsh J.C. Clinical Effects and Pharmacokinetics of Different Dosage Schedules of Adriamycin. Cancer Res, 1976, 36, pp:216-221.
8. Bachur N.R.. Adriamycin (NS-123127) Pharmacology. Cancer Chemother Rep, 1975, 6, 153-158.
9. Ambudkar S.V., Dey S., Hrycyna C.A., Ramachandra M., Pastan I., Gottesman M.M.; "Biochemical, cellular, and pharmacological aspects of the multidrug transporter." Annu. Rev. Pharmacol. Toxicol. 1999, vol. 39, pp. 361-398,
10. Zunino F., Pratesi G., Perego P. Role of the Sugar Moiety in the Pharmacological Activity of Anthracyclines: Development of a Novel Series of Disaccharide Analogs. Biochem. Pharmacol., v. 61, 933-938, 2001.
11. DuVernay V.H. Molecular Pharmacology of Anthracycline Antitumor Antibiotics. In: Canver and Chemotherapy, vol. III. New York: Academic Press, 1981, p. 233-271
12. Gonzalez-Paz O., Polizzi D., De Cesare M., Zunino F., Bigioni M., Maggi C.A., Manzini S., Pratesi G. Tissue Distribution, Antitumor Activity and *in vivo* Apoptosis Induction by MEN10755 in Nude Mice. Eur. J. Cancer 37, 431-437, 2001
13. Horton D., Priebe W., Sznaidman M.L., Varela O. Synthesis and antitumor activity of anthracycline disaccharide glycosides containing daunosamine. J. Antibiot., 46, 1720-1730, 1993.

14. Animati F., Arcamone F., Giannini G., Lombardi P., Monteagudo E. S-fluoro-anthracyclines, processes for their preparation and pharmaceutical compositions containing them. U.S. Patent No. 5,814,608, September 29, 1998
15. Animati F., Berettoni M., Cipollone A., Franciotti M., Lombardi P., Monteagudo E., Arcamone F., New Anthracycline Disaccharides Synthesis of L-daunosaminyl-α-(1→4)-2-deoxy-L-rhamnosyl and of L-daunosaminyl-α-(1→4)-2-deoxy-L-fucosyl daunorubicin analogues. J. Chem. Soc. Perkin Trans I, 1327-1329, 1996.
16. Arcamone F., Animati F., Bigioni M., Capranico G., Caserini C., Cipollone A., De Cesare M., Ettorre A., Guano F., Manzini S., Monteagudo E., Pratesi G., Salvatore C., Supino R., Zunino F., Configurational Requirements of the Sugar Moiety for the Pharmacological Activity of Anthracycline Disaccharides. Biochemical Pharmacology 57, 1133-1139, 1999.
17. Arcamone F., Animati F., Berettoni M., Bigioni M., Capranico G., Casazza A.M., Caserini C., Cipollone A., De Cesare M., Franciotti M.,Lombardi P., Madami A., Manzini S., Monteagudo E., Polizzi D., Pratesi G., Righetti S.C., Salvatore C., Supino R., Zunino F. Doxorubicin Disaccharide Analogue: Apoptosis-Related Improvement of Efficacy *in vivo*. J. Natl. Cancer Inst. 89, 1217-1223, 1997.
18. Pratesi G., Monestiroli S.V. Preclinical evaluation of new anthracyclines. Curr. Med.Chem., 8, 9-13, 2001.
19. Pratesi G., De Cesare M., Caserini C., Perego P., Dal Bo L., Polizzi D., Supino R., Bigioni M., Manzini S., Iafrate E., Salvatore C., Casazza A., Arcamone F., Zunino F. Improved Efficacy and Enlarged Spectrum of Activity of a Novel Anthracycline Disaccharide Analogue of Doxorubicin Against Human Tumor Xenografts. Clin. Cancer Res. 4, 2833-2839, 1998.
20. Leenders R.G.G., Gerrits K.A.A., Ruijtenbeek R., Scheeren H.W., Haisma H.J., Boven E. β-Glucuronyl Carbamate Based Pro-moieties Designed for Prodrugs in ADEPT. Tetrahedron Letters, 36, 1701-1704, 1995
21. Kenten J., Simpson D.M. Compounds and methods for the selective treatment of cancer and bacterial infections. U.S. Patent No. 6,218,519, April 17, 2001.
22. Ghosh A.K., Khan S., Marini F., Nelson A., Farquhar D. A Daunorubicin β-Galactoside Prodrug for Use in Conjunction with Gene-Directed Enzyme Prodrug Therapy. Tetrahedron Letters 41, 4871-4874, 2000.
23. Bakina E., Farquhar D. Intensely cytotoxic anthracycline prodrugs: galactosides. Anticacer Drug Des., *14*, 607-515, 1999.
24. Domb A.J., Benita S., Polacheck I., Linden G. "Conjugates of biologically active substances", United States Patent, patent number 6,011,008; Jan. 4, 2000
25. Barondes, S.H.; Castronovo, V.; Cooper, D.N.W.; Cummings, R.D.; Drickamer, K.; Feizi, T.; Gitt, M.A.; Hirabayashi, J.; Hudges, C.; Kasai, K.-I.; Leffler, H.; Liu, F.-T.; Lotan, R.; Mercurio, A.M.; Monsigny, M.; Pillai, S.; Poirer, F.; Raz, A.; Rigby, P.W.J.; Rini, J.M.; Wang, J.L. Galectins: A

Family of Animal beta-galactoside-binding Lectins *Cell,* **1994**, *76*, 597-598.

26. Lotan, R.; Ito, H.; Yasui, W.; Yokozaki, H.; Lotan, D.; Tahara, E. Expression of a 31-kDa Lactoside-binding Lectin in Normal Human Gastric Mucosa and in Primary and Metastatic Gastric Carcinomas. *Int. J. Cancer* **1994**, *56,* 474-480 .
27. Klyosov, A.A., Platt, D., and Zomer, E. Preclinical Studies of Anticancer Efficacy of 5-Fluorouracil when Co-Administered with the 1,4-β-D-Galactomannan. Preclinica, 2003, 1, No. 4, 175-186.
28. Eric WP Damen, Franciscus MH de Groot, Hans W. Scheeren; "Novel Anthracycline prodrugs." Exp. Opin. Ther. Patents, 2001, vol. 11 (4), pp 651-666
29. Ahmad S., Tester R. "Polysaccharides as drug carries", 22nd International carbohydrate Symposium, Glasgow, UK, 23-27 July, 2004, C37.
30. Houba, P.H.J.; Boven, E., Erkelens, C.A.M.; Leenders, R.G.G.; Scheren, J.W.; Pinedo, H.M., Haisma, H.J. The efficacy of the anthracycline prodrug daunorubicin-GA3 in human overian cancer xenoggragts. Br. J. Cancer, 1998, 78 (12), 1600-1606
31. Olsufyeva E.N., Tevyashova A.N., Trestchalin I.D., Preobrazhenskaya M.N., Platt D., Klyosov A.; "Synthesis and antitumor activity of new D-galactose-containing derivatives of doxorubicin." Carbohydrate Research, 2003, vol. 338, pp. 1359-1367.
32. T. Ouchi., M. Matsumoto, K. Ihara, and Y. Ohya; "Synthesis and Cytotoxicity of Oxidized Galactomannan/ADR Conjugate", Pure Appl. Chem., 1997, A34(6), pp. 975-989.
33. Klyosov, A.A. and Platt, D. U.S. Patent No. 6,645,946. Delivery of a therapeutic agent in a formulation for reduced toxicity, November 11, 2003.

Human Immunodeficiency Virus

Chapter 6

Carbohydrate-Based Vaccines against HIV/AIDS

Lai-Xi Wang

Institute of Human Virology, University of Maryland Biotechnology Institute, University of Maryland, 725 West Lombard Street, Baltimore, MD 21201

The human immunodeficiency virus (HIV) is the cause of AIDS. Despite tremendous efforts in the past two decades, an effective HIV/AIDS vaccine capable of inducing broadly neutralizing antibody response in humans is still not on the horizon. HIV has evolved a number of mechanisms to escape the host immune surveillance, which account for the difficulties in HIV vaccine development. The outer envelope glycoprotein gp120 is heavily glycosylated. The carbohydrates cover a major area of the protein surface and constitute a strong defense mechanism for the virus to evade host immune attacks. Accumulating evidence has implicated that the HIV-1 carbohydrates themselves could also serve as targets for vaccines. A notable example is the broadly neutralizing antibody 2G12 that exerts its anti-HIV activity by targeting a multivalent oligomannose cluster on gp120. The present review provides an overview on the structure and function of HIV-1 carbohydrates as well as their relevance to HIV vaccine design. Recent work on the synthesis and antigenicity of HIV-1 carbohydrate epitopes and their mimics is highlighted.

Introduction

The acquired immunodeficiency syndrome (AIDS) is caused by the infection of human immunodeficiency virus (HIV) (*1-3*). HIV type 1 (HIV-1) is the human retrovirus that causes the global epidemic of HIV/AIDS. Since the identification of the first case in 1981, AIDS has become one of the deadliest diseases in human medical history. According to the statistics from the World Health Organization, some 25 million people had already died of the disease within the past two decades and over 40 million people are currently living with HIV/AIDS (*4*). The epidemic is still expanding. Most people agree that the best hope to stop the worldwide HIV/AIDS epidemic is the development of an effective HIV vaccine (*5,6*).

Vaccination has been successful in controlling some of the worst infectious diseases in human history. For example, vaccines have virtually eradicated polio (*7,8*) and smallpox (*9,10*), which once took millions of lives on the earth. Vaccination has also played a key role in combating common flu. Therefore, it was once expected that a vaccine would follow closely behind upon the identification of HIV as the cause of AIDS. However, developing an effective HIV vaccine has proven to be much more difficult and complicated than anticipated.

The large-scale, phase III clinical trial (a 5000-person, 3-year trial) of a candidate vaccine, AIDSVAX (a genetically engineered version of monomeric gp120), was completed in 2003 but yielded only disappointing results. The vaccine raised very weak neutralizing antibodies and did not show protective effects in humans (*11*). The failure of the clinical trials with gp120 was not a surprise to many who have been working in the HIV/AIDS field. In order to achieve persistent infection, HIV has evolved strong defense mechanisms to evade immune recognition, including frequent mutation of neutralizing epitopes, heavy glycosylation of the envelope, switch of conformations, and formation of oligomeric envelope spikes (*5,6,12,13*). Each level of the HIV's defenses provides an additional dimension of complexity in vaccine design.

A number of excellent reviews have been published to address general scientific obstacles and strategies in HIV-1 vaccine development (*5,6,12-20*). We have recently written a comprehensive review on the bioorganic aspects of HIV-1 vaccine design (*14*). For an effective HIV-1 vaccine, at least one of the following two types of immune effectors must be elicited: the neutralizing antibodies that bind and clean up the virus (humoral immunity) (*12,19*), or the HIV-specific cytotoxic T lymphocytes (CTL) that recognize and kill infected cells (cellular immunity) (*21,22*). While CTL alone may not be sufficient to prevent HIV infection, experiments in animal models have proved that sufficient levels of broadly neutralizing antibodies can protect animals against viral challenge (*23,24*). Therefore, the design of an immunogen capable of inducing broadly neutralizing antibody responses remains a major goal in HIV vaccine development.

Work on antibody-based HIV-1 vaccines has hitherto focused on the protein backbone and related peptide fragments of the viral envelope glycoproteins. However, frequent sequence mutation of the neutralizing epitopes poses a major challenge for the conventional vaccine approaches. New concepts need to be developed in HIV vaccine design. Accumulating data have suggested that the carbohydrate portions of HIV-1 envelope can also serve as attractive targets for HIV-1 vaccine development. Carbohydrates account for about half of the molecular weight of the outer envelope glycoprotein gp120, which cover a large area of the surface of the envelope and play a major protective role for viral immune evasion. Immunological and biochemical studies have indeed implicated the existence of novel carbohydrate epitopes on HIV-1 envelope. For example, the epitope of the broadly neutralizing antibody 2G12 has been characterized as a unique cluster of high-mannose type oligosaccharides located on HIV-1 gp120 (*25-27*). The present review intends to provide an overview on our understanding of the structure and biological functions of the HIV-1 carbohydrates, and on how the information might be explored for developing carbohydrate-based vaccines against HIV/AIDS. Recent work on the design, synthesis, and antigenicity of various HIV-1 carbohydrate antigens and their mimics is highlighted.

Structural Features of Glycosylation of HIV-1 Gp120

HIV-1 has two envelope glycoproteins, gp120 and gp41. A typical gp120 molecule contains about 24 consensus N-glycosylation sites (NXS/T). Some early immunochemical studies also implicated the existence of O-linked glycosylations in gp120 (*28-30*). But direct structural evidence for the existence of O-glycans on HIV-1 gp120 is lacking. Basics of glycobiology and AIDS were discussed in two excellent earlier reviews (*31,32*). Discussed here are some major findings and recent progresses on the structural studies.

Several groups have investigated the structures of HIV-1 gp120 N-glycans from both virus-derived and recombinant gp120, which were found to be highly heterogeneous and diverse (*33-37*). On the other hand, the transmembrane envelope glycoprotein gp41 has four conserved N-glycosylation sites that are normally occupied by N-glycans (*38,39*). However, to the best of our knowledge, the types of the N-glycans on each of the glycosylation sites of gp41 have not been fully characterized.

Analysis of the glycosylation pattern of the HIV-1_{IIIB} gp120 expressed in Chinese hamster ovary (CHO) cells has revealed that all the 24 potential N-glycosylation sites are utilized for carbohydrate attachment. These include 13 sites for complex type oligosaccharides and 11 sites for high-mannose type or hybrid type oligosaccharides (*35*). The significantly high numbers of high-mannose type oligosaccharides are unusual for a mammalian glycoprotein. A high density of high-mannose type oligosaccharides was also observed for free, virus-derived gp120, as well as the gp120 associated with HIV-1 infected lymphoblastoid (H9) cells. In one study, high-mannose type structures (Man_5 to Man_9) was shown to be approximate 60% of the total sugar chains (*36*); in another investigation, the high-mannose type structures (Man_7 to Man_9) account for 80% of the total N-glycans for T cell-associated gp120, and 50% for the free virus-derived gp120 with Man_8 and Man_9 as the main components (*34*).

As for the complex type N-glycans on gp120, bi-, tri-, and tetra-antennary structures , with varied degree of sialylations, were characterized (*33,34,36,37*). In addition, a significant portion of the complex type oligosaccharides of the virus-derived gp120 was found to bear bisecting N-acetylglucosamine residues (*36*). The structural diversity of HIV-1 N-glycans is also dependent on the types of infected cells where the virus was derived, and on the expression system where the recombinant gp120 was produced. For example, the N-glycans of the recombinant gp120 produced in insect cells were found to be of high-mannose type but not complex type (*40,41*). There was also evidence for the presence of Fuc-α-(1-3) or α-(1-4)-GlcNAc moieties beyond the chitobiose core in the N-glycans of HIV-1 gp120 (*42*). Moreover, sulfation of GlcNAc residues of complex type N-glycans was characterized on both HIV-1 gp120 and gp160 produced in the human lymphoblastoid cell line Molt-4 (*43*).

Clustering of oligosaccharides on the viral surface is another feature of HIV-1 glycosylation. The three-dimension structure of de-glycosylated gp120 was resolved (*44,45*). The core gp120 structure allowed an investigation on how the carbohydrates were globally distributed on the surface of the HIV-1 gp120. Remodeling of N-glycans on the de-glycosylated core revealed that a great deal of gp120 surface was covered by a carbohydrate coat (*13,44,45*). These N-glycans shield some receptor-binding regions of the protein backbone, and limit the access of neutralizing antibodies.

Since the HIV-1 carbohydrates are produced by the host glycosylation machinery and may appear as "self" to the immune system, the dense glycosylation would reduce the potential of the underneath protein backbone to serve as immunological targets. The dense glycosylation region on gp120 is thus called immunologically "silent" surface (*45*). Interestingly, when the types of oligosaccharides of N-glycans are sorted out at each glycosylation site, the high-mannose type N-glycans appear to be clustered together on one surface, while the complex type N-glycans are clustered as a distinct domain on another area of the surface, with little structural overlap of the two domains (*37*). The presence of the two distinct oligosaccharide clusters, together with some "unusual" structures in the subunit oligosaccharides (e.g., fucosylation, sulfation, and bisecting structures), may provide new carbohydrate targets for vaccine development.

Biological Functions of HIV-1 N-Glycosylation

Effects of Glycosylation on the Antigenicity and Immunogenicity of the Viral Envelope

Early studies have demonstrated that N-glycosylation of HIV-1 affects the processing and maturation of the envelope glycoproteins and the infectivity of the virus (*46-48*). Glycosylation is also necessary for the correct folding of HIV-1 gp120 (*49*). Meanwhile, compelling data have shown that N-glycans of gp120 exert profound effects on the antigenicity and immunogenicity of the viral envelope glycoproteins. The dense carbohydrate shield forms a strong barrier to help protect HIV from immune recognition and limit effectiveness of antibody neutralization.

Strong evidence has come from the studies in rhesus monkeys with simian immunodeficiency virus (SIV) mutants lacking N-glycans in the V1 region of the envelope (*50*). When glycosylation sites were removed from the V1 loop of the SIV gp120, not only did the resulting mutants become much more sensitive to antibody neutralization, but the glycosylation mutants also exhibited greatly enhanced immunogenicity in the animals, eliciting much better immune responses that are reactive to both mutants and the wild-type virus (*50*). Similarly, it was also reported that deleting the V1/V2 loops from HIV-1 gp120, as well as removing other specific N-glycosylation sites on the HIV-1 envelope glycoprotein, rendered the underneath protein domains more vulnerable to antibody binding and dramatically increased viral sensitivity to antibody neutralizations (*51-54*).

On the other hand, a recent analysis of the mutations in the envelope glycoproteins by sequencing the gp160 genes of multiple escape viruses revealed a "dynamic glycosylation" as a new mechanism for HIV to evade neutralizing antibodies (*55*). It was found that a high frequency of mutations occurred at the consensus N-glycosylation sites (NXS/T) during chronic and persistent infections. Strikingly, a mutation at the old glycosylation site was usually companied with an insertion of an alternative N-glycosylation site nearby, resulting in a reposition or shifting of the N-glycans. The glycosylation mutants became more resistant to antibody neutralization. That is, HIV-1 escapes neutralizing antibodies by selected changes in N-glycan packing, but maintaining the global glycosylation, e.g., the total number of N-glycans, relatively constant. The so-called "evolving glycan shield" represents a new strategy for HIV-1 to maintain persistent replication in the face of an evolving antibody repertoire.

Carbohydrates as Ligands for HIV-1 Infection and Transmission

In addition to a strong defensive role for HIV immune evasion, the viral surface N-glycans was also implicated to play an active role in HIV-1 infection and transmission. It was proposed that immature dendritic cells, which are primary antigen-presenting cells and are richly expressed in the skin and at the mucosal surfaces, are the first cells targeted by HIV-1 during transmission (*56,57*). A recently identified, dendritic cell specific C-type lectin, DC-SIGN, was found to be able to bind to HIV-1 envelope gp120 and to mediate the migration of HIV from the mucosal infection sites to secondary lymphoid organs, where the virus infects T cells (*58*).

More detailed studies on molecular levels suggested that the interaction between HIV-1 gp120 and dendritic cells was carbohydrate dependent, and the high-mannose type oligosaccharides on HIV-1 gp120 served as ligands for DC-SIGN being expressed on dendritic cells (*59-62*). In addition, HIV-1 oligosaccharides such as the high-mannose type oligosaccharides and those N-glycans with terminal fucose and N-acetylglucosamine residues may serve as ligands for mannose receptors richly expressed on macrophages. The specific carbohydrate-receptor interactions may facilitate the viral infection of macrophages and may also have an influence on the viral tropism (*63-65*).

Carbohydrate antigens on HIV-1

The Epitope of the Broadly Neutralizing Antibody 2G12

The identification of conserved and accessible epitopes on HIV-1 is an essential step for vaccine development. The characterization of the epitopes of some broadly neutralizing antibodies provides important clues on rational immunogen design. So far, only a few monoclonal antibodies, including 2F5, 4E10, b12, and 2G12, were characterized that showed broadly neutralizing activities against both laboratory-adapted and primary HIV isolates (*12-14*). The epitopes of 2F5 and 4E10 were mapped to the C-terminal ectodomain of the transmembrane envelope glycoprotein gp41 (*66-69*), whereas the epitopes for the neutralizing antibodies b12 and 2G12 were characterized on the outer envelope glycoprotein gp120 (*25-27,70,71*).

Among the broadly HIV-neutralizing antibodies so far identified, the human monoclonal antibody 2G12 is the only one that directly targets the surface carbohydrate antigen of HIV-1. Several pieces of evidence suggest that the epitope of 2G12 is a unique cluster of high-mannose type oligosaccharides (oligomannose) on HIV-1 gp120. Initial mutational studies indicated that the oligomannose sugar chains at the *N*-glycosylation sites N295, N332, N339, N386, N392, and N448 might be involved in 2G12 recognition (*25*). Two recent studies further proposed that the epitope of 2G12 might consist of several Manα1-2Man moieties contributed by the oligomannose sugar chains at sites N295, N332, and N392 that form a unique cluster on gp120 (*26,27*).

Systematic mutational studies suggested that peptide portions of gp120 are not directly involved in the binding of 2G12, but serve primarily as a rigid scaffold to hold the oligomannose sugars in proximity to form a unique cluster (*25-27*). The X-ray crystal structures of Fab 2G12 and its complexes with disaccharide Manα1,2Man and high-mannose oligosaccharide $Man_9GlcNAc_2$ were recently solved (*72*).

The X-ray structural study revealed an unusual domain-swapped structure, in which the V_H domains of 2G12 exchange in its two Fab regions so that an extended multivalent binding surface is created to accommodate an oligomannose cluster. Based on the X-ray structure, a model was proposed for the recognition between 2G12 and the N-glycans on gp120. The model implicated the binding of the N-glycans of N332 and N392 in the primary combining sites, with a potential binding of the N-glycan at N339 to the secondary binding site at the V_H/V_H' interface. All the evidence pointed to the fact that the epitope of broadly neutralizing antibody 2G12 is a novel oligomannose cluster on gp120. Although high-mannose oligosaccharide moiety exists in some human glycoproteins, such a high-density, clustering oligomannose structure as the 2G12 epitope has not been found in normal human glycoproteins so far. Therefore, the unique carbohydrate antigenic structure on HIV-1 gp120 provides an ideal template for designing a vaccine that may generate HIV-neutralizing antibodies but will not raise cross-reactivity or auto immune reactions in humans.

Other Potential Carbohydrate Epitopes on HIV-1 Envelope

Besides antibody 2G12, some other anti-carbohydrate antibodies were also reported to exhibit anti-HIV properties (*73-76*). For example, natural IgM anti-carbohydrate antibodies in baby rabbit serum that are specific for high-mannose oligosaccharides are able to effectively lyse virus-coated CD4+ T cells, regardless the virus strains (*73*). The observations implicated the existence of new potential carbohydrate epitopes on HIV-1 envelope glycoproteins other than the 2G12 epitope that was characterized as a novel oligomannose cluster. The structural analysis of HIV-1 glycosylations also revealed some unusual structures, such as the notable Fucα1,3 or α(1,4)GlcNAc moieties beyond the chitobiose core and the sulfated GlcNAc structure in certain context of the N-glycans. These structures, when coming together to form a unique configuration (clustering) on the envelope, may generate novel antigenic structures that are not expressed in normal human glycoproteins. In analogy to the clustering domain of high-mannose type oligosaccharides, a clustering domain of some complex type N-glycans was also present on the HIV-1 gp120 (*37*), which may also provide potential antigenic structures. Moreover, when an extended HIV-1 peptide portion is considered in the context, certain HIV-1 glycopeptides would be expected to constitute novel epitopes. All these carbohydrate antigenic structures may have the potentials for HIV-1 vaccine development.

Towards Carbohydrate-Based Vaccines against HIV/AIDS

The frustrating situation with conventional protein-based HIV vaccines urges the exploration of new vaccine approaches. As described above, the dense carbohydrate coat on HIV-1 envelope provides a strong defense for the virus to evade humoral immune recognition. However, this strong shield is not completely seamless. Accumulating evidence suggested that the HIV-1 surface carbohydrates themselves could also be targets for developing anti-HIV strategy. For example, a potent anti-HIV protein cyanovirin-N, isolated from a cyanobacterium, exerts its potent and broad-spectrum inhibitory activity against HIV-1 through tight binding to the high-mannose type oligosaccharides (Man_9 and Man_8) of gp120 (*77,78*). Some lectins (carbohydrate-binding proteins) specific for oligomannose structures also demonstrated anti-HIV-1 activities in *in vitro* assays (*79-84*).

From the vaccine development perspective, the most notable example is the discovery of the carbohydrate-specific neutralizing antibody 2G12. Thus, the carbohydrate epitope of 2G12, a unique oligomannose cluster, provides an excellent template for vaccine design. Several groups, including ours, have already begun to explore its potential in the hope of developing a carbohydrate-based vaccine against HIV/AIDS.

Binding of High-Mannose Type Oligosaccharide Subunits to 2G12

Structural analysis indicated that HIV-1 expressed an array of high-mannose type oligosaccharides, ranging from Man_5, Man_6, Man_7, Man_8, to Man_9 (*34,36,37*). Although it was demonstrated that 2G12 recognizes an oligosaccharide cluster consisting of several high-mannose type oligosaccharides on gp120, and biochemical analysis suggested that Manα1,2Man residues were crucial for the recognition (*26,27*), it was hitherto not clear how the distinct oligomannose subunits (Man5-Man9) behave individually in the binding to 2G12. The information is important for selecting the right 'building block" for making epitope mimics. Accordingly, we prepared three typical HIV-1 high-mannose type oligosaccharides, namely $Man_5GlcNAc$, $Man_6GlcNAc$, and $Man_9GlcNAc$ (Figure 1) with high-purity, and examined their affinity by competitive inhibition of 2G12 binding to immobilized gp120 in an enzyme-linked immunosorbent assay (ELISA).

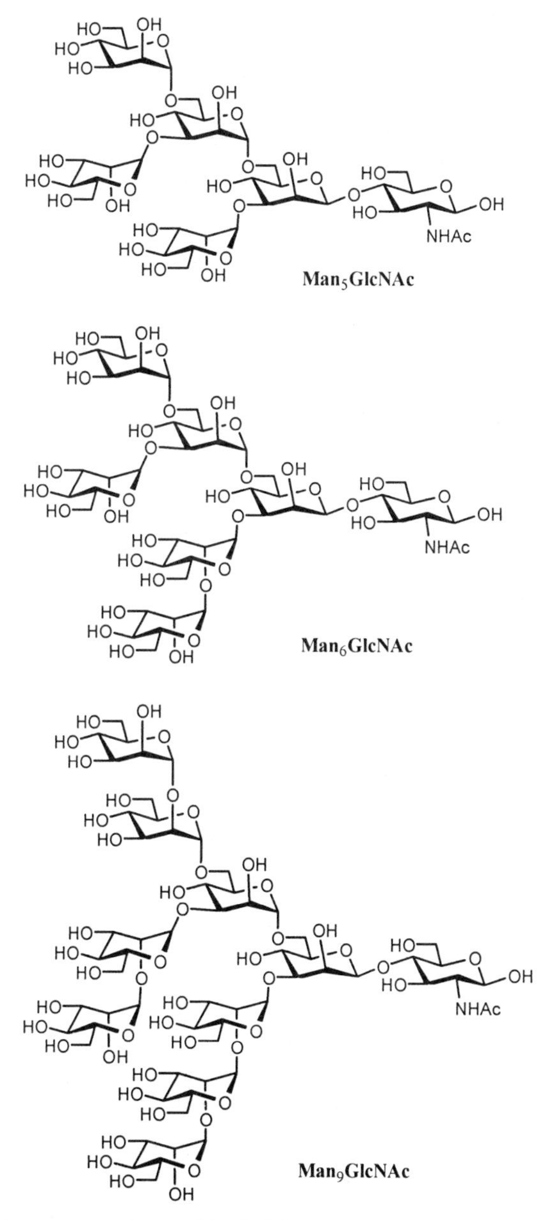

Figure 1. Structures of typical HIV-1 high-mannose type oligosaccharides.

The inhibition potency was demonstrated in Figure 2. Because of the unexpected low solubility of $Man_5GlcNAc$ and $Man_6GlcNAc$ in the assay buffer, a more accurate inhibition curve could not be achieved. The IC_{50} (concentration for 50% inhibition) data were deduced to be 0.96 mM, 70 mM, and 200 mM for $Man_9GlcNAc$, $Man_6GlcNAc$, and $Man_5GlcNAc$, respectively.

The results suggest that Man_9GlNAc is 74-fold and 210-fold more effective in inhibition of 2G12 binding than $Man_6GlcNAc$ and $Man_5GlcNAc$, respectively (*85*). The much higher affinity of $Man_9GlcNAc$ to 2G12 than that of $Man_5GlcNAc$ or $Man_6GlcNAc$ provides direct evidence that terminal Manα1,2Man linkages are essential for the antibody-epitope recognition. In comparison, $Man_9GlcNAc$ contains three terminal Manα1,2Man linkages, $Man_6GlcNAc$ contains one, whereas $Man_5GlcNAc$ does not have any terminal Manα1,2Man moiety. The results are consistent with previous observations from the binding of 2G12 to glycosidase-treated gp120. For example, treatment of gp120 with *Aspergillus saitoi* mannosidase, which selectively removes terminal Manα1,2Man linkages without touching the Manα1,3Man or Manα1,6Man residues on gp120, significantly reduced the binding of gp120 to 2G12 (*27*).

Our direct binding studies further suggest that the larger high-mannose oligosaccharide such as $Man_9GlcNAc$ is the favorable subunit on gp120 for 2G12 recognition. Therefore, if native high-mannose type oligosaccharides would be considered for incorporation into an immunogen design, the larger oligosaccharide such as Man_9 should be the choice as the favorable "building block".

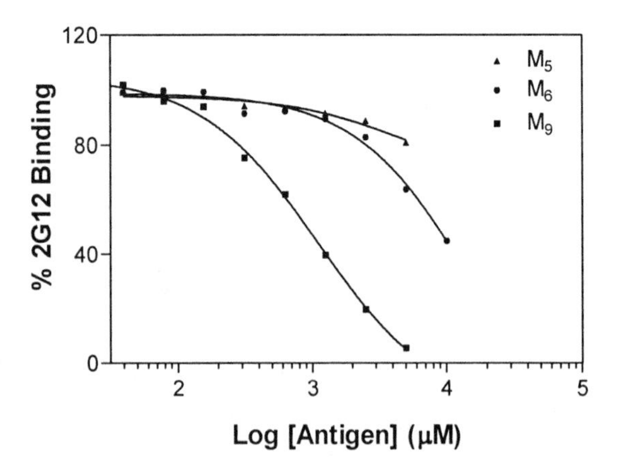

Figure 2. Inhibition of 2G12 binding to gp120 by high-mannose type oligosaccharides. 2G12 binding was plotted against the log of competing carbohydrate concentrations in μM units (Adapted from Wang et al, Chem & Biol., 2004, ***11****, 127).*

In another development related to 2G12 epitope, Wong *et al* synthesized a series of oligosaccharides with varied number and configuration of Manα1,2Man residues, using a programmable, reactivity-based one-pot synthetic method (*86*). The synthetic oligosaccharides, together with the standard oligosaccharide $Man_9GlcNAc_2$, were evaluated for their ability to inhibit the binding of 2G12 to immobilized gp120 in an ELISA assay. The structures and relative inhibitory activities (at 2 mM concentration) of the synthetic oligosaccharides are listed in Figure 3.

Comparison between the trimannose **2** (15% inhibition at 2 mM) and tetramannose **3** (79% inhibition at 2 mM) suggested that adding up of an extra α1,2-linked mannose at an appropriate location could significantly enhance its binding efficiency. Pentamannose **4** (79% inhibition at 2 mM), which possesses a divalent Manα1,2Man configuration, shows similar activity to the tetramannose **3** that has two sequential Manα1,2Man units. Interestingly, the heptamannose **5** (65% inhibition at 2 mM), which contains an additional Manα1,2Manα1,2Man branch at the 6-position of the core mannose in **3**, does not show further increased affinity over **3**. It is interesting to observe that the tetramannose **3** exhibited similar affinity to antibody 2G12 as the native $Man_9GlcNAc_2$, which showed 71% inhibition at 2mM in the competitive assay.

Synthesis and Antigenicity of Template-Assembled Oligomannose Clusters as the Mimics of 2G12 Epitope

Although Man_9 and the synthetic oligosaccharides, tetramannose (**3**) and pentamannose (**4**), showed enhanced affinity to 2G12 over other oligosaccharides so far tested, the affinity is still too weak. The mM-range exhibited by these oligosaccharide subunits are far away from the nM-range exhibited by the binding of HIV-1 gp120 to 2G12. We reasoned that assembly of oligomannose such as Man_9 on a suitable scaffold molecule should provide novel oligosaccharide clusters that may mimic or capture 2G12 epitope as it is presented on HIV-1 gp120.

To test the hypothesis, we synthesized bi-, tri- and tetra-valent Man_9 clusters based on a galactopyranoside scaffold (Figure 4). The choice of a galactopyranoside as a template was based on the following considerations. When a galactopyranoside is used as the scaffold to present the oligosaccharides, the oligosaccharide chains being installed at C-3, 4, and 6 positions will face above the galactose ring to form a cluster, while the oligomannose sugar chain at position C-2 is likely to be located on the flank of the cluster. We expect that this arrangement will at least partially mimic the spatial orientation of the carbohydrate epitope of antibody 2G12, which most likely include the N-glycans at N295, N332, and N392 (*26,27*). The distances

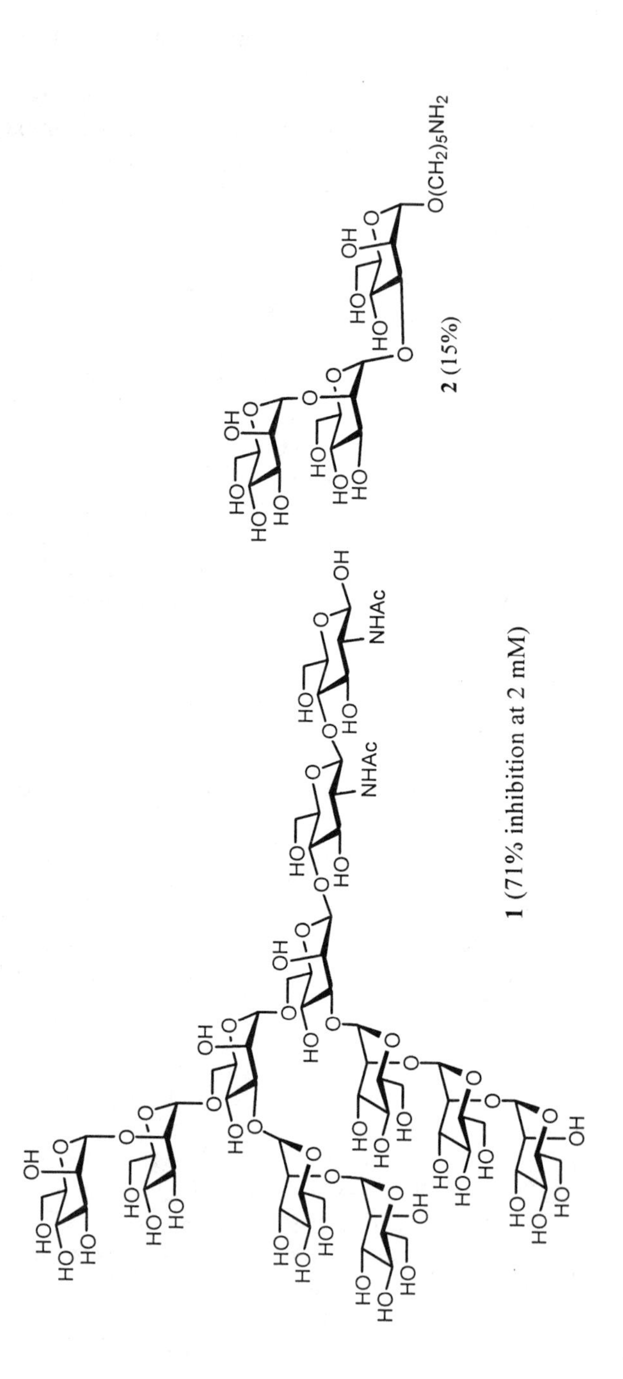

1 (71% inhibition at 2 mM)

2 (15%)

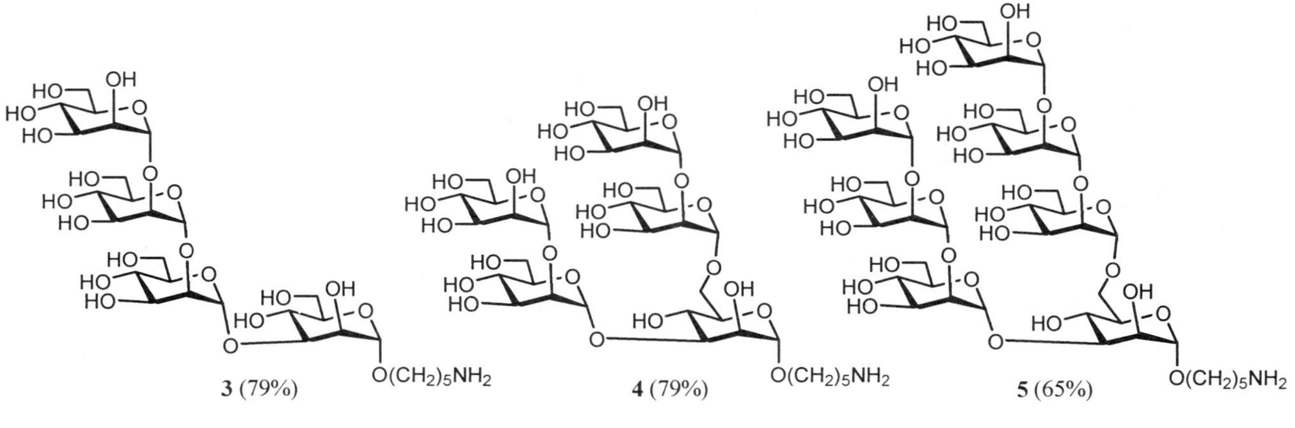

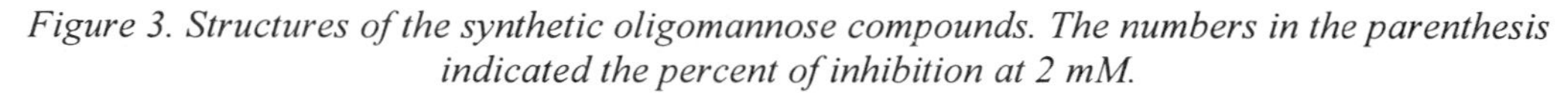

Figure 3. Structures of the synthetic oligomannose compounds. The numbers in the parenthesis indicated the percent of inhibition at 2 mM.

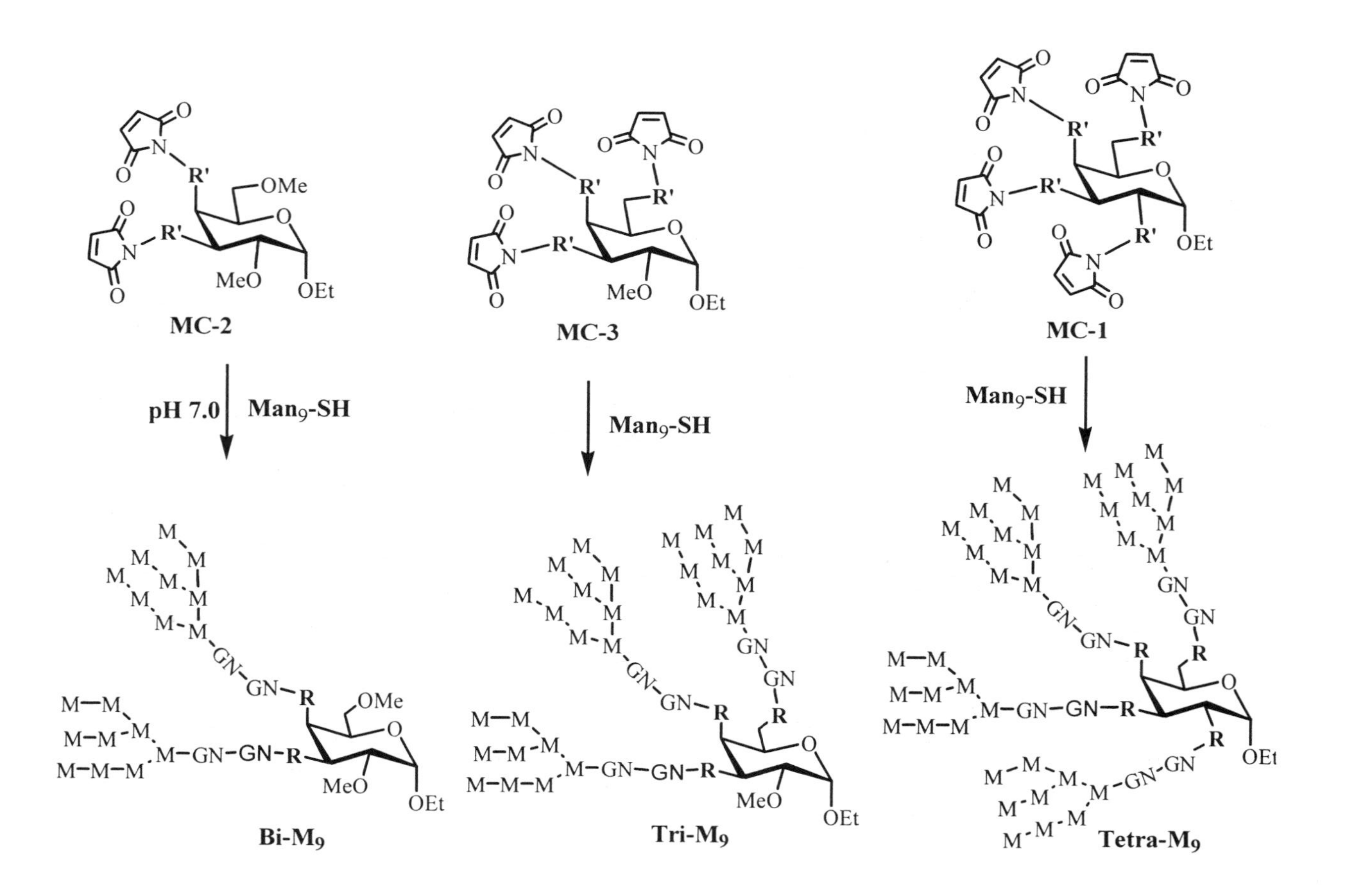
MC-2
pH 7.0
Man$_9$-SH
Bi-M$_9$
MC-3
Man$_9$-SH
Tri-M$_9$
MC-1
Man$_9$-SH
Tetra-M$_9$

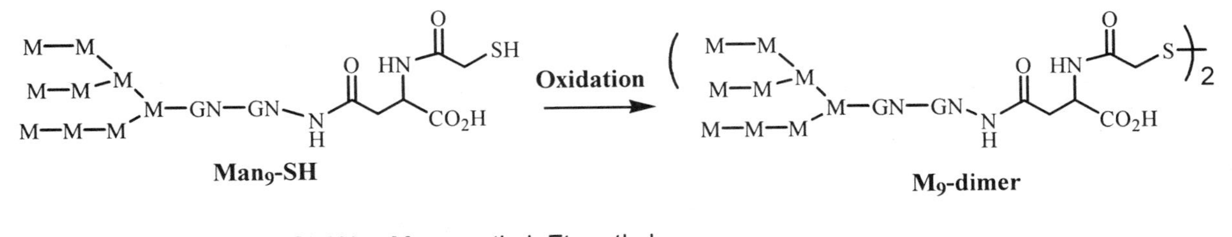

M = mannose; GN = GlcNAc; Me = methyl; Et = ethyl

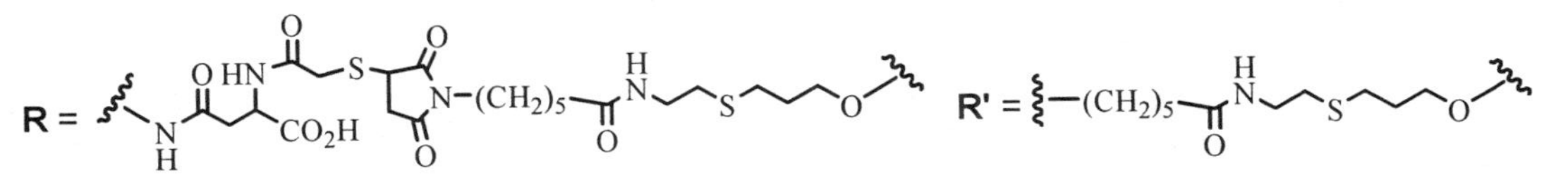

Figure 4. Structures and synthesis of a set of synthetic oligomannose clusters.

between each pair of the Asn residues, N295-N332, N332-N392, and N295-N392 , were measured to be 5.8, 20.3, and 23.6 Å, respectively, based on the reported structure of gp120 core (*45*). By adjusting the length of the spacer, we expect that the distances among the conjugating sites in the synthetic cluster could fall in the range as measured on the gp120 core.

The key step for the oligosaccharide cluster synthesis was the chemoselective ligation between a thiol-tagged oligosaccharide and a maleimide-functionalized template (*85*). We have previously used this type of ligation for preparing large and complex multivalent peptides and glycoconjugates (*87-90*). The SH-tagged high-mannose oligosaccharide, Man_9-SH, was prepared in two steps: the selective N-acylation of the free amino group in $Man_9GlcNAc_2Asn$ with N-succinimidyl S-acetylthioacetate (SATA) (*91*) and subsequent removal of the S-acetyl group by treatment with hydroxylamine. The bi-, tri-, and tetravalent maleimide cluster, MC-1, MC-2, and MC-3 were prepared through functional group manipulations on ethyl α-D-galactopyranoside (*87,90*).

Chemoselective ligation between the Man_9-SH and the maleimide clusters proceeded very efficiently in a phosphate buffer (pH 6.6) containing acetonitrile to give the corresponding oligosaccharide clusters in high yields. On the other hand, a dimer of $Man_9GlcNAc_2Asn$ was prepared through oxidation of Man_9-SH (Figure 4). All the products were purified by reverse- phase HPLC and characterized by electron spray ionization-mass spectroscopy (ESI-MS) (*85*).

The affinity of the synthetic clusters was evaluated by a competitive inhibition of 2G12 binding to the immobilized HIV-1 gp120 in an ELISA assay (Figure 5). The IC50 data were deduced to be 0.4, 0.13, 0.044, and 0.013 mM for Man9-dimer, Bi-Man9, Tri-Man9, and Tetra-Man9, respectively. If IC_{50} was taken as an indication for relative affinity, the affinity of the tetravalent cluster (Tetra-Man_9) to antibody 2G12 would be 73-fold higher than that of subunit $Man_9GlcNAc_2Asn$. On the other hand, the tri- and bi-valent oligomannose clusters are 22- and 7-fold more effective than $Man_9GlcNAc_2Asn$ in binding to 2G12. Another interesting finding from our binding studies came from the two bivalent oligomannose compounds, Bi-Man_9 and Man_9-dimer. The Bi-Man_9, which has a longer distance between the two Man_9 subunits, is 3-fold more effective than the Man_9-dimer in inhibition of 2G12 binding to gp120. The results suggest that control of the geometry and the distance between the oligosaccharide subunits is important to achieve a tight binding to antibody 2G12. It should be pointed out that although the synthetic oligosaccharide clusters reached a *micromolar* level in binding to 2G12, the affinity is still much lower than that of HIV-1 gp120, which binds to 2G12 at a *nanomolar* level. Therefore, further optimization of the epitope mimics is necessary.

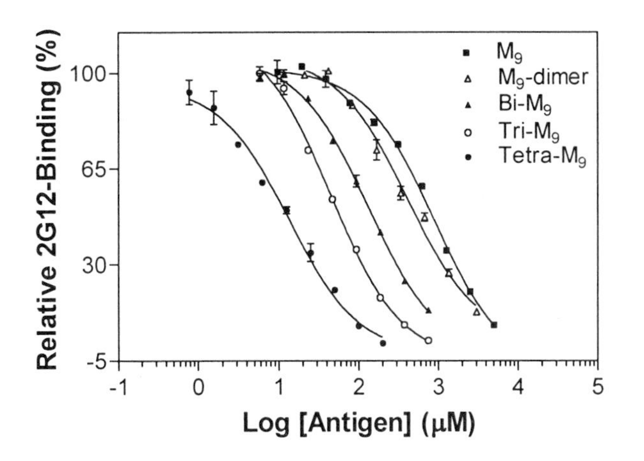

Figure 5. Inhibition of 2G12 binding to gp120 by synthetic oligomannose clusters. 2G12 binding was plotted against the log of competing ligand in μM units (Adapted from Wang et al, Chem & Biol., 2004, ***11****, 127).*

In exploring new scaffold for assembling mimics of 2G12 epitope, we recently synthesized a trivalent oligomannose cluster using cholic acid as the scaffold molecule (*92*) (Figure 6). The design of such an oligosaccharide cluster was based on the proposed model for the recognition between 2G12 and HIV-1 glycans, which suggested that the N-glycans at N295 and N332 of gp120 would bind to the primary combining sites of 2G12, while the N-glycan at N339 might interact with the secondary binding site of 2G12 (*72*). Therefore, the cholic acid-based oligomannose cluster was made to mimic the oligosaccharide cluster formed by the N-glycans from the N295, N332, and N339 sites.

The cluster was synthesized in two key steps: introduction of maleimide functionality at the 3α, 7α, and 12α-positions of cholic acid and chemoselective ligation of the maleimide cluster with Man_9-SH. Binding studies revealed that the synthetic oligomannose cluster was 46-fold more effective than subunit $Man_9GlcNAc_2Asn$ in inhibiting 2G12-binding to HIV-1 gp120. This cholic acid-based mimic showed a moderate enhancement in the 2G12-affinity when compared to the trivalent, galactoside-scaffolded mimic. A free amino group introduced at the C24 position of the cluster would facilitate its coupling to a carrier protein to provide a functional immunogen.

The scaffold approach described here provided a new avenue for epitope mimic design. Thus, better mimics of 2G12 epitope might be achieved by optimizing

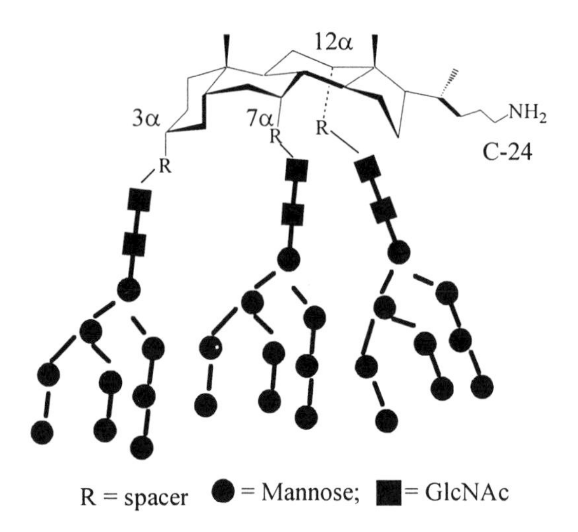

Figure 6. Structure of the synthetic, cholic acid-based oligomannose cluster.

the length of the spacers, and the configuration and rigidity of the scaffold molecule. It should be pointed out that carbohydrates themselves are generally poorly immunogenic. To improve their immunogenicity, the optimal epitope mimics thus generated should be conjugated to a carrier protein such as KLH to fulfill a functional immunogen. It has been demonstrated that KLH-conjugate vaccines of tumor-associated oligosaccharide antigens, which are otherwise barely immunogenic, can raise moderate titers of both IgG and IgM antibodies that are specific for the carbohydrate antigens and are able to neutralize cancer cells (*93-96*). The development of carbohydrate-based HIV vaccines may follow a similar path.

HIV-1 Glycopeptides as Potential Targets for HIV-1 Vaccine Design

While novel HIV-1 carbohydrates, such as the epitope of 2G12, should be further investigated for HIV vaccines, certain HIV-1 glycopeptides may have a greater potential to form unique epitopes (antigenic determinants) on HIV-1 than sugar chains alone. To explore this potential, first of all, it is necessary to develop efficient methods for constructing the HIV-1 glycopeptides of interest, which are hitherto not easy to synthesize (*97-100*). We have initiated a project aiming to construct various HIV-1 glycopeptides by a chemoenzymatic approach, and to apply them as components for HIV-1 vaccines. The first, native HIV-1 glycopeptides that we synthesized were the gp120 fragment derived from the sequence 336-342, which contains the N339 N-glycan as part of the 2G12 epitope (*72*) (Figure 7).

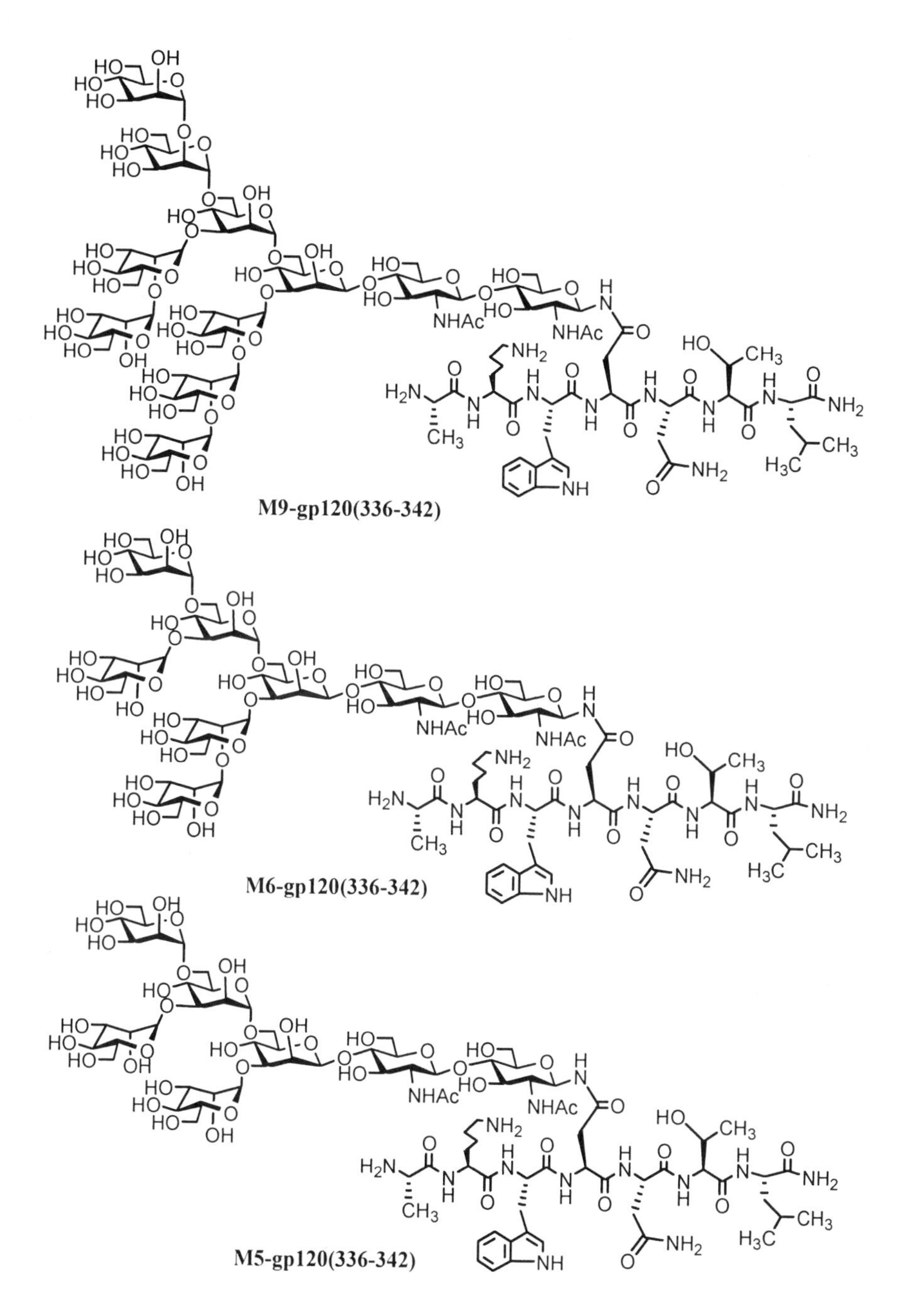

Figure 7. Structures of the synthetic gp120 glycopeptides.

Endoglycanase (ENGase)-çatalyzed oligosaccharide transfer was used as the key step for the glycopeptide synthesis (*101-104*). First, the acceptor GlcNAc-peptide was synthesized by automatic peptide synthesis. Homogeneous glycoforms of the gp120 peptide containing Man_9, Man_6, and Man_5 N-glycans, respectively, were then constructed using endo-β-*N*-acetylglucosaminidase from *Arthrobacter* (Endo-A) as the key enzyme. Using $Man_9GlcNAc_2Asn$ as the glycosyl donor and the chemically synthesized GlcNAc-peptide as the glycosyl acceptor, the Endo-A catalyzed transglycosylation gave the Man_9-glycopeptide in 28% isolated yield. Similarly, Transglycosylation using $Man_5GlcNAc_2Asn$ and $Man_6GlcNAc_2Asn$ as the glycosyl donors, which were prepared through pronase digestion of chicken ovalbumin, gave the Man_5-and Man_6-glycopeptides in 25% and 27% yields, respectively. Interestingly, it was observed that Endo-A was equally efficient for transferring all high-mannose type oligosaccharides, making it particularly useful for constructing an array of high-mannose type N-glycopeptides (*105*).

Next, we attempted to synthesize a large glycopeptide from the gp120 sequence 293-334 that carries two N-glycans at the glycosylation sites N295 and N332, which are components of the 2G12 epitope (*27*). To carry out a pilot experiment, we replaced the 32 amino acid residues in the V3 loop with a proline residue. The synthesis of the double-glycosylated gp120 glycopeptide was summarized in Figure 8.

The undeca-peptide containing two GlcNAc moieties at the putative N295 and N332 sites was synthesized by solid phase peptide synthesis. After de-*O*-acetylation, the $(GlcNAc)_2$-peptide was purified by HPLC and characterized by mass spectrometry (*104*). The Acm-protecting group was selectively removed by treatment with $Hg(OAc)_2$ and then H_2S to give the free cysteine-containing peptide that was subsequently cyclized to form the cyclic peptide [ESI-MS: 1676.71 $(M+H)^+$, 838.84 $(M+2H)^{2+}$]. Endo-A catalyzed transglycosylation to the cyclic peptide was performed in a 30% aqueous acetone using $Man_9GlcNAc_2Asn$ as the oligosaccharide donor. HPLC indicated the formation of three transglycosylation products, which were purified by HPLC and characterized by ESI-MS.

The major products isolated are the mono-glycosylated glycopeptides in about 1:1 ratio in a 32% total yield, which were temporally assigned as A and B [ESI-MS: glycopeptide A, 1680.63 $(M+H+Na)^{2+}$, 1669.56 $(M+2H)^{2+}$; glycopeptide B, 1680.72 $(M+H+Na)^{2+}$, 1669.95 $(M+2H)^{2+}$]. The desired, double-glycosylated glycopeptide (C) was isolated in only 2.5% yield [ESI-MS: 1675.87 $(M+2H+Na)^{3+}$, 1667.65 $(M+3H)^{3+}$] (Wang LX, *et al*, unpublished results). The low yield of double-glycosylated glycopeptide may be partially due to the steric hindrance for the addition of the second oligosaccharide moiety. Nevertheless, the experiment provides proof-of-concept data showing that the ENGase-catalyzed approach is applicable to the assembling of very complex

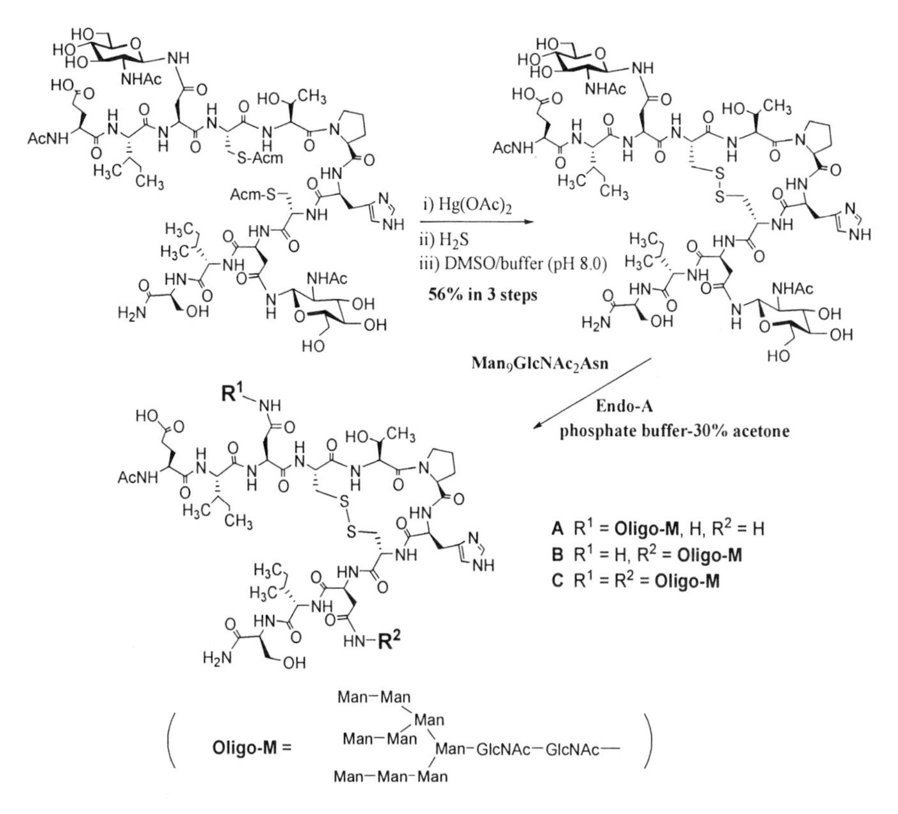

Figure 8. Synthesis of the cyclic gp120 glycopeptide mimics carrying two native N-glycans.

glycopeptides carrying more than one multiple N-glycans in a single step. This is the first example of ENGase-catalyzed synthesis of a cyclic N-glycopeptide carrying two N-glycans.

Recently, Danishefsky *et al* launched a total chemical synthesis of HIV-1 gp120 glycopeptides carrying either a hybrid type or a high-mannose type N-glycans (*106,107*). The total synthesis consists of the pre-assembly of the full-size N-glycans by multiple step chemical synthesis, and the convergent coupling of the N-glycans with a selectively protected peptide containing a free aspartic acid. The structures of the two gp120 glycopeptides synthesized were shown in Figure 9. The synthetic glycopeptides contains a hybrid or high-mannose type N-glycan at the N332 glycosylation site, which was part of the 2G12 epitope. However, for the vaccine perspective, it is not clear why the particular sequence

(316-335) was chosen. The successful synthesis of various HIV-1 glycopeptides, either by the total chemical synthesis or the chemoenzymatic approach, paved the way toward testing this new class of HIV-1 antigenic structures for vaccine development.

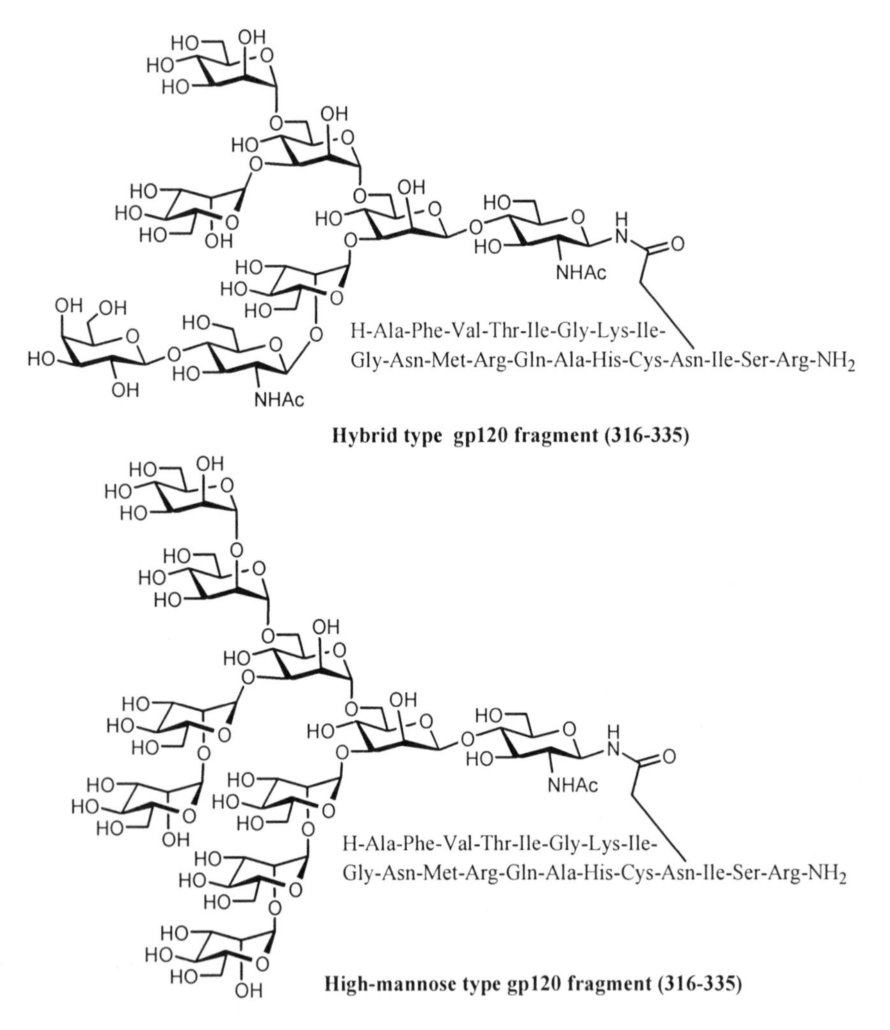

Figure 9. Structures of the two gp120 glycopeptides synthesized by total chemical synthesis.

Conclusion

Carbohydrate-based conjugate vaccines have been developed for eliciting protective immunity against many pathogens such as bacteria (*108*). Carbohydrate-based cancer vaccines have also been under extensive research and some candidates have moved to clinical trials (*93-96*). However, carbohydrate antigens have not been adequately exploited for HIV-1 vaccine design, despite their abundance on HIV-1 surface. The existence of the broadly neutralizing antibody 2G12 and the characterization of its epitope as a novel oligosaccharide cluster on HIV-1 gp120 raised a possibility of developing a carbohydrate-based vaccine against HIV/AIDS. The 2G12-binding studies with natural and synthetic high-mannose oligosaccharides described in this article have yielded useful information that defines the structural requirement for the antibody recognition of the oligosaccharide subunits. The design and synthesis of template-assembled oligosaccharide clusters as mimics of 2G12 epitope has made it one step closer towards an HIV vaccine. On the other hand, the exploration of various HIV-1 glycopeptides as new and unique epitopes may open a new avenue for vaccine development. It is our hope that continuing studies will eventually lead to an effective, carbohydrate-based vaccine to conquer HIV/AIDS.

Acknowledgment

The HIV vaccine work in my group was supported in part by the Institute of Human Virology, University of Maryland Biotechnology Institute, and the National Institutes of Health (NIH grants AI54354 and AI51235 to LXW).

References

1. Gallo, R. C. *Science* **2002**, *298*, 1728-30.
2. Montagnier, L. *Science* **2002**, *298*, 1727-8.
3. Gallo, R. C.; Montagnier, L. *N. Engl. J. Med.* **2003**, *349*, 2283-5.
4. UNAIDS "AIDS epidemic update: December 2003 (www.unaids.org)," the World Health Organization, 2003.
5. Calarota, S. A.; Weiner, D. B. *AIDS* **2003**, *17 Suppl 4*, S73-84.
6. Nabel, G. J. *Nature* **2001**, *410*, 1002-7.
7. Melnick, J. L. *Clin. Microbiol. Rev.* **1996**, *9*, 293-300.
8. Sutter, R. W.; Tangermann, R. H.; Aylward, R. B.; Cochi, S. L. *Infect. Dis. Clin. North Am.* **2001**, *15*, 41-64.
9. Gross, C. P.; Sepkowitz, K. A. *Int. J. Infect. Dis.* **1998**, *3*, 54-60.
10. Hilleman, M. R. *Vaccine* **2000**, *18*, 1436-47.

11. Cohen, J. *Science* **2003**, *300*, 28-9.
12. Burton, D. R. *Nat. Rev. Immunol.* **2002**, *2*, 706-13.
13. Wyatt, R.; Sodroski, J. *Science* **1998**, *280*, 1884-8.
14. Wang, L. X. *Curr. Pharm. Des.* **2003**, *9*, 1771-87.
15. Ho, D. D.; Huang, Y. *Cell* **2002**, *110*, 135-8.
16. Bojak, A.; Deml, L.; Wagner, R. *Drug Discov. Today* **2002**, *7*, 36-46.
17. Hone, D. M.; DeVico, A. L.; Fouts, T. R.; Onyabe, D. Y.; Agwale, S. M.; Wambebe, C. O.; Blattner, W. A.; Gallo, R. C.; Lewis, G. K. *J. Hum. Virol.* **2002**, *5*, 17-23.
18. Graham, B. S. *Annu. Rev. Med.* **2002**, *53*, 207-21.
19. Moore, J. P.; Burton, D. R. *Nat. Med.* **1999**, *5*, 142-4.
20. Sattentau, Q. J.; Moulard, M.; Brivet, B.; Botto, F.; Guillemot, J. C.; Mondor, I.; Poignard, P.; Ugolini, S. *Immunol. Lett.* **1999**, *66*, 143-9.
21. McMichael, A. J.; Rowland-Jones, S. L. *Nature* **2001**, *410*, 980-7.
22. Barouch, D. H.; Santra, S.; Schmitz, J. E.; Kuroda, M. J.; Fu, T. M.; Wagner, W.; Bilska, M.; Craiu, A.; Zheng, X. X.; Krivulka, G. R.; Beaudry, K.; Lifton, M. A.; Nickerson, C. E.; Trigona, W. L.; Punt, K.; Freed, D. C.; Guan, L.; Dubey, S.; Casimiro, D.; Simon, A.; Davies, M. E.; Chastain, M.; Strom, T. B.; Gelman, R. S.; Montefiori, D. C.; Lewis, M. G.; Emini, E. A.; Shiver, J. W.; Letvin, N. L. *Science* **2000**, *290*, 486-92.
23. Mascola, J. R.; Stiegler, G.; VanCott, T. C.; Katinger, H.; Carpenter, C. B.; Hanson, C. E.; Beary, H.; Hayes, D.; Frankel, S. S.; Birx, D. L.; Lewis, M. G. *Nat. Med.* **2000**, *6*, 207-10.
24. Baba, T. W.; Liska, V.; Hofmann-Lehmann, R.; Vlasak, J.; Xu, W.; Ayehunie, S.; Cavacini, L. A.; Posner, M. R.; Katinger, H.; Stiegler, G.; Bernacky, B. J.; Rizvi, T. A.; Schmidt, R.; Hill, L. R.; Keeling, M. E.; Lu, Y.; Wright, J. E.; Chou, T. C.; Ruprecht, R. M. *Nat. Med.* **2000**, *6*, 200-6.
25. Trkola, A.; Purtscher, M.; Muster, T.; Ballaun, C.; Buchacher, A.; Sullivan, N.; Srinivasan, K.; Sodroski, J.; Moore, J. P.; Katinger, H. *J. Virol.* **1996**, *70*, 1100-8.
26. Sanders, R. W.; Venturi, M.; Schiffner, L.; Kalyanaraman, R.; Katinger, H.; Lloyd, K. O.; Kwong, P. D.; Moore, J. P. *J. Virol.* **2002**, *76*, 7293-305.
27. Scanlan, C. N.; Pantophlet, R.; Wormald, M. R.; Ollmann Saphire, E.; Stanfield, R.; Wilson, I. A.; Katinger, H.; Dwek, R. A.; Rudd, P. M.; Burton, D. R. *J. Virol.* **2002**, *76*, 7306-21.
28. Hansen, J. E.; Clausen, H.; Nielsen, C.; Teglbjaerg, L. S.; Hansen, L. L.; Nielsen, C. M.; Dabelsteen, E.; Mathiesen, L.; Hakomori, S. I.; Nielsen, J. O. *J. Virol.* **1990**, *64*, 2833-40.
29. Hansen, J. E.; Nielsen, C.; Arendrup, M.; Olofsson, S.; Mathiesen, L.; Nielsen, J. O.; Clausen, H. *J. Virol.* **1991**, *65*, 6461-7.
30. Hansen, J. E.; Jansson, B.; Gram, G. J.; Clausen, H.; Nielsen, J. O.; Olofsson, S. *Arch. Virol.* **1996**, *141*, 291-300.
31. Feizi, T.; Larkin, M. *Glycobiology* **1990**, *1*, 17-23.
32. Feizi, T. In *Carbohydrates in Chemistry and Biology*; Wiley-VCH, 2000; Vol. 4, pp 851-63.

33. Mizuochi, T.; Spellman, M. W.; Larkin, M.; Solomon, J.; Basa, L. J.; Feizi, T. *Biochem. J.* **1988**, *254*, 599-603.
34. Geyer, H.; Holschbach, C.; Hunsmann, G.; Schneider, J. *J. Biol. Chem.* **1988**, *263*, 11760-7.
35. Leonard, C. K.; Spellman, M. W.; Riddle, L.; Harris, R. J.; Thomas, J. N.; Gregory, T. J. *J. Biol. Chem.* **1990**, *265*, 10373-82.
36. Mizuochi, T.; Matthews, T. J.; Kato, M.; Hamako, J.; Titani, K.; Solomon, J.; Feizi, T. *J. Biol. Chem.* **1990**, *265*, 8519-24.
37. Zhu, X.; Borchers, C.; Bienstock, R. J.; Tomer, K. B. *Biochemistry* **2000**, *39*, 11194-204.
38. Perrin, C.; Fenouillet, E.; Jones, I. M. *Virology* **1998**, *242*, 338-45.
39. Lee, W. R.; Yu, X. F.; Syu, W. J.; Essex, M.; Lee, T. H. *J. Virol.* **1992**, *66*, 1799-803.
40. Yeh, J. C.; Seals, J. R.; Murphy, C. I.; van Halbeek, H.; Cummings, R. D. *Biochemistry* **1993**, *32*, 11087-99.
41. Butters, T. D.; Yudkin, B.; Jacob, G. S.; Jones, I. M. *Glycoconj. J.* **1998**, *15*, 83-8.
42. Bolmstedt, A.; Biller, M.; Hansen, J. E.; Moore, J. P.; Olofsson, S. *Arch. Virol.* **1997**, *142*, 2465-81.
43. Shilatifard, A.; Merkle, R. K.; Helland, D. E.; Welles, J. L.; Haseltine, W. A.; Cummings, R. D. *J. Virol.* **1993**, *67*, 943-52.
44. Kwong, P. D.; Wyatt, R.; Robinson, J.; Sweet, R. W.; Sodroski, J.; Hendrickson, W. A. *Nature* **1998**, *393*, 648-59.
45. Wyatt, R.; Kwong, P. D.; Desjardins, E.; Sweet, R. W.; Robinson, J.; Hendrickson, W. A.; Sodroski, J. G. *Nature* **1998**, *393*, 705-11.
46. Pal, R.; Hoke, G. M.; Sarngadharan, M. G. *Proc. Natl. Acad. Sci. USA* **1989**, *86*, 3384-8.
47. Lee, W. R.; Syu, W. J.; Du, B.; Matsuda, M.; Tan, S.; Wolf, A.; Essex, M.; Lee, T. H. *Proc. Natl. Acad. Sci. USA* **1992**, *89*, 2213-7.
48. Johnson, W. E.; Sauvron, J. M.; Desrosiers, R. C. *J. Virol.* **2001**, *75*, 11426-36.
49. Li, Y.; Luo, L.; Rasool, N.; Kang, C. Y. *J. Virol.* **1993**, *67*, 584-8.
50. Reitter, J. N.; Means, R. E.; Desrosiers, R. C. *Nat. Med.* **1998**, *4*, 679-84.
51. Wyatt, R.; Moore, J.; Accola, M.; Desjardin, E.; Robinson, J.; Sodroski, J. *J. Virol.* **1995**, *69*, 5723-33.
52. Cao, J.; Sullivan, N.; Desjardin, E.; Parolin, C.; Robinson, J.; Wyatt, R.; Sodroski, J. *J. Virol.* **1997**, *71*, 9808-12.
53. Kolchinsky, P.; Kiprilov, E.; Bartley, P.; Rubinstein, R.; Sodroski, J. *J. Virol.* **2001**, *75*, 3435-43.
54. Kolchinsky, P.; Kiprilov, E.; Sodroski, J. *J. Virol.* **2001**, *75*, 2041-50.
55. Wei, X.; Decker, J. M.; Wang, S.; Hui, H.; Kappes, J. C.; Wu, X.; Salazar-Gonzalez, J. F.; Salazar, M. G.; Kilby, J. M.; Saag, M. S.; Komarova, N. L.; Nowak, M. A.; Hahn, B. H.; Kwong, P. D.; Shaw, G. M. *Nature* **2003**, *422*, 307-12.
56. Rowland-Jones, S. L. *Curr. Biol.* **1999**, *9*, R248-50.

57. Grouard, G.; Clark, E. A. *Curr. Opin. Immunol.* **1997**, *9*, 563-7.
58. Geijtenbeek, T. B.; Kwon, D. S.; Torensma, R.; van Vliet, S. J.; van Duijnhoven, G. C.; Middel, J.; Cornelissen, I. L.; Nottet, H. S.; KewalRamani, V. N.; Littman, D. R.; Figdor, C. G.; van Kooyk, Y. *Cell* **2000**, *100*, 587-97.
59. Mitchell, D. A.; Fadden, A. J.; Drickamer, K. *J. Biol. Chem.* **2001**, *276*, 28939-45.
60. Feinberg, H.; Mitchell, D. A.; Drickamer, K.; Weis, W. I. *Science* **2001**, *294*, 2163-6.
61. Hong, P. W.; Flummerfelt, K. B.; de Parseval, A.; Gurney, K.; Elder, J. H.; Lee, B. *J. Virol.* **2002**, *76*, 12855-65.
62. Su, S. V.; Hong, P.; Baik, S.; Negrete, O. A.; Gurney, K. B.; Lee, B. *J. Biol. Chem.* **2004**, *279*, 19122-32.
63. Shepherd, V. L.; Lee, Y. C.; Schlesinger, P. H.; Stahl, P. D. *Proc. Natl. Acad. Sci. USA* **1981**, *78*, 1019-22.
64. Stahl, P. D.; Rodman, J. S.; Miller, M. J.; Schlesinger, P. H. *Proc. Natl. Acad. Sci. USA* **1978**, *75*, 1399-403.
65. Larkin, M.; Childs, R. A.; Matthews, T. J.; Thiel, S.; Mizuochi, T.; Lawson, A. M.; Savill, J. S.; Haslett, C.; Diaz, R.; Feizi, T. AIDS **1989**, *3*, 793-8.
66. Conley, A. J.; Kessler, J. A., II; Boots, L. J.; McKenna, P. M.; Schleif, W. A.; Emini, E. A.; Mark, G. E., III; Katinger, H.; Cobb, E. K.; Lunceford, S. M.; Rouse, S. R.; Murthy, K. K. *J. Virol.* **1996**, *70*, 6751-8.
67. Parker, C. E.; Deterding, L. J.; Hager-Braun, C.; Binley, J. M.; Schulke, N.; Katinger, H.; Moore, J. P.; Tomer, K. B. *J. Virol.* **2001**, *75*, 10906-11.
68. Zwick, M. B.; Labrijn, A. F.; Wang, M.; Spenlehauer, C.; Saphire, E. O.; Binley, J. M.; Moore, J. P.; Stiegler, G.; Katinger, H.; Burton, D. R.; Parren, P. W. *J. Virol.* **2001**, *75*, 10892-905.
69. Stiegler, G.; Kunert, R.; Purtscher, M.; Wolbank, S.; Voglauer, R.; Steindl, F.; Katinger, H. *AIDS Res. Hum. Retroviruses* **2001**, *17*, 1757-65.
70. Burton, D. R.; Pyati, J.; Koduri, R.; Sharp, S. J.; Thornton, G. B.; Parren, P. W.; Sawyer, L. S.; Hendry, R. M.; Dunlop, N.; Nara, P. L.; Lamacchia, M.; Garratty, E.; Stiehm, E. R.; Bryson, Y. T.; Moore, J. P.; Ho, D. D.; Barbas, C. F. *Science* **1994**, *266*, 1024-7.
71. Saphire, E. O.; Parren, P. W.; Pantophlet, R.; Zwick, M. B.; Morris, G. M.; Rudd, P. M.; Dwek, R. A.; Stanfield, R. L.; Burton, D. R.; Wilson, I. A. *Science* **2001**, *293*, 1155-9.
72. Calarese, D. A.; Scanlan, C. N.; Zwick, M. B.; Deechongkit, S.; Mimura, Y.; Kunert, R.; Zhu, P.; Wormald, M. R.; Stanfield, R. L.; Roux, K. H.; Kelly, J. W.; Rudd, P. M.; Dwek, R. A.; Katinger, H.; Burton, D. R.; Wilson, I. A. *Science* **2003**, *300*, 2065-71.
73. Gerencer, M.; Barrett, P. N.; Kistner, O.; Mitterer, A.; Dorner, F. *AIDS Res. Hum. Retroviruses* **1998**, *14*, 599-605.
74. Arendrup, M.; Sonnerborg, A.; Svennerholm, B.; Akerblom, L.; Nielsen, C.; Clausen, H.; Olofsson, S.; Nielsen, J. O.; Hansen, J. E. *J. Gen. Virol.* **1993**, *74*, 855-63.

75. Hansen, J. E.; Nielsen, C.; Clausen, H.; Mathiesen, L. R.; Nielsen, J. O. *Antiviral Res.* **1991**, *16*, 233-42.
76. Tomiyama, T.; Lake, D.; Masuho, Y.; Hersh, E. M. *Biochem. Biophys. Res. Commun.* **1991**, *177*, 279-85.
77. Bolmstedt, A. J.; O'Keefe, B. R.; Shenoy, S. R.; McMahon, J. B.; Boyd, M. R. *Mol. Pharmacol.* **2001**, *59*, 949-54.
78. Shenoy, S. R.; Barrientos, L. G.; Ratner, D. M.; O'Keefe, B. R.; Seeberger, P. H.; Gronenborn, A. M.; Boyd, M. R. *Chem. Biol.* **2002**, *9*, 1109-18.
79. Ezekowitz, R. A.; Kuhlman, M.; Groopman, J. E.; Byrn, R. A. *J. Exp. Med.* **1989**, *169*, 185-96.
80. Hansen, J. E.; Nielsen, C. M.; Nielsen, C.; Heegaard, P.; Mathiesen, L. R.; Nielsen, J. O. *AIDS* **1989**, *3*, 635-41.
81. Balzarini, J.; Schols, D.; Neyts, J.; Van Damme, E.; Peumans, W.; De Clercq, E. *Antimicrob. Agents Chemother.* **1991**, *35*, 410-6.
82. Gattegno, L.; Ramdani, A.; Jouault, T.; Saffar, L.; Gluckman, J. C. *AIDS Res. Hum. Retroviruses* **1992**, *8*, 27-37.
83. Hammar, L.; Hirsch, I.; Machado, A. A.; De Mareuil, J.; Baillon, J. G.; Bolmont, C.; Chermann, J. C. *AIDS Res. Hum. Retroviruses* **1995**, *11*, 87-95.
84. Saifuddin, M.; Hart, M. L.; Gewurz, H.; Zhang, Y.; Spear, G. T. *J. Gen. Virol.* **2000**, *81*, 949-55.
85. Wang, L. X.; Ni, J.; Singh, S.; Li, H. *Chem. Biol.* **2004**, *11*, 127-34.
86. Lee, H. K.; Scanlan, C. N.; Huang, C. Y.; Chang, A. Y.; Calarese, D. A.; Dwek, R. A.; Rudd, P. M.; Burton, D. R.; Wilson, I. A.; Wong, C. H. *Angew. Chem. Int. Ed. Engl.* **2004**, *43*, 1000-3.
87. Wang, L. X.; Ni, J.; Singh, S. *Bioorg. Med. Chem.* **2003**, *11*, 129-36.
88. Li, H.; Wang, L. X. *Org. Biomol. Chem.* **2003**, *1*, 3507-13.
89. Ni, J.; Singh, S.; Wang, L. X. *Bioconjug. Chem.* **2003**, *14*, 232-8.
90. Ni, J.; Powell, R.; Baskakov, I. V.; DeVico, A.; Lewis, G. K.; Wang, L. X. *Bioorg. Med. Chem.* **2004**, *12*, 3141-8.
91. Duncan, R. J.; Weston, P. D.; Wrigglesworth, R. *Anal. Biochem.* **1983**, *132*, 68-73.
92. Li, H.; Wang, L. X. *Org. Biomol. Chem.* **2004**, *2*, 483-8.
93. Danishefsky, S. J.; Allen, J. R. *Angew. Chem. Int. Ed. Engl.* **2000**, *39*, 836-63.
94. Kudryashov, V.; Glunz, P. W.; Williams, L. J.; Hintermann, S.; Danishefsky, S. J.; Lloyd, K. O. *Proc. Natl. Acad. Sci. USA* **2001**, *98*, 3264-9.
95. Gilewski, T.; Ragupathi, G.; Bhuta, S.; Williams, L. J.; Musselli, C.; Zhang, X. F.; Bencsath, K. P.; Panageas, K. S.; Chin, J.; Hudis, C. A.; Norton, L.; Houghton, A. N.; Livingston, P. O.; Danishefsky, S. J. *Proc. Natl. Acad. Sci. USA* **2001**, *98*, 3270-5.
96. Ragupathi, G.; Cappello, S.; Yi, S. S.; Canter, D.; Spassova, M.; Bornmann, W. G.; Danishefsky, S. J.; Livingston, P. O. *Vaccine* **2002**, *20*, 1030-8.

97. Grogan, M. J.; Pratt, M. R.; Marcaurelle, L. A.; Bertozzi, C. R. *Annu. Rev. Biochem.* **2002**, *71*, 593-634.
98. Sears, P.; Wong, C. H. *Science* **2001**, *291*, 2344-50.
99. Herzner, H.; Reipen, T.; Schultz, M.; Kunz, H. *Chem. Rev.* **2000**, *100*, 4495-538.
100. Arsequell, G.; Valencia, G. *Tetrahedron: Asymmetry* **1999**, *10*, 3045-94.
101. Wang, L. X.; Fan, J. Q.; Lee, Y. C. *Tetrahedron Lett.* **1996**, *37*, 1975-78.
102. Wang, L.-X.; Tang, M.; Suzuki, T.; Kitajima, K.; Inoue, Y.; Inoue, S.; Fan, J.-Q.; Lee, Y. C. *J. Am. Chem. Soc.* **1997**, *119*, 11137-46.
103. Mizuno, M.; Haneda, K.; Iguchi, R.; Muramoto, I.; Kawakami, T.; Aimoto, S.; Yamamoto, K.; Inazu, T. *J. Am. Chem. Soc.* **1999**, *121*, 284-90.
104. Wang, L. X.; Singh, S.; Ni, J. In "*Synthesis of Carbohydrates through Biotechnology*"; Wang, P. G., Ichikawa, Y., Eds.; *ACS Symposium Series 873*, Chapter 6, pp 73-92. American Chemical Society: Washington, D. C., 2004.
105. Singh, S.; Ni, J.; Wang, L. X. *Bioorg. Med. Chem. Lett.* **2003**, *13*, 327-30.
106. Mandal, M.; Dudkin, V. Y.; Geng, X.; Danishefsky, S. J. *Angew. Chem. Int. Ed. Engl.* **2004**, *43*, 2557-61.
107. Geng, X.; Dudkin, V. Y.; Mandal, M.; Danishefsky, S. J. *Angew. Chem. Int. Ed. Engl.* **2004**, *43*, 2562-65.
108. Morley, S. L.; Pollard, A. J. *Vaccine* **2001**, *20*, 666-87.

Chapter 7

Toward a Carbohydrate-Based HIV-1 Vaccine

Daniel Calarese[1], Christopher Scanlan[2,4], Hing-Ken Lee[3], Pauline Rudd[4], Chi-Huey Wong[3,5], Raymond Dwek[4], Dennis Burton[2], and Ian A. Wilson[1,5]

Departments of [1]Molecular Biology, [2]Immunology, and [3]Chemistry, The Scripps Research Institute, 10550 North Torrey Pines Road, La Jolla, CA 92037
[4]The Glycobiology Institute, Department of Biochemistry, University of Oxford, South Parks Road, Oxford OX1 3QU, United Kingdom
[5]Skaggs Institute for Chemical Biology, The Scripps Research Institute, 10550 North Torrey Pines Road, La Jolla CA 92037

1. Glycosylation and Immunogenicity

Pathogen glycosylation is often perceived as a barrier to immune recognition and, by extension, to vaccine design. This notion is perhaps best illustrated in the case of HIV-1, where the glycosylation of the viral surface glycoproteins (gp120 and gp41) appears to profoundly affect the antibody response to the virus and likely contributes to viral immune evasion [1-3]. The carbohydrate chains, which cover a substantial portion of the antigenic surface of HIV-1 (Figure 1), are poorly immunogenic and act as a shield to prevent antibody recognition of the viral particle. For this reason, most efforts aimed at eliciting a protective immunity to HIV, have focused on the protein components of the virus.

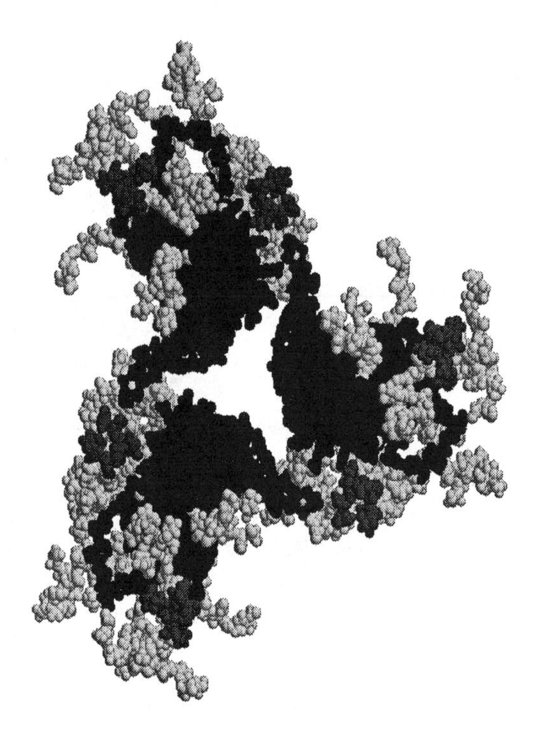

Figure 1. Proposed model for trimeric gp120, based on information in [57], with the protein core shown in blue and N-linked glycan chains in yellow. The glycans attached to N332 and N392, which form the primary attachment sites for 2G12, are indicated in red. This figure depicts a view of the trimeric gp120 as seen from the host cell membrane. The V1/V2 loops and attached carbohydrates are not indicated as no structural data are available on these moieties. (See page 1 of color inserts.)

Although immune recognition of carbohydrates is generally reduced compared to protein antigens, robust antibody responses to some certain carbohydrates on pathogens can be elicited, particularly when the sugars are pathogen-specific and do not resemble carbohydrates of the host. Moreover, some of these carbohydrate antigens have formed the basis of successful prophylactic vaccines [4]. Therefore, in principle, the glycosylation of HIV-1 should be considered a target for vaccine design.

Recently, structural studies of a neutralizing anti-HIV-1 antibody and its carbohydrate epitope have allowed the first steps to be taken towards the design of a carbohydrate-based vaccine for HIV.

Reasons for the Poor Immunogenicity of Carbohydrates

Several factors limit the immune response to carbohydrates. The *low affinity* of protein-carbohydrate interactions places a physical constraint on the degree of antibody affinity. Many carbohydrates on pathogens are similar to those on self-proteins (proteins from the host organism), preventing essential *self/non-self discrimination* needed for immune surveillance, and potentially lead to immunological tolerance. Self/non-self discrimination is necessary in order for the immune system to distinguish between the host and the invading pathogen. Finally, a given glycoprotein may carry a wide *heterogeneity* of sugar types at a particular location, thus presenting multiple distinct isoforms of a potentially immunogenic surface that dilutes the immune response.

i) Low affinity of Antibody-Glycan Interactions.

The affinity of proteins for carbohydrates is generally low, where the Ka for interaction is typically in the mid-micromolar range. This physical constraint severely limits the ability of the B-cell receptors to respond to a glycan epitope and, hence, limits the evolution of soluble antibodies (IgG, IgA) that could potently neutralize a pathogen. One way in which the affinity of protein-carbohydrate binding can be increased is through multivalency of the ligand-receptor interaction [5-7]. In fact, the majority of naturally occurring lectins bind their sugar ligands through multiple low-affinity interactions, giving rise to a high avidity interaction. For an antibody, the receptor valency depends on the class of Ig; IgM can display up to ten combining sites, whereas IgG has only two.

The avidity derived from two combining sites (i.e. two Fab molecules, which are the antigen binding portions of an immunoglobulin) may approach the product of the individual binding constants of the Fabs. This can give an apparent affinity for a glycan ligand, comparable to that of a protein epitope (for example a 10^3 dissociation constant for each Fab could maximally give rise to a 10^6 Ka for the IgG directed against a polysaccharide epitope (Figure 2)).

However, for optimal multivalent binding to occur, the epitope(s) must be orientated in such a way that allows simultaneous interaction of two or more Fabs with the antigen surface [8]. Therefore, clustered glycans or repetitive polysaccharides are much more likely to elicit high affinity antibodies than dispersed, monovalent glycan epitopes.

ii) Similarity of Carbohydrates on Pathogens to Self Antigens

Many pathogens cover their surface with glycans that resemble the structures present in the host. Some bacterial glycosyl transferases appear to have evolved,

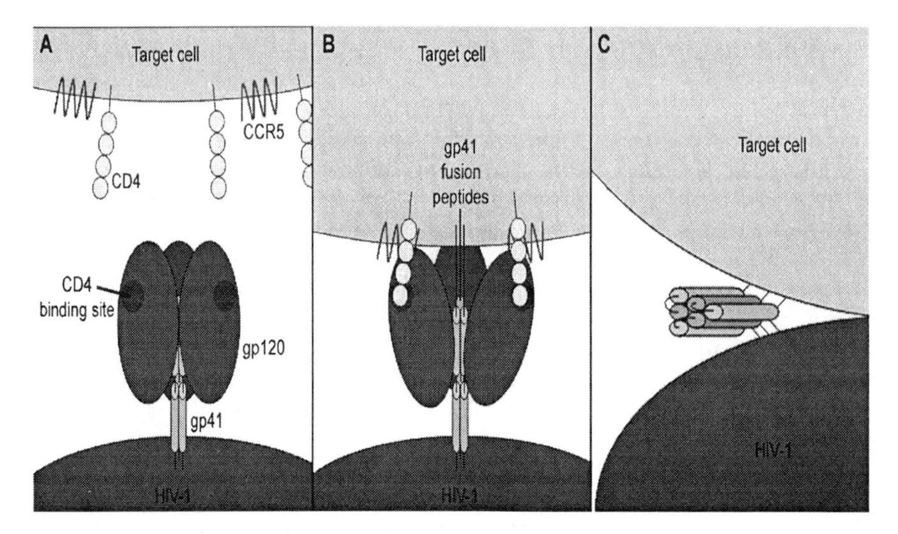

Figure 2. HIV-1 entry into the host cell. A) Components of HIV-1 entry into the host cell. On the surface of the virus is a homotrimeric pre-fusion complex made up of gp41/gp120 heterodimers. Gp120 has a recessed CD4 binding pocket (labeled). B) Gp120 binds to CD4 and the CCR5 coreceptor, which signals gp41 to change conformation and insert its fusion peptides into the host cell. C) Gp120 is subsequently shed from gp41, and gp41 undergoes a large conformational change causing it to snap shut into a six helix coiled-coil bundle. This activation of the spring-loaded mechanism brings the viral membrane and host cell membrane in close contact and allows the virus to fuse with the host cell. Figure adapted from [58]. (See page 2 of color inserts.)

convergently, with their host, to produce glycans that mimic host carbohydrates. For example, the glycosyl transferases of some streptococcal pathogens appear to have evolved a convergent functionality with their host to produce glycosaminoglycans similar to that of the host [9].

Alternatively, most surface proteins from membrane viruses are glycosylated by the addition of 'self'-glycans to viral proteins by the host glycosylation apparatus itself. Many highly pathogenic viruses are heavily glycosylated, including Influenza [10], Ebola [11-13], HIV [14-18] and Hepatitis [14-18]. The extent to which the self glycan coating limits the humoral immune response is unknown, but the decoration of pathogen surfaces with self glycans is a widespread phenotype, suggesting a functional role in immune evasion.

iii) Heterogeneity of Pathogen Glycosylation

Unlike proteins, which are a direct product of the encoding DNA, glycans are post-translational secondary modifications. The cell type where a protein is

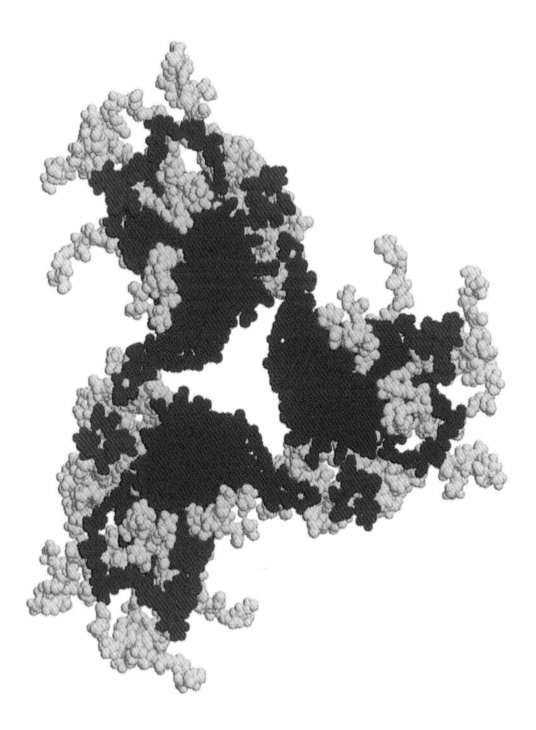

Plate 7.1. Proposed model for trimeric gp120, based on information in [57], with the protein core shown in blue and N-linked glycan chains in yellow. The glycans attached to N332 and N392, which form the primary attachment sites for 2G12, are indicated in red. This figure depicts a view of the trimeric gp120 as seen from the host cell membrane. The V1/V2 loops and attached carbohydrates are not indicated as no structural data are available on these moieties.

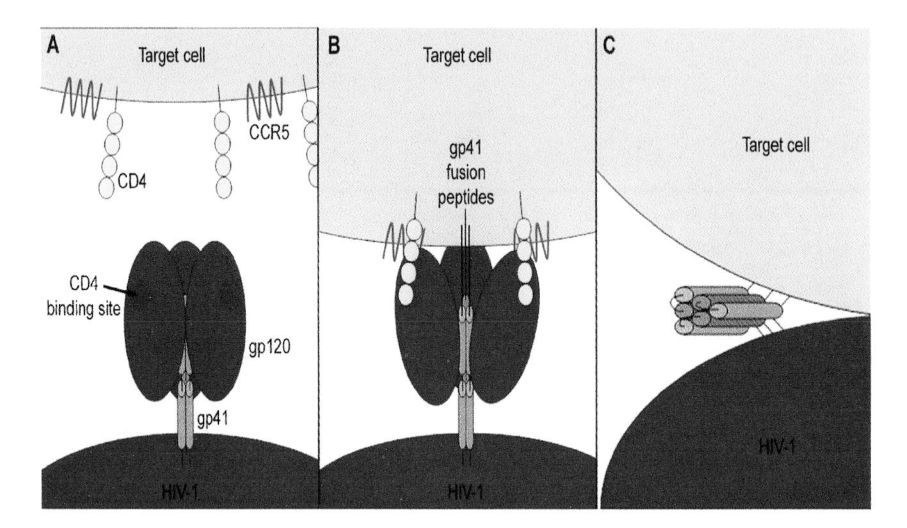

Plate 7.2. HIV-1 entry into the host cell. A) Components of HIV-1 entry into the host cell. On the surface of the virus is a homotrimeric pre-fusion complex made up of gp41/gp120 heterodimers. Gp120 has a recessed CD4 binding pocket (labeled). B) Gp120 binds to CD4 and the CCR5 coreceptor, which signals gp41 to change conformation and insert its fusion peptides into the host cell. C) Gp120 is subsequently shed from gp41, and gp41 undergoes a large conformational change causing it to snap shut into a six helix coiled-coil bundle. This activation of the spring-loaded mechanism brings the viral membrane and host cell membrane in close contact and allows the virus to fuse with the host cell. Figure adapted from [58].

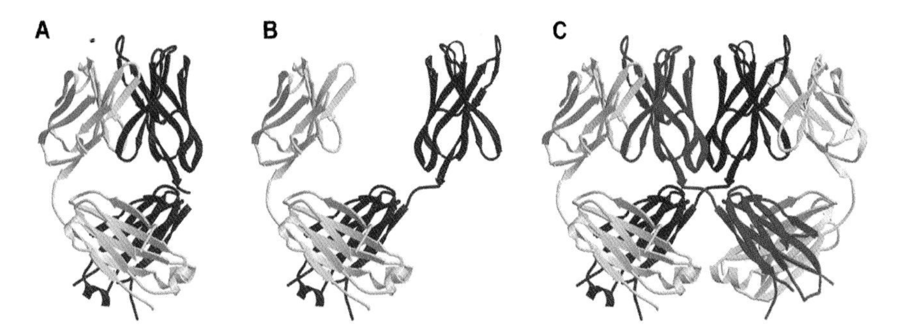

Plate 7.3. Comparison of a typical and domain-swapped antibody. A) Structure of a typical Fab molecule, with the heavy chain shown in purple and light chain in silver. B) Structure of a Fab 2G12 monomer, as seen in the crystal structures. The variable heavy domain is clearly separated away from the rest of the Fab molecule. C) Novel structure of the Fab 2G12 dimer, with variable heavy domains swapped between the two Fab molecules [52]. The heavy chains of the two Fabs are shown in purple and blue, and the light chains are colored silver. Figure was generated using Molscript [59] and Raster3D [60].

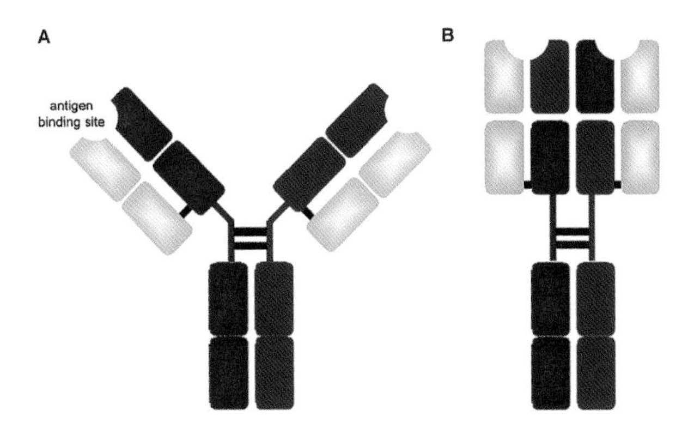

Plate 7.4. Consequence of domain swapping to overall IgG1 conformation. A) The structure of a typical Y-shaped IgG1. B) Structure of 2G12 IgG1. The swapping of the variable heavy domains between the two Fab molecules causes the IgG to adopt a linear conformation. This unusual linear conformation of 2G12 IgG has been observed by cryo-electron microscopy [52, 61]. Figure was generated using Molscript [59] and Raster3D [60].

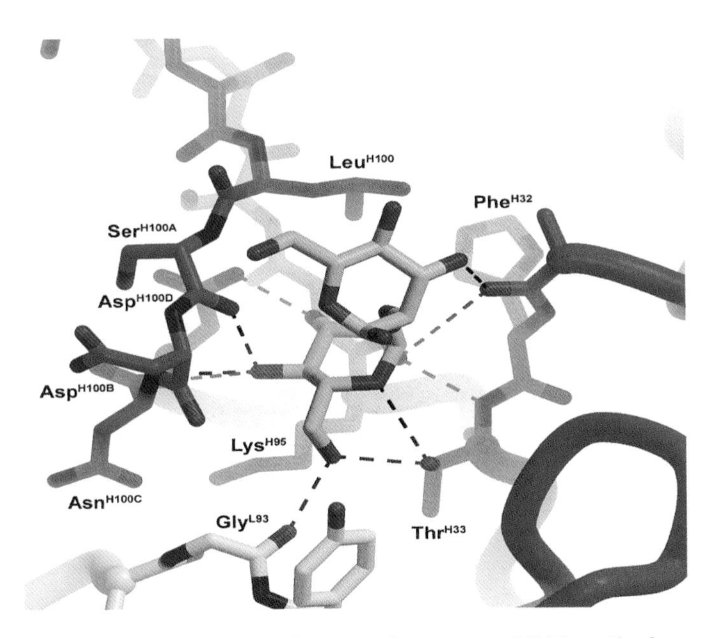

Plate 7.5. Detailed interactions between the primary 2G12 antibody combining site and Manα1-2Man [52]. The Fab heavy and light chains are shown in blue and silver, respectively. The Manα1-2Man is bound in a pocket made up of the L3, H1, H2, and H3 CDR loops of Fab 2G12. Potential hydrogen bonds between Fab 2G12 and Manα1-2Man are shown with dashed lines. Figure was generated using Molscript [59] and Raster3D [60].

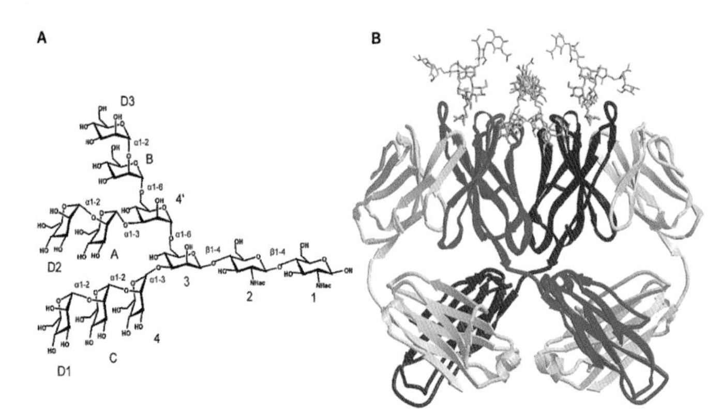

Plate 7.6. A) Structure of Man9GlcNAc2. B) Structure of Fab 2G12 with Man9GlcNAc2 [52]. The carbohydrate is shown in yellow ball-and-stick. A Man9GlcNAc2 moiety is bound in the primary combining sites of both Fabs. In addition, two noncrystallographically unique Man9GlcNAc2 moeities also bind at the interface between the variable heavy domains of the two Fabs. Figure was generated using Molscript [59] and Raster3D [60].

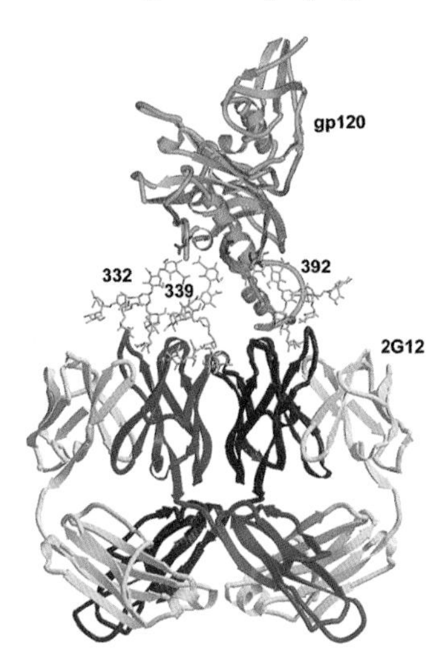

Plate 7.7. Model of 2G12 glycan recognition of gp120 [52]. The Fab 2G12 + Man9GlcNAc2 (shown in yellow ball-and-stick) structure docks onto the previously determined structure for gp120 [62] (shown in red), almost perfectly spanning the N-linked glycosylation sites implicated as most critical for 2G12 binding. In this model, the Man9GlcNAc2 moieties in the primary combining sites are unchanged from the determined crystal structure, and the Man9 moiety at the interface between the two Fabs has been slightly modified to show the potential for its origin to be Asn339. This model shows that 2G12 has attained a high level of affinity and specificity to gp120 by adopting a conformation to recognize and bind the carbohydrate cluster. Figure was generated using Molscript [59] and Raster3D [60].

synthesized is often the dominant factor in determining the type of glycans added to a protein. Therefore, the exact carbohydrate structures present on virus may depend on the tissue(s) infected, rather than being an inherent property of the pathogen. Bacteria and fungi generally synthesize their own glycans, and are, therefore, less influenced by the host.

Furthermore, the synthesis of carbohydrate conjugates occurs though sequential enzymatic pathways, with many, often competing, parallel reactions. Therefore, the same amino acid, in the same protein, from the same cell line, can often be derivatized by many diverse carbohydrate moieties. Such microheterogeneity may well dilute the immunogenicity of a particular epitope [19].

Current Carbohydrate Vaccines

Some current vaccines, based on the carbohydrate structures from bacterial pathogens, have been successful in providing protective immunity, or are in advanced clinical trials. These targets of these immunogens include *Neisseria meningitidesm, Haemophilus Influenzae B, Staphylococcus aureus, Salmonella typhi,* and *Streptococcus pneumoniae.*

The reasons why all of the existing carbohydrate vaccines are based on bacterial surface polysaccharides can be rationalized in the context of carbohydrate immunogenicity:

i) The immunogens are based on repetitive or clustered epitopes which are ideal for multivalent recognition by B-cell receptors and immunoglobulins.

ii) The oligosaccharide units that form the basis for Fab recognition are distinct from self carbohydrates. These differences can be attributed to the unique bacterial enzymes which synthesize the capsular polysaccharides. (Perhaps for this reason, the *N. meningitides* B strain has proved harder to combat because the structural linkage of polysialic acid is present in neuronal tissue (Table 1)). The species and sub-type specificity of glycosylation is important. For example, immunity raised by C-type meningococcal vaccines is not generally reactive against the B-type polysaccharide antigen. Thus, differences in glycosylation allow self to be discriminated from non-self and, conversely, small differences in polysaccharide antigens, within a bacterial species, allow the persistence of vaccine-resistant sub-types.

Table 1. Carbohydrate-based vaccines against bacterial pathogens.

Pathogen	Carbohydrate Epitope	Ref
Neisseria meningitides A	(1-6)- N-acetyl-D-mannosamine-1-phosphate	[63]
Neisseria meningitides B	(2-8) polysialic acid*	[64]
Neisseria meningitides C	(2-9) polysialic acid	[64]
Haemophilus Influenzae B	polyribosyl ribitol-phosphate (PRP)	[8, 65]
Staphylococcus aureus	Type 5: [-4-3-O-Ac-β-D-ManNac(1-4)-α-L-FucNAc-(1-3)-β-D-FucNAc-(1-)]n Type 8: [-3-4-O-Ac-β-D-ManNac(1-3)-α-L-FucNAc-(1-3)-β-D-FucNAc-(1-)]n	[66-68]
Salmonella typhi	[-4) 3-O-Ac-α-D-GalNAc-(1-]n	[69, 70]
Streptococcus pneumonia	β-D-Glcp-(1-6)-[α-D-NeupNAc-(2-3)-β-D-Galp-(1-4)]-β-D-GlcpNAc-(1-3)-β-D-Galp-(1-4)	[71, 72]

*also present in mammalian neuronal polysialic acid

iii) Finally, the diversity of glycosylation, found on the exterior of a eukaryotic cell (and hence on the surface of most glycosylated viruses) may not be found within the capsid antigens of a single bacterium. Generally, a few repeating polysaccharide units can be isolated as a major neutralizing determinant on a bacterial surface (Table 1), as evidenced by their utility in vaccine design. Thus, glycan microheterogeneity probably does not significantly undermine the antibody response to bacterial antigens to the same extent as in eukaryotic glycosylation.

2. A Carbohydrate Vaccine for HIV-1?

It is generally accepted that a prophylactic vaccine against HIV is required to control the AIDS epidemic. The immune response to HIV is generally characterized by low levels of neutralizing antibodies. An immunogen capable of eliciting potent neutralizing antibodies against a broad range of HIV-1 isolates has, to date, not been found. A neutralizing antibody response is essential for a successful vaccine, as the neutralizing activity of antibodies may be directed towards both free virus particles and infected cells. First, antibodies are capable of neutralizing a virus by binding to the virus and sterically hindering it from host cell contact. Second, the antibodies can target the virus for complement-mediated effector functions, such as virolysis or phagocytosis. In addition, antibodies can target infected cells for clearance and, in certain cases, may inhibit new virion particles from being released from these cells (reviewed in [20]).

Antibodies are typically directed against the large envelope spike proteins of viruses and, therefore, HIV-1 virus envelope proteins gp41 and gp120 are the key targets for HIV-1 vaccine design. These proteins are originally made as a single protein (gp160), which is then cleaved by cellular proteases into two proteins (gp41 and gp120) that remain non-covalently bound as a heterodimer. The final product on the viral surface is a homotrimeric complex made of gp41/gp120 heterodimers. Viral infection occurs when gp120 binds the CD4 receptor and a coreceptor (typically the CCR5 or CXCR4 membrane protein) of a host cell (such as $CD4^+$ T cells or macrophages). Once gp120 is bound, it is believed that conformational changes take place that involve gp120 monomers being shed from the gp41 trimer in conjunction with the fusion peptides from the gp41 trimer inserting themself into the lipid bilayer of the host cell. This is then followed by a "snapping shut" of the gp41 trimer into a compact six stranded coiled-coil complex that juxtaposes the host and viral membranes, allowing fusion to occur (Figure 2) (reviewed in [21]).

As expected, the virus has evolved various modes of defense against the host immune system (including the direct offense against the host immune system).

The gp120 monomers which are shed during the fusion process can act as decoys to dampen antibody binding and also present immunogenic epitopes exposed only in the monomer state, but hidden in the trimeric envelope complex. Also, in the trimer complex, the gp41 protein is mainly sequestered by gp120 and only transiently exposed during the fusion process. This partial exposure of gp41 in the pre-fusogenic state (gp120 bound) results in low antigenicity, as it is sterically difficult for an antibody bind to gp41. This allows the important fusion machinery of gp41 to remain hidden until the moment of infection.

The gp120 glycoprotein is also well evolved to evade host adapted immunity. First, gp120 displays immunodominant loops, which are susceptible to antigenic variation without undermining overall fitness. These immunodominant variable loops lead to the creation of antibodies that can only recognize a small subset of viral isolates. Second, the conserved regions of gp120 are difficult to access and are, therefore, poorly antigenic. The CD4 binding site is found in a deep pocket of gp120, and the co-receptor binding site is believed to be only exposed after CD4 binding (reviewed in [20]).

Finally, the gp120 protein is heavily glycosylated. The carbohydrates on gp120 are perhaps HIV-1's greatest defense against the host immune system. Gp120 contains an average of twenty four N-linked glycosylation sites. For the reasons discussed above, these carbohydrates are poorly immunogenic because they are self oligosaccharides. Coating the gp120 surface with self carbohydrates hides many potential immunogenic epitopes of the protein. This highly glycosylated surface of gp120 has been called the "silent face", as it was believed to be immunologically silent [22]. The high-mannose oligosaccharides are also believed to facilitate viral infection by binding to the lectin DC-SIGN on dendritic cells, which then transport the virus to the host lymph nodes where they may easily infect $CD4^+$ T cells [23].

This combination of defense mechanisms the HIV-1 virus uses to avoid the immune system result in a poor, non-neutralizing host immune response. However, a few potent, broadly neutralizing antibodies have been discovered from assaying combinatorial antibody phage display libraries [24] and cell lines [25] from infected individuals. These antibodies include b12, which uses a long H3 loop to gain access to the recessed CD4 binding site in gp120 [26], and 4E10 and 2F5 which recognize highly conserved linear epitopes on gp41 [27-30].

Of particular relevance here is the antibody 2G12, a potent, broadly neutralizing antibody which protects against viral challenge in animal models (reviewed in [31]). 2G12 is the only known anti-HIV-1 antibody which binds to carbohydrates on HIV-1 [32, 33]. As described in the next sections, the structure of both the epitope and the Fab of 2G12 have been characterized.

These studies have made possible rational HIV-1 vaccine design based on the 2G12 epitope.

3. HIV gp120 glycosylation

A remarkable feature of HIV is the dense sugar array surrounding the exposed envelope proteins gp120 and gp41. This extensive glycosylation is known to impact on every aspect of virus-host biology. The folding of viral proteins [34-37], the transmission of the virus [38-41] and the efficiency of the immune response to infection [3] are all profoundly affected by these carbohydrate structures.

Like other viruses, HIV proteins are glycosylated entirely by cellular enzymes. In the endoplasmic reticulum (ER), glycan precursors are added to newly translated proteins in a tightly regulated (and evolutionarily conserved) set of reactions. Following exit from the ER the glycoprotein is transported to the Golgi where immature glycans are processed, by multiple glycosidases and transferases, to produce a broad spectrum of complex glycans. The location of a glycan is encoded by the linear glycosylation squence (NxS/T); however, the enzymatic and physical environment determines the type of carbohydrate produced. Because the expression pattern of Golgi enzymes differs greatly between cell-types, gp120 glycosylation is cell-type specific. Furthermore, the accessibility of a carbohydrate for a particular enzyme determines its probability for further modification [19, 42-44].

Several studies have investigated the types of carbohydrates found on gp120 and showed the presence of oligomannose, complex and hybrid carbohydrates [14-18]. Significantly, about half of the glycans on gp120 are oligomannose structures. Site-specific analysis showed that these glycans are clustered together on the outer face of gp120. The complex glycans are located around the receptor binding sites and on the variable loops. The dense packing of the oligomannose glycans and the more exposed positioning of the complex glycans is, therefore, consistent with their level of enzymatic processing [14].

Despite the high mutation rate of HIV, the locations of the N-linked glycosylation sites in gp120 are generally well conserved between different isolates and clades [45]. This suggests that positive selection acts to maintain these sites. What exactly that selection pressure might be is unclear; no one site is found in all isolates and even the most conserved glycosylation motifs display significant heterogeneity. The dispensible nature of individual sites was highlighted by a study where each of the 24 N-linked sites of gp120 was individually deleted [46]. The resulting mutant viruses generally showed little or no decrease in activity.

4. Antibody recognition of gp120 glycans

Several reports have indicated that the presence of a glycosylation site can influence the binding of antibodies. For example, the presence or absence of glycans on the V3 loop of gp120 has been shown to modulate the binding of monoclonal antibodies. This appears to be a result of steric occlusion of a protein epitope rather than direct antibody-protein binding [2, 10, 47-49]. Thus, the acquisition (or loss) of a glycosylation site can have a dramatic effect on the immunogenicity of the surrounding protein surface. This has led to the concept of the "evolving glycan shield" of gp120 [3], with changes in glycosylation enchancing the rate of immune evasion during antibody-antigen co-evolution.

Some data have indicated that (sialyl)-Tn O-linked epitopes, thought to be present on gp120, can be recognized by antibodies. These sugars can act as immunogens [50] and this epitope is, therefore, a potential carbohydrate site for antibody neutralization.

Because gp120 and gp41 are synthesized by host cells, their glycosylation is identical in many ways to that of host glycoproteins. Thus gp120 and gp41 have no intrinsically non-self glycan structures and exhibit significant microheterogeneity. For all these reasons, any anti-HIV-1 carbohydrate antibodies should be expected to be fundamentally different from existing bacterial anti- polysaccharide antibodies.

Only one specific glycan epitope has been determined on gp120. This cluster of oligomannose glycans on the outer domain of gp120 is recognized by the neutralizing antibody 2G12. Investigation of the 2G12 epitope has illuminated surprising structural and immunological similarities between the outer domain of gp120 and the immunogenic carbohydrate surfaces of other pathogens [45, 51, 52].

Structural basis of 2G12 recognition of gp120

Mutagenesis studies on gp120 revealed that only substitutions that remove the N-linked glycosylation sites of gp120 had a significant effect on 2G12 affinity. Substitutions removing the sequences for some N-linked glycosylation motifs (NxS/T) on the silent face of gp120 were sufficient to prevent 2G12 attachment. Comparison of mutagenesis data with 2G12 escape mutants and sequence alignments of sequences known to be bound by 2G12 identified a small cluster of gp120 glycans as the basis of the 2G12 epitope.

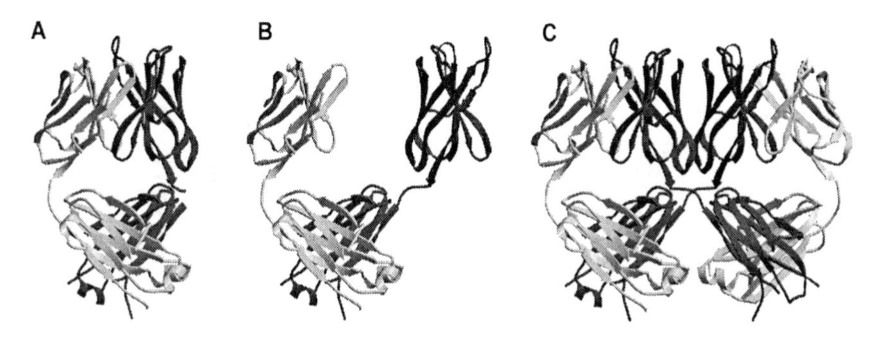

Figure 3. Comparison of a typical and domain-swapped antibody. A) Structure of a typical Fab molecule, with the heavy chain shown in purple and light chain in silver. B) Structure of a Fab 2G12 monomer, as seen in the crystal structures. The variable heavy domain is clearly separated away from the rest of the Fab molecule. C) Novel structure of the Fab 2G12 dimer, with variable heavy domains swapped between the two Fab molecules [52]. The heavy chains of the two Fabs are shown in purple and blue, and the light chains are colored silver. Figure was generated using Molscript [59] and Raster3D [60]. (See Page 2 of color insert.)

Treatment of gp120 with mannosidases (exoglycosidases, which remove terminal, sugars from gp120) demonstrated that the removal of non-reducing Manα1-2Man sugars from gp120 was sufficient to abrogate 2G12 binding. Consistent with these data, the 2G12-gp120 interaction could be inhibited by D-mannose. Additionally, 2G12 binding to gp120 could be inhibited by Manα1-2Man, but not by Manα1-3, 4 or 6 di-mannose [45, 53].

Further research has elucicated the specific interactions between 2G12 and its carbohydrate ligand. The crystal structures of the Fab fragments of 2G12 with $Man_9GlcNAc_2$ and with Manα1-2Man showed an extraordinary configuration where the variable heavy domains are domain-swapped between two Fabs (Figure 3)[52]. Further biochemical and biophysical evidence (such as cryo-electron microscopy, analytical ultracentrifugation, gel filtration, and alanine mutagenesis) suggested that this domain-swapped dimer of Fabs is the natural, functional form of the 2G12 antibody. The consequence of this unusual configuration creates an entirely novel architecture for an IgG. As opposed to a classical Y- or T-shaped IgG, the 2G12 IgG would have to be linear in order to accommodate the domain swap of the variable heavy domains (Figure 4).

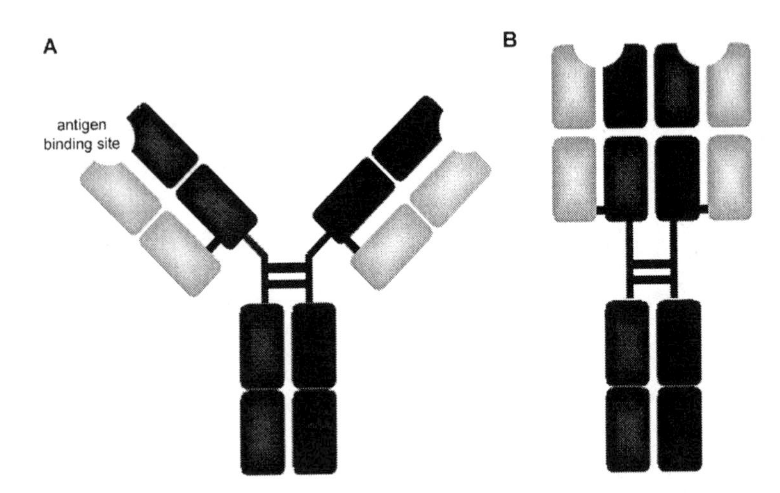

Figure 4. Consequence of domain swapping to overall IgG1 conformation. A) The structure of a typical Y-shaped IgG1. B) Structure of 2G12 IgG1. The swapping of the variable heavy domains between the two Fab molecules causes the IgG to adopt a linear conformation. This unusual linear conformation of 2G12 IgG has been observed by cryo-electron microscopy [52, 61]. Figure was generated using Molscript [59] and Raster3D [60]. (See Page 3 of color insert.)

The crystal structures give us a detailed look at the molecular interactions between 2G12 and its carbohydrate ligands. The primary combining site is a pocket which binds mainly to the Manα1-2Man at the tip of the D1 arm of a $Man_9GlcNAc_2$ moiety (Figure 5).

Additional interactions are also found between 2G12 and the mannose 3 and 4 sugars of the $Man_9GlcNAc_2$, which explains why 2G12 has higher affinity for $Man_9GlcNAc_2$ than Manα1-2Man. These additional interactions also give 2G12 a specificity for the Manα1-2Man of the D1 arm of a $Man_9GlcNAc_2$. In addition to the primary combining sites, a novel potential binding surface was created at the newly created interface between the two variable heavy domains. At this interface, non-crystallographically unique $Man_9GlcNAc_2$ moieties can be found binding to 2G12 via their D2 arms (Figure 6). These protein-carbohydrate interactions are not observed in Manα1-2Man crystal structure and, therefore, may only represent a secondary low affinity binding site that has a different specificity than Manα1-2Man.

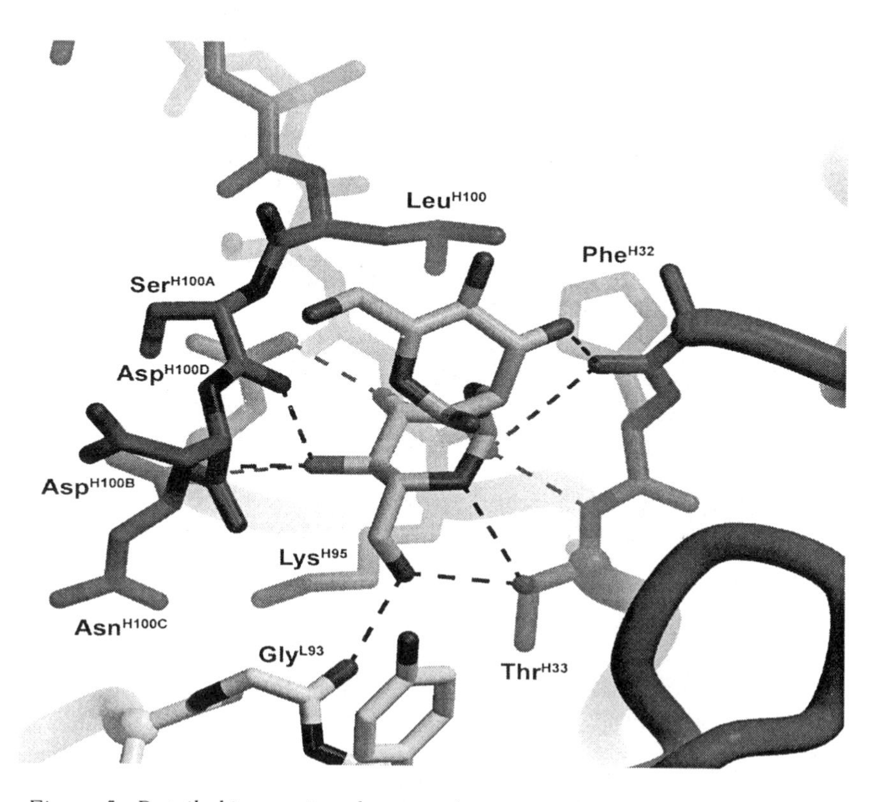

Figure 5. Detailed interactions between the primary 2G12 antibody combining site and Manα1-2Man [52]. The Fab heavy and light chains are shown in blue and silver, respectively. The Manα1-2Man is bound in a pocket made up of the L3, H1, H2, and H3 CDR loops of Fab 2G12. Potential hydrogen bonds between Fab 2G12 and Manα1-2Man are shown with dashed lines. Figure was generated using Molscript [59] and Raster3D [60]. (See Page 3 of color insert.)

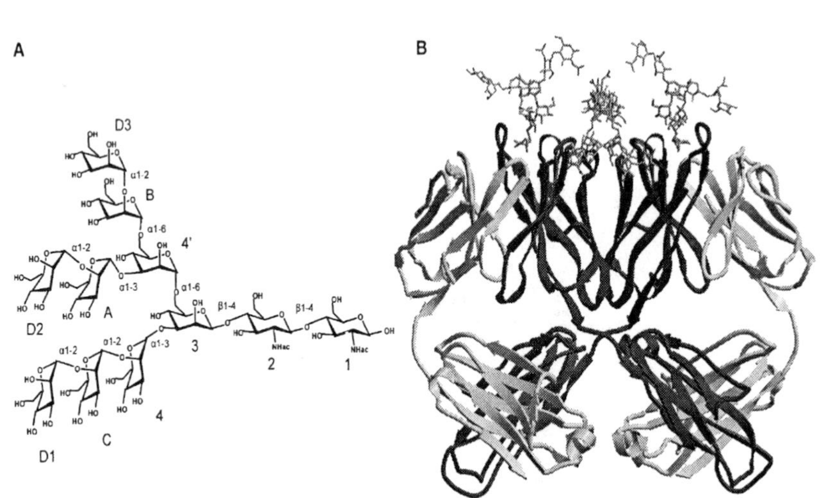

Figure 6. A) Structure of $Man_9GlcNAc_2$. B) Structure of Fab 2G12 with $Man_9GlcNAc_2$ [52]. The carbohydrate is shown in yellow ball-and-stick. A $Man_9GlcNAc_2$ moiety is bound in the primary combining sites of both Fabs. In addition, two noncrystallographically unique $Man_9GlcNAc_2$ moeities also bind at the interface between the variable heavy domains of the two Fabs. Figure was generated using Molscript [59] and Raster3D [60].
(See Page 4 of color insert.)

The novel architecture of the 2G12 antibody is an unexpected immunological solution to the problem of specific binding a carbohydrate cluster. The 2G12 antibody is able to bind with high affinity to gp120 because it binds divalently, with the geometrical spacing between the two primary combining sites (~35 Å) precisely aligned to the spacing of the carbohydrate ligands on gp120. In fact, docking of the 2G12+ $Man_9GlcNAc_2$ structure right onto gp120 show that the $Man_9GlcNAc_2$ moieties found in the primary combining sites perfectly span the previously implicated N-linked glycosylation sites (specifically Asn^{332} and Asn^{392}) on gp120 (Figure 7).

Further structural dynamics have shown that this conformation of $Man_9GlcNAc_2$ is energetically favorable and changes of less than half an angstrom in the crystal structure can cause it to attach at those N-linked sites. Most importantly, 2G12 shows that the immune system is capable of directing a high affinity antibody response specific against a self glycan that is arranged as a non-self cluster.

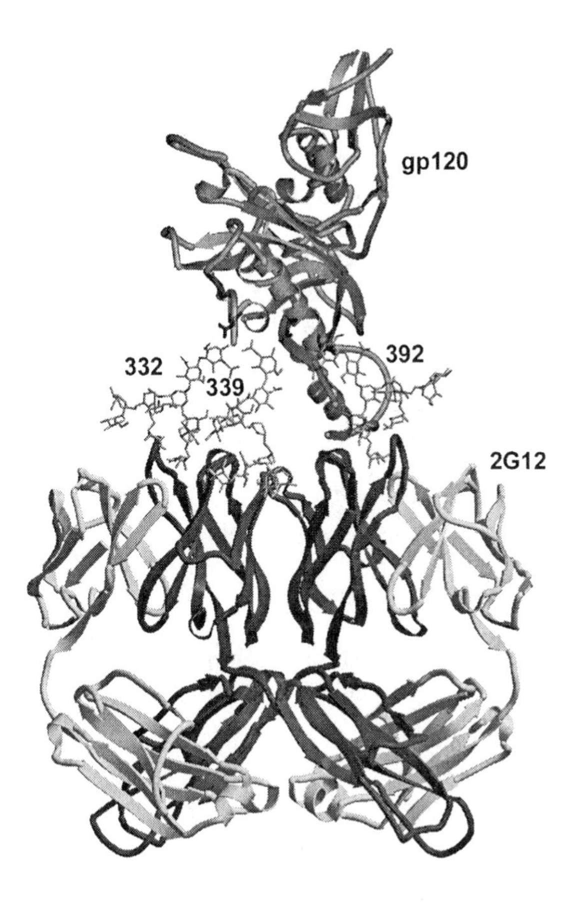

Figure 7. Model of 2G12 glycan recognition of gp120 [52]. The Fab 2G12 + $Man_9GlcNAc_2$ (shown in yellow ball-and-stick) structure docks onto the previously determined structure for gp120 [62] (shown in red), almost perfectly spanning the N-linked glycosylation sites implicated as most critical for 2G12 binding. In this model, the $Man_9GlcNAc_2$ moieties in the primary combining sites are unchanged from the determined crystal structure, and the Man9 moiety at the interface between the two Fabs has been slightly modified to show the potential for its origin to be Asn^{339}. This model shows that 2G12 has attained a high level of affinity and specificity to gp120 by adopting a conformation to recognize and bind the carbohydrate cluster. Figure was generated using Molscript [59] and Raster3D [60]. *(See Page 4 of color insert.)*

Immunological properties of the 2G12 epitope

The glycans on gp120 are somewhat unusual and shows some similarities to the types of carbohydrates which have previously elicited protective antibody responses:

i) The location of the N-linked sites are fairly well conserved, explaining the ability of 2G12 to recognize and neutralize a wide range HIV-1 isolates. The close packing of these conserved sites limits the accessibility of the ER/golgi enzymes to modify the basic oligomannose structure. Thus, the glycan microheterogeneicity found elsewhere on gp120 is significantly reduced.

ii) The dense clustering of oligomannoses is conceptually, if not chemically, similar to the repetitive glycans on many bacterial surfaces. This clustering is highly unusual and not previously observed in any mammalian (self) glycoproteins. Therefore, the recognition of this particular cluster of self glycans (which would normally be tolerated) by 2G12 can be explained by the close arrangement of the individual self glycans into a novel non-self cluster.

iii) The high avidity required for strong protein-carbohydrate interactions can be supported by the multivalent presentation of α1-2 linked mannose termini on gp120.

These features of gp120 glycosylation then open the door for design of carbohydrate-based HIV vaccines which utilize clustered carbohydrates as the immunogens to elicit a 2G12-like antibody response.

5. Glycan Immunogens

Recently, a few laboratories have begun designing immunogens that present a self glycan in a non-self cluster. This immunogen design has mainly been focused on presenting $Man_9GlcNAc_2$ (or a similar, cross-reactive carbohydrate) as a cluster with a geometry similar to that found on the gp120 envelope glycoprotein of HIV. One approach investigators have taken is to synthesize Man9-containing glycopeptides [54, 55]. The creation of these glycopeptides provides a protein scaffold in which to display the carbohydrate as a multivalent array.

Another group has used a reactivity-based one-pot synthesis approach to synthesize various Manα1-2Man containing oligomannoses [53]. The different oligomannoses were then screened for their ability to inhibit 2G12 binding to gp120. This approach provides non-self carbohydrates that are cross-reactive with $Man_9GlcNAc_2$, in addition to further elucidating the minimal template needed for binding 2G12. Two oligomannose candidates were discovered: a tetrasaccharide which mimics the D1 arm of $Man_9GlcNAc_2$ and a pentasaccharide which mimics the D2+D3 arms of $Man_9GlcNAc_2$ (Figure 8). The D1 and D2+D3 arm oligomannoses are capable of inhibiting 2G12 binding to gp120 at levels nearly equivalent to that of Man9. The next step in the design of a potential vaccine would be the creation of multivalent constructs with these oligomannoses in order to further mimic the carbohydrate cluster epitope of gp120.

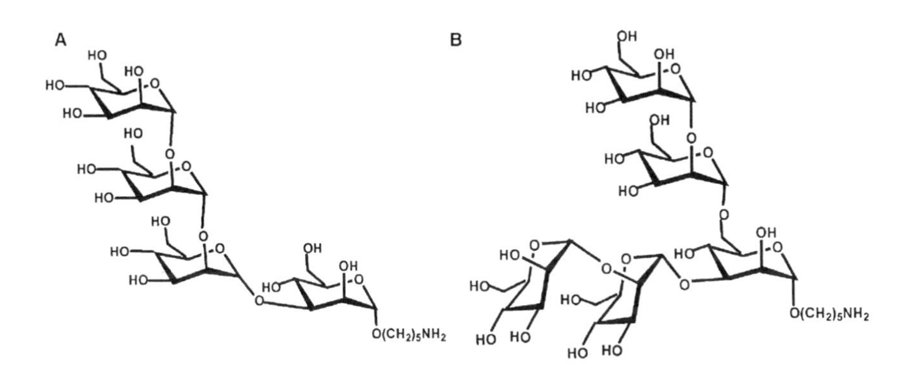

Figure 8. A) Structure of the tetrasaccharide, a mimic of the D1 arm of $Man_9GlcNAc_2$. B) Structure of the pentasaccharide, a mimic of the D2+D3 arms of $Man_9GlcNAc_2$. [53]

Finally, a third group has synthesized a template-assembled oligomannose cluster in an attempt to mimic the $Man_9GlcNAc_2$ cluster on gp120 [56]. Using the model of 2G12 binding to gp120 as a guideline, they attached three $Man_9GlcNAc_2$ moieties on a maleimide cluster which is anchored to a cholic acid molecule.

The final product is a molecule which displays three $Man_9GlcNAc_2$ moieties at a geometry similar to that of the N332, N339, and N392 carbohydrates of gp120 (Figure 9). Indeed, this template-assembled Man9 cluster is capable of binding 2G12 46-fold better than a single $Man_9GlcNAc_2$ molecule. Investigations into improving the geometry, through changing the size and rigidity of the spacer between the carbohydrate and the cholic acid, should improve the mimicry of the gp120 carbohydrate epitope even further.

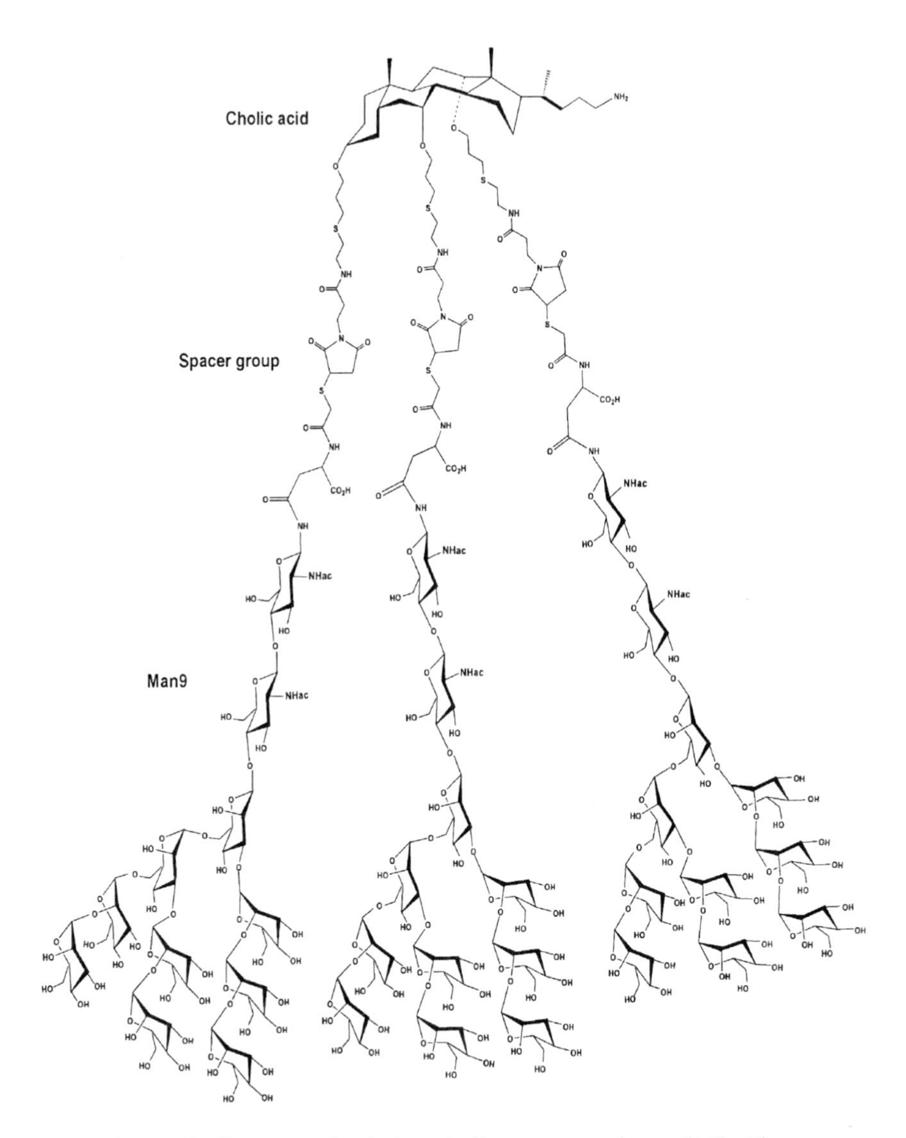

Figure 9. Structure of a designed oligomannose cluster [56]. Three Man$_9$GlcNAc$_2$ moieties were attached to a maleimide cluster. This cluster is attached via a spacer group onto a rigid template, cholic acid. The final product is a molecule which displays a Man9GlcNAc2 cluster like that of gp120.

Acknowledgements

Support is acknowledged from NIH GM46192 (IAW), AI33292 (DRB), the International Aids Vaccine Initiative (IAVI), and a joint TSRI/Oxford University graduate program (CS). This is publication #16877-MB from The Scripps Research Institute.

References

1. Reitter, J. N., Means, R. E. & Desrosiers, R. C. (1998). A role for carbohydrates in immune evasion in AIDS. *Nat Med* 4, 679-84.
2. Cole, K. S., Steckbeck, J. D., Rowles, J. L., Desrosiers, R. C. & Montelaro, R. C. (2004). Removal of N-linked glycosylation sites in the V1 region of simian immunodeficiency virus gp120 results in redirection of B-cell responses to V3. *J Virol* 78, 1525-39.
3. Wei, X., Decker, J. M., Wang, S., Hui, H., Kappes, J. C., Wu, X., Salazar-Gonzalez, J. F., Salazar, M. G., Kilby, J. M., Saag, M. S., Komarova, N. L., Nowak, M. A., Hahn, B. H., Kwong, P. D. & Shaw, G. M. (2003). Antibody neutralization and escape by HIV-1. *Nature* 422, 307-12.
4. Nyame, A. K., Kawar, Z. S. & Cummings, R. D. (2004). Antigenic glycans in parasitic infections: implications for vaccines and diagnostics. *Arch Biochem Biophys* 426, 182-200.
5. Dam, T. K., Roy, R., Das, S. K., Oscarson, S. & Brewer, C. F. (2000). Binding of multivalent carbohydrates to concanavalin A and Dioclea grandiflora lectin. Thermodynamic analysis of the "multivalency effect". *J Biol Chem* 275, 14223-30.
6. Dam, T. K., Roy, R., Page, D. & Brewer, C. F. (2002). Negative cooperativity associated with binding of multivalent carbohydrates to lectins. Thermodynamic analysis of the "multivalency effect". *Biochemistry* 41, 1351-8.
7. Monsigny, M., Mayer, R. & Roche, A. C. (2000). Sugar-lectin interactions: sugar clusters, lectin multivalency and avidity. *Carbohydr Lett* 4, 35-52.
8. Mammen, M., Choi, S.K., Whitesides, G.M. (1998). Polyvalent Interactions in Biological Systems: Implications for Design and Use of Multivalent Ligands and Inhibitors. *Angew. Chem. Int. Ed. Engl.* 37, 2754-2794.
9. DeAngelis, P. L. (2002). Evolution of glycosaminoglycans and their glycosyltransferases: Implications for the extracellular matrices of animals and the capsules of pathogenic bacteria. *Anat Rec* 268, 317-26.
10. Wilson, I.A., Skehel, J.J., Wiley, D.C. (1981). Structure of the haemagglutinin membrane glycoprotein of influenza virus at 3 A resolution. *Nature* 289, 366-73.

11. Feldmann, H., Nichol, S. T., Klenk, H. D., Peters, C. J. & Sanchez, A. (1994). Characterization of filoviruses based on differences in structure and antigenicity of the virion glycoprotein. *Virology* 199, 469-73.
12. Jeffers, S. A., Sanders, D. A. & Sanchez, A. (2002). Covalent modifications of the ebola virus glycoprotein. *J Virol* 76, 12463-72.
13. Lin, G., Simmons, G., Pohlmann, S., Baribaud, F., Ni, H., Leslie, G. J., Haggarty, B. S., Bates, P., Weissman, D., Hoxie, J. A. & Doms, R. W. (2003). Differential N-linked glycosylation of human immunodeficiency virus and Ebola virus envelope glycoproteins modulates interactions with DC-SIGN and DC-SIGNR. *J Virol* 77, 1337-46.
14. Zhu, X., Borchers, C., Bienstock, R. J. & Tomer, K. B. (2000). Mass spectrometric characterization of the glycosylation pattern of HIV-gp120 expressed in CHO cells. *Biochemistry* 39, 11194-204.
15. Kuster, B., Wheeler, S. F., Hunter, A. P., Dwek, R. A. & Harvey, D. J. (1997). Sequencing of N-linked oligosaccharides directly from protein gels: in-gel deglycosylation followed by matrix-assisted laser desorption/ionization mass spectrometry and normal-phase high-performance liquid chromatography. *Anal Biochem* 250, 82-101.
16. Mizuochi, T., Spellman, M. W., Larkin, M., Solomon, J., Basa, L. J. & Feizi, T. (1988). Carbohydrate structures of the human-immunodeficiency-virus (HIV) recombinant envelope glycoprotein gp120 produced in Chinese-hamster ovary cells. *Biochem J* 254, 599-603.
17. Feizi, T. & Larkin, M. (1990). AIDS and glycosylation. *Glycobiology* 1, 17-23.
18. Mizuochi, T., Matthews, T. J., Kato, M., Hamako, J., Titani, K., Solomon, J. & Feizi, T. (1990). Diversity of oligosaccharide structures on the envelope glycoprotein gp 120 of human immunodeficiency virus 1 from the lymphoblastoid cell line H9. Presence of complex-type oligosaccharides with bisecting N-acetylglucosamine residues. *J Biol Chem* 265, 8519-24.
19. Rudd, P. M. & Dwek, R. A. (1997). Glycosylation: heterogeneity and the 3D structure of proteins. *Crit Rev Biochem Mol Biol* 32, 1-100.
20. Burton, D. R. (2002). Antibodies, viruses and vaccines. *Nat Rev Immunol* 2, 706-13.
21. Burton, D. R., Desrosiers, R. C., Doms, R. W., Koff, W. C., Kwong, P. D., Moore, J. P., Nabel, G. J., Sodroski, J., Wilson, I. A. & Wyatt, R. T. (2004). HIV vaccine design and the neutralizing antibody problem. *Nat Immunol* 5, 233-6.
22. Moore, J. P. & Sodroski, J. (1996). Antibody cross-competition analysis of the human immunodeficiency virus type 1 gp120 exterior envelope glycoprotein. *J Virol* 70, 1863-72.
23. Geijtenbeek, T. B., Kwon, D. S., Torensma, R., van Vliet, S. J., van Duijnhoven, G. C., Middel, J., Cornelissen, I. L., Nottet, H. S., KewalRamani, V. N., Littman, D. R., Figdor, C. G. & van Kooyk, Y. (2000). DC-SIGN, a dendritic cell-specific HIV-1-binding protein that enhances trans-infection of T cells. *Cell* 100, 587-97.

24. Burton, D. R., Pyati, J., Koduri, R., Sharp, S. J., Thornton, G. B., Parren, P. W., Sawyer, L. S., Hendry, R. M., Dunlop, N., Nara, P. L. & et al. (1994). Efficient neutralization of primary isolates of HIV-1 by a recombinant human monoclonal antibody. *Science* 266, 1024-7.
25. Buchacher, A., Predl, R., Strutzenberger, K., Steinfellner, W., Trkola, A., Purtscher, M., Gruber, G., Tauer, C., Steindl, F., Jungbauer, A. & et al. (1994). Generation of human monoclonal antibodies against HIV-1 proteins; electrofusion and Epstein-Barr virus transformation for peripheral blood lymphocyte immortalization. *AIDS Res Hum Retroviruses* 10, 359-69.
26. Saphire, E. O., Parren, P. W., Pantophlet, R., Zwick, M. B., Morris, G. M., Rudd, P. M., Dwek, R. A., Stanfield, R. L., Burton, D. R. & Wilson, I. A. (2001). Crystal structure of a neutralizing human IgG against HIV-1: a template for vaccine design. *Science* 293, 1155-9.
27. Conley, A. J., Kessler, J. A., 2nd, Boots, L. J., Tung, J. S., Arnold, B. A., Keller, P. M., Shaw, A. R. & Emini, E. A. (1994). Neutralization of divergent human immunodeficiency virus type 1 variants and primary isolates by IAM-41-2F5, an anti-gp41 human monoclonal antibody. *Proc Natl Acad Sci U S A* 91, 3348-52.
28. Parker, C. E., Deterding, L. J., Hager-Braun, C., Binley, J. M., Schulke, N., Katinger, H., Moore, J. P. & Tomer, K. B. (2001). Fine definition of the epitope on the gp41 glycoprotein of human immunodeficiency virus type 1 for the neutralizing monoclonal antibody 2F5. *J Virol* 75, 10906-11.
29. Zwick, M. B., Labrijn, A. F., Wang, M., Spenlehauer, C., Saphire, E. O., Binley, J. M., Moore, J. P., Stiegler, G., Katinger, H., Burton, D. R. & Parren, P. W. (2001). Broadly neutralizing antibodies targeted to the membrane-proximal external region of human immunodeficiency virus type 1 glycoprotein gp41. *J Virol* 75, 10892-905.
30. Stiegler, G., Kunert, R., Purtscher, M., Wolbank, S., Voglauer, R., Steindl, F. & Katinger, H. (2001). A potent cross-clade neutralizing human monoclonal antibody against a novel epitope on gp41 of human immunodeficiency virus type 1. *AIDS Res Hum Retroviruses* 17, 1757-65.
31. Ferrantelli, F. & Ruprecht, R. M. (2002). Neutralizing antibodies against HIV -- back in the major leagues? *Curr Opin Immunol* 14, 495-502.
32. Trkola, A., Pomales, A. B., Yuan, H., Korber, B., Maddon, P. J., Allaway, G. P., Katinger, H., Barbas, C. F., 3rd, Burton, D. R., Ho, D. D. & et al. (1995). Cross-clade neutralization of primary isolates of human immunodeficiency virus type 1 by human monoclonal antibodies and tetrameric CD4-IgG. *J Virol* 69, 6609-17.
33. Trkola, A., Purtscher, M., Muster, T., Ballaun, C., Buchacher, A., Sullivan, N., Srinivasan, K., Sodroski, J., Moore, J. P. & Katinger, H. (1996). Human monoclonal antibody 2G12 defines a distinctive neutralization epitope on the gp120 glycoprotein of human immunodeficiency virus type 1. *J Virol* 70, 1100-8.

34. Fenouillet, E. & Gluckman, J. C. (1991). Effect of a glucosidase inhibitor on the bioactivity and immunoreactivity of human immunodeficiency virus type 1 envelope glycoprotein. *J Gen Virol* 72 (Pt 8), 1919-26.
35. Fennie, C. & Lasky, L. A. (1989). Model for intracellular folding of the human immunodeficiency virus type 1 gp120. *J Virol* 63, 639-46.
36. Otteken, A. & Moss, B. (1996). Calreticulin interacts with newly synthesized human immunodeficiency virus type 1 envelope glycoprotein, suggesting a chaperone function similar to that of calnexin. *J Biol Chem* 271, 97-103.
37. Fischer, P. B., Collin, M., Karlsson, G. B., James, W., Butters, T. D., Davis, S. J., Gordon, S., Dwek, R. A. & Platt, F. M. (1995). The alpha-glucosidase inhibitor N-butyldeoxynojirimycin inhibits human immunodeficiency virus entry at the level of post-CD4 binding. *J Virol* 69, 5791-7.
38. Gu, R., Westervelt, P. & Ratner, L. (1993). Role of HIV-1 envelope V3 loop cleavage in cell tropism. *AIDS Res Hum Retroviruses* 9, 1007-15.
39. Dumonceaux, J., Goujon, C., Joliot, V., Briand, P. & Hazan, U. (2001). Determination of essential amino acids involved in the CD4-independent tropism of the X4 human immunodeficiency virus type 1 m7NDK isolate: role of potential N glycosylations in the C2 and V3 regions of gp120. *J Virol* 75, 5425-8.
40. Ly, A. & Stamatatos, L. (2000). V2 loop glycosylation of the human immunodeficiency virus type 1 SF162 envelope facilitates interaction of this protein with CD4 and CCR5 receptors and protects the virus from neutralization by anti-V3 loop and anti-CD4 binding site antibodies. *J Virol* 74, 6769-76.
41. Li, Y., Rey-Cuille, M. A. & Hu, S. L. (2001). N-linked glycosylation in the V3 region of HIV type 1 surface antigen modulates coreceptor usage in viral infection. *AIDS Res Hum Retroviruses* 17, 1473-9.
42. Opdenakker, G., Rudd, P. M., Ponting, C. P. & Dwek, R. A. (1993). Concepts and principles of glycobiology. *Faseb J* 7, 1330-7.
43. Van den Steen, P., Rudd, P. M., Dwek, R. A. & Opdenakker, G. (1998). Concepts and principles of O-linked glycosylation. *Crit Rev Biochem Mol Biol* 33, 151-208.
44. Rudd, P. M., Elliott, T., Cresswell, P., Wilson, I. A. & Dwek, R. A. (2001). Glycosylation and the immune system. *Science* 291, 2370-6.
45. Scanlan, C. N., Pantophlet, R., Wormald, M. R., Ollmann Saphire, E., Stanfield, R., Wilson, I. A., Katinger, H., Dwek, R. A., Rudd, P. M. & Burton, D. R. (2002). The broadly neutralizing anti-human immunodeficiency virus type 1 antibody 2G12 recognizes a cluster of alpha1-2 mannose residues on the outer face of gp120. *J Virol* 76, 7306-21.
46. Ohgimoto, S., Shioda, T., Mori, K., Nakayama, E. E., Hu, H. & Nagai, Y. (1998). Location-specific, unequal contribution of the N glycans in simian immunodeficiency virus gp120 to viral infectivity and removal of multiple glycans without disturbing infectivity. *J Virol* 72, 8365-70.

47. Koito, A., Stamatatos, L. & Cheng-Mayer, C. (1995). Small amino acid sequence changes within the V2 domain can affect the function of a T-cell line-tropic human immunodeficiency virus type 1 envelope gp120. *Virology* 206, 878-84.
48. Bolmstedt, A., Sjolander, S., Hansen, J. E., Akerblom, L., Hemming, A., Hu, S. L., Morein, B. & Olofsson, S. (1996). Influence of N-linked glycans in V4-V5 region of human immunodeficiency virus type 1 glycoprotein gp160 on induction of a virus-neutralizing humoral response. *J Acquir Immune Defic Syndr Hum Retrovirol* 12, 213-20.
49. Losman, B., Bolmstedt, A., Schonning, K., Bjorndal, A., Westin, C., Fenyo, E. M. & Olofsson, S. (2001). Protection of neutralization epitopes in the V3 loop of oligomeric human immunodeficiency virus type 1 glycoprotein 120 by N-linked oligosaccharides in the V1 region. *AIDS Res Hum Retroviruses* 17, 1067-76.
50. Hansen, J. E., Jansson, B., Gram, G. J., Clausen, H., Nielsen, J. O. & Olofsson, S. (1996). Sensitivity of HIV-1 to neutralization by antibodies against O-linked carbohydrate epitopes despite deletion of O-glycosylation signals in the V3 loop. *Arch Virol* 141, 291-300.
51. Sanders, R. W., Venturi, M., Schiffner, L., Kalyanaraman, R., Katinger, H., Lloyd, K. O., Kwong, P. D. & Moore, J. P. (2002). The mannose-dependent epitope for neutralizing antibody 2G12 on human immunodeficiency virus type 1 glycoprotein gp120. *J Virol* 76, 7293-305.
52. Calarese, D. A., Scanlan, C. N., Zwick, M. B., Deechongkit, S., Mimura, Y., Kunert, R., Zhu, P., Wormald, M. R., Stanfield, R. L., Roux, K. H., Kelly, J. W., Rudd, P. M., Dwek, R. A., Katinger, H., Burton, D. R. & Wilson, I. A. (2003). Antibody domain exchange is an immunological solution to carbohydrate cluster recognition. *Science* 300, 2065-71.
53. Lee, H. K., Scanlan, C. N., Huang, C. Y., Chang, A. Y., Calarese, D. A., Dwek, R. A., Rudd, P. M., Burton, D. R., Wilson, I. A. & Wong, C. H. (2004). Reactivity-based one-pot synthesis of oligomannoses: defining antigens recognized by 2G12, a broadly neutralizing anti-HIV-1 antibody. *Angew Chem Int Ed Engl* 43, 1000-3.
54. Mandal, M., Dudkin, V. Y., Geng, X. & Danishefsky, S. J. (2004). In pursuit of carbohydrate-based HIV vaccines, part 1: The total synthesis of hybrid-type gp120 fragments. *Angew Chem Int Ed Engl* 43, 2557-61.
55. Geng, X., Dudkin, V. Y., Mandal, M. & Danishefsky, S. J. (2004). In pursuit of carbohydrate-based HIV vaccines, part 2: The total synthesis of high-mannose-type gp120 fragments. *Angew Chem Int Ed Engl* 43, 2562-5.
56. Li, H. & Wang, L. X. (2004). Design and synthesis of a template-assembled oligomannose cluster as an epitope mimic for human HIV-neutralizing antibody 2G12. *Org Biomol Chem* 2, 483-8.
57. Kwong, P. D., Wyatt, R., Sattentau, Q. J., Sodroski, J., Hendrickson, W. A. (2000). Oligomeric modeling and electrostatic analysis of the gp120 envelope glycoprotein of human immunodeficiency virus. *J Virol.* 74:1961-72).

58. Montefiori, D. & Moore, J. P. (1999). HIV vaccines. Magic of the occult? *Science* 283, 336-7.
59. Kraulis, P. J. (1991). MOLSCRIPT: A Program to Produce Both Detailed and Schematic Plots of Protein Structures. *J of Appl. Cryst.* 24, 946-950.
60. Merritt, E. A. B., D.J. (1997). Raster3D Photorealistic Molecular Graphics'. *Methods in Enz.* 277, 505-524.
61. Schulke, N., Vesanen, M. S., Sanders, R. W., Zhu, P., Lu, M., Anselma, D. J., Villa, A. R., Parren, P. W., Binley, J. M., Roux, K. H., Maddon, P. J., Moore, J. P. & Olson, W. C. (2002). Oligomeric and conformational properties of a proteolytically mature, disulfide-stabilized human immunodeficiency virus type 1 gp140 envelope glycoprotein. *J Virol* 76, 7760-76.
62. Kwong, P. D., Wyatt, R., Majeed, S., Robinson, J., Sweet, R. W., Sodroski, J. & Hendrickson, W. A. (2000). Structures of HIV-1 gp120 envelope glycoproteins from laboratory-adapted and primary isolates. *Structure Fold Des* 8, 1329-39.
63. Liu, T. Y., Gotschlich, E. C., Jonssen, E. K. & Wysocki, J. R. (1971). Studies on the meningococcal polysaccharides. I. Composition and chemical properties of the group A polysaccharide. *J Biol Chem* 246, 2849-58.
64. Bhattacharjee, A. K., Jennings, H. J., Kenny, C. P., Martin, A. & Smith, I. C. (1975). Structural determination of the sialic acid polysaccharide antigens of Neisseria meningitidis serogroups B and C with carbon 13 nuclear magnetic resonance. *J Biol Chem* 250, 1926-32.
65. Ip, C. C., Manam, V., Hepler, R. & Hennessey, J. P., Jr. (1992). Carbohydrate composition analysis of bacterial polysaccharides: optimized acid hydrolysis conditions for HPAEC-PAD analysis. *Anal Biochem* 201, 343-9.
66. Robbins, J. B., Schneerson, R., Horwith, G., Naso, R. & Fattom, A. (2004). Staphylococcus aureus types 5 and 8 capsular polysaccharide-protein conjugate vaccines. *Am Heart J* 147, 593-8.
67. Fournier, J. M., Vann, W. F. & Karakawa, W. W. (1984). Purification and characterization of Staphylococcus aureus type 8 capsular polysaccharide. *Infect Immun* 45, 87-93.
68. Moreau, M., Richards, J. C., Fournier, J. M., Byrd, R. A., Karakawa, W. W. & Vann, W. F. (1990). Structure of the type 5 capsular polysaccharide of Staphylococcus aureus. *Carbohydr Res* 201, 285-97.
69. Szu, S. C., Li, X. R., Stone, A. L. & Robbins, J. B. (1991). Relation between structure and immunologic properties of the Vi capsular polysaccharide. *Infect Immun* 59, 4555-61.
70. Szu, S. C., Stone, A. L., Robbins, J. D., Schneerson, R. & Robbins, J. B. (1987). Vi capsular polysaccharide-protein conjugates for prevention of typhoid fever. Preparation, characterization, and immunogenicity in laboratory animals. *J Exp Med* 166, 1510-24.

71. Brisson, J. R., Uhrinova, S., Woods, R. J., van der Zwan, M., Jarrell, H. C., Paoletti, L. C., Kasper, D. L. & Jennings, H. J. (1997). NMR and molecular dynamics studies of the conformational epitope of the type III group B Streptococcus capsular polysaccharide and derivatives. *Biochemistry* 36, 3278-92.
72. Wessels, M. R., Pozsgay, V., Kasper, D. L. & Jennings, H. J. (1987). Structure and immunochemistry of an oligosaccharide repeating unit of the capsular polysaccharide of type III group B Streptococcus. A revised structure for the type III group B streptococcal polysaccharide antigen. *J Biol Chem* 262, 8262-7.

Pathogen Management

Chapter 8

Cationic Polysaccharides in the Treatment of Pathogenic *Candida* Infections

Avital Mazar Ben-Josef[1], David Platt[2], and Eliezer Zomer[2]

[1]Huntington Woods, MI 48070
[2]Pro-Pharmaceuticals, 189 Well Avenue, Newton, MA 02459

Introduction

Commensally microorganisms considered to be the normal microflora of our body surfaces can become opportunistic pathogens responsible for severe and often fatal infections in humans (1-3). The expansion in numbers of fungal infections is also due to an expanding population of immunocompromised patients from diseases like AIDS, or many therapeutics such as steroid and chemo therapies, and advances in organ transplantation that facilitate fungal invasion. The prevalence of hospital acquired infections has almost doubled in recent years and now accounts for 10-15%, of which nosocomial infections with pathogenic yeast are responsible for up to 70% of the cases (4).

The increased frequency, severity, and number of fungal species identified as pathogens have further created a critical need for new, safe, anti-fungal drugs. Today, the most commonly used drugs for systemic and local fungal infections are amphotericin B (AMB) and azole agents (5). AMB is still the drug of choice in many systemic mycoses due to its broad spectrum and fungicidal activity. However, AMB is nephro-toxic and is administered intravenously only (6). The triazoles, fluconazole and itraconazole, have a broad spectrum of activity, can be administered orally, and are considered less toxic than AMB. However, azoles are fungistatic, and drug resistance has become a significant problem (7 - 12).

The Compounds

Cationic Polysaccharide from Mucor rouxii

The polysaccharide purified from the cell wall of the fungus *Mucor rouxii* (CPM), is a heat stable complex carbohydrate with an estimated molecular weight of 4.3 Kd. The purified polysaccharide is composed of glucosamine residues with the majority connected through a 1,4 beta glucosidic bond, and the minority has either a 3,4 or 4,6 glucosidic bond with about 5% *N*-acetylated residues.

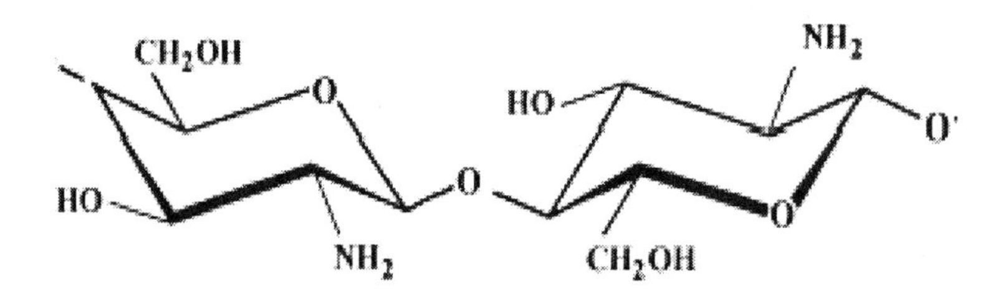

Structural units of polyglucosamine consist mostly of D-glucosamine units connected in beta-1,4 bond configuration.

Cationic Polysaccharide From Crab' Chitin

This polysaccharide is available commercially as deacetylated chitin (chitosan) with over 95% 2-amino-2-deoxy-D-glucopyranose. The polymer' monomeric units are bound through a 1,4 beta glycosidic bond. A 1% solution of this polymer was prepared by dissolving it in mild hydrochloric acid at room temperature. This soluble preparation was combined with sodium pyrithionate (CPP) at acidic pH at 5 to 1 ratio (w/w) to study potential synergistic effect.

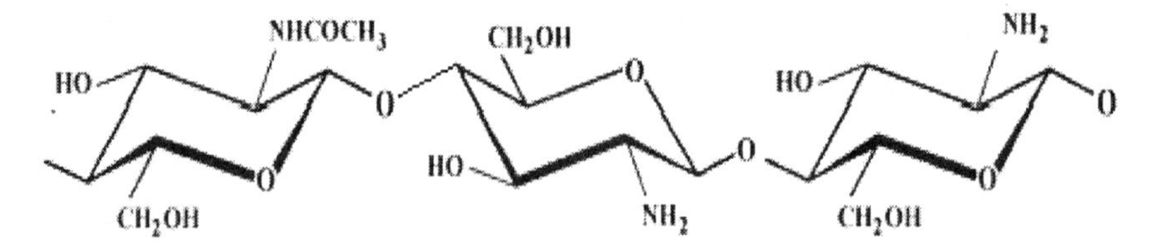

Structural units of chitosan consist mostly of D-glucosamine units connected in beta 1-4 bond configuration with about 5% N acetyl-D-glucosamine.

In Vitro Activity

CPM was proven to possess a remarkable in vitro fungicidal activity against a wide spectrum of azole-resistant *Candida* with apparently no cross-resistance with any currently used antifungal agents (13). Its broad anti-*Candida* activity is well comparable to that of AMB; however, the fungicidal activity of CPM is faster than that of AMB. Use of CPM in concentrations of 8 and 16 times fold greater than the Minimum Inhibitory Concentration (MIC) (5 and 10 μg/ml), resulted in the killing of 99.6% and 99.9%, respectively, of *Candida albicans* within 15 minutes of exposure. Using AMB in similar concentrations resulted in 99.9% killing after 4 hours for *C. albicans and C. glabrata* (Fig. 1 & 2). CPM possess a very narrow therapeutic range (MIC = 0.078 – 0.321 μg/ml) in all 112 *Candida* species (Table 1) that were tested including Azole–resistant species (13).

Table 1. In vitro susceptibilities of *C. albicans* and non-*albican*s Candida to Cationic Polysaccharide from *Mucor rouxii* (CPM).

Organisms	*Antifungal agent*	*MIC Range (μg/ml)*
C. albicans (29 tested)	CPM	0.156-0.312
C. parapsilosis (10)	CPM Fluconazole	0.078-0.312 0.16-20
C. glabrata (15)	CPM Ketoconazole Fluconazole	0.078-10.00 0.01-6.3 1.25-40
C. tropicalis (10)	CPM Fluconazole	0.039-0.312 0.63-80
C. lusitaniae (10)	CPM Fluconazole	0.156-0.312 0.31-20
C. krusie (12)	CPM Fluconazole	0.039-0.312 0.16-80
c. guillermondii (10)	CPM Fluconazole	0.078-0.312 5-10
C. kefyr (5)	CPM	0.156-0.312
C. stellatoidea (4)	CPM	0.312
C. rugosa (2)	CPM Fluconazole	0.312 5
C. lambica (1)	CPM Fluconazole	0.078 20
C. paratropicalis (1)	CPM	0.312
C. lipolitica (1)	CPM	0.312
Saccharomyces cerevisiae (2)	CPM	0.156

Like CPM, CPP was evaluated for in-vitro anti-*Candida* activity and demonstrated superior *in vitro* fungicidal activity against a wide spectrum of pathogenic yeasts, including azole-resistant isolates. The MIC's of CPP for *C. albicans* were found to be similar or one dilution lower than that of CPM, with a MIC of 0.156-0.312 μg/ml

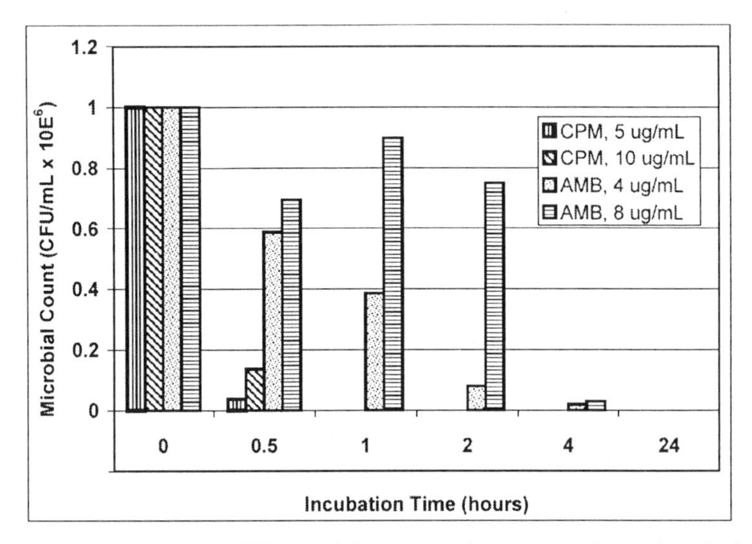

Figure 1. Comparison of fungicidal activity of cationic polysaccharide from Mucor rouxii *(CPM) and Amphotericin b (AMB) – for* C. albicans

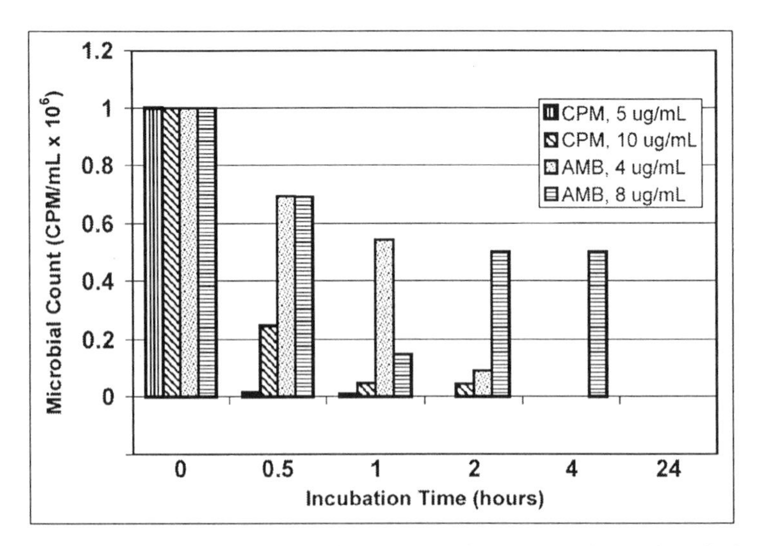

Figure 2. Comparison of fungicidal activity of cationic polysaccharide from Mucor rouxii *(CPM) and amphotericin b (AMB) – for* C. glabrata

Mode of Action

In recent years, the role of carbohydrates in biological interactions has been increasingly recognized (14). Carbohydrate-binding receptors, initiating diverse signal transduction pathways, have been described in hepatocytes (15), alveolar macrophages (16) and other systems (17). Many animal lectins contain carbohydrate-recognition domains (CRDs) (18), which specifically recognize a variety of sugars. The mannose-binding protein, for instance, is a serum lectin that binds to the surface of yeasts and bacteria and mediates an innate immune response (15, 19). It recognizes and binds to mannose, N-acetyl-glucosamine and L-fucose with approximately equal affinities (20). This process is achieved, in part, by clustering of several CRDs, each with its own sugar binding site.

In the case of the macrophage mannose receptor, the full affinity for the yeast mannan can only be attained when five of the eight CRDs are present (21). Many of these lectins require calcium for binding activity (C-type lectins). At the binding site of these C-type lectins, specific oligosaccharide equatorial hydroxyl groups act as coordination ligands and bind to Ca^{2+}. An example of the latter would be the binding of the antifungal agents from the *pradimicins* family, that bind to the yeast cell wall through the mannopyranoside-binding site in the presence of Ca^{2+} (22). On the other hand, calcium and magnesium are also known to inhibit a number of aminoglycoside-membrane interactions including those with bacterial, plasma and subcellular membranes (23, 24).

The high-affinity sites for aminoglycoside-membrane interaction were identified as the acidic phospholipids of the membrane (25, 26), and the binding is due to a charge interaction between those polycationic antibiotics and the anionic head of the acidic phospholipids. Divalent cations can interfere with these reactions.

Although the exact mechanism of action of CPM is not fully understood, it seems that a few steps are involved that affect critical functions within the yeast cell. These steps take place concurrently and rapidly.

The first step involves the aggregation of the yeast cells. The aggregate formation reaches a plateau within 60 minutes. This aggregation process can be inhibited with the addition of high concentration of calcium (1-5 g/L). The effect of calcium on CPM is a direct one. Calcium will bind directly to CPM and will inhibit its ability to bind to the yeast cell. However, the binding of calcium to CPM is loose enough and the calcium can be removed by dialysis.

Simultaneously, the second step involves the binding of CPM to the cell wall by means of carbohydrate-recognition domains. CPM is a relatively large size molecule. It is very unlikely that a compound of that size that also holds highly charged residues would be able to penetrate the yeast cell membrane and accumulate inside the cell. However, it may easily penetrate the relatively nonselective fungal cell wall and gain access to the cytoplasmic membrane. This binding of CPM to the cell wall is very tight (Fig. 3), as was shown by the lack of displacement of radioactive labeled CPM when washed with excess of unlabeled CPM (27).

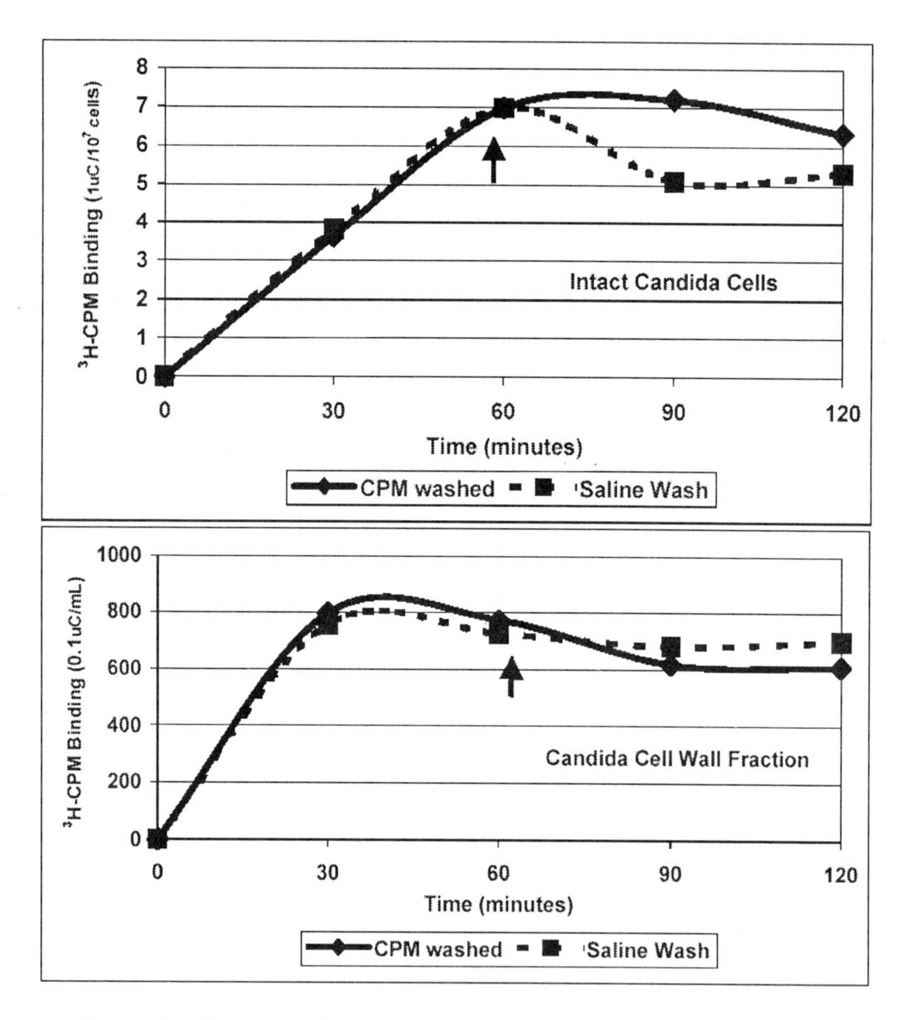

Figure 3. Effect of washing with unlabelled cationic polysaccharide from Mucor rouxii (CPM) on the uptake/binding of [^{3}H]cpm in intact cells (a) and in cell-wall fraction (b). The arrow indicates the time of addition of a 150-fold excess of unlabelled CPM.

♦ *Washed with unlabelled CPM;* ■ *unwashed*

The rapid irreversible action of the compound suggests the accessibility of a cellular target(s). The purified membrane fractions from *C. albicans* contain 10-12 membrane proteins. Binding studies reveal that radioactive CPM failed to bind to purified membrane fraction under denaturing conditions. Thus it is likely that the compound may be acting on a non-protein component(s) of the plasma membrane.

Further proof of this theory was demonstrated when *C. albicans* protoplasts were exposed to inhibitory concentrations of CPM. In the absence of CPM, *C. albicans* protoplasts were able to regenerate and produce colony forming units (CFUs). However, exposure of the proto-plasts to CPM in concentration of 10μg/ml resulted in 99% inhibition of the production of CFUs within 30minutes (Fig. 4). These results corresponded well with the percentage of killing observed when whole cells were exposed to CPM at the same concentration. Furthermore, when examining the physical appearance of the exposed protoplasts, complete lysis of the protoplasts was observed within 60 min of exposure, even in the presence of 1 M sorbitol. (28).

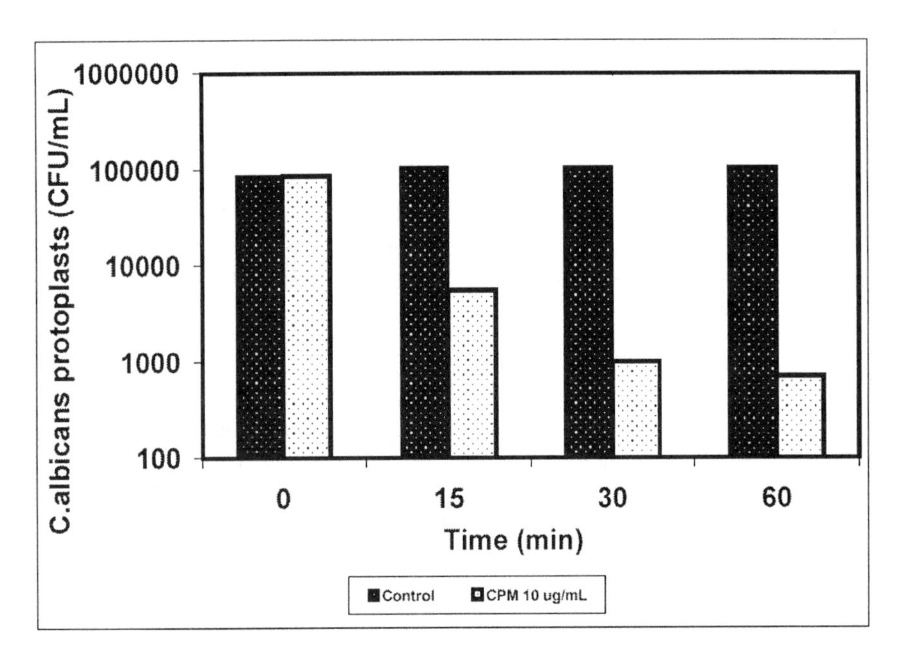

Figure 4. Effect of cationic polysaccharide from Mucor rouxii *(CPM) on* C. albicans *Protoplasts.*

Although it is not clear which component(s) of the plasma membrane interact with CPM, this indirect interaction affects the action of the proton translocating ATPase pump (H^+ -ATPase). The H^+ -ATPase of fungi is a plasma membrane located, ATP-driven proton pump belonging to the p-type ATPase superfamily. To date, 211 members (ranging from 646 to 1956 amino acids in size) of the P-type ATPase have been identified in a wide spectrum of organisms ranging from archeabacteria to man (29, 30).

The charged substrates, the P-type ATPase translocate includes Na^+, K^+, Ca^+, Mg^{2+}, Cd^+, H^+ and phospholipids. Based on their substrate specificity, the ion translocating P-type ATPases are grouped into five (types I - V) families (30). The distinguishing feature of the P-type ATPases is the formation of a phosphorylated enzyme intermediate during the reaction cycle (hence they are called P-type). The phosphorylation of the enzyme invariably involves an aspartic acid residue of a highly conserved motif which consists of DKTGT (29-31).

The number of P-type ATPases present in an organism is highly variable, ranging from just a few (pathogenic bacteria), seven to nine (free living bacteria), as many as 16 in *Saccharomyces cerevisiae*, to more than 30 members in plants and animals. Studies with the *S. cerevisiae* H^+ -ATPase revealed that it is an integral protein of which greater than 80% is exposed to the cytoplasmic side of the cell, 15% is estimated to be associated with the lipid bilayer forming 10 membrane spanning helical regions while the remaining 5% of the protein is exposed to the extra cytoplasmic side of the cell (32).

The plasma membrane H^+ -ATPase plays an essential role in fungal cell physiology (31). This ion translocating enzyme is mainly responsible for maintaining the electrochemical proton gradient necessary for nutrient uptake and the regulation of the intracellular pH of fungal cells (31). Interference with its function will lead to cell death.

Glucose-induced acidification of the external medium by carbon-starved yeast cells is a convenient measure of H^+ -ATPase-mediated proton pumping. CPM was found to inhibit the acidification of external medium by *Candida* species in a concentration dependent manner (Fig. 5). It is important to note the inhibition of H^+ -ATPase function was achieved at the MIC concentrations of CPM. AMB, which is known to interact with the sterol component of the membrane, had no significant effect on H^+ -ATPase.

However, studies of purified fractions of *C. albicans* plasma membrane, which measured the enzymatic activity specifically, showed that the inhibition of the pump activity by CPM is an indirect one. The enzymatic activity was measured by the liberation of inorganic phosphate via the hydrolysis of ATP. The ATP hydrolysis was not inhibited by CPM (Fig. 6A) (28). On the other hand, 5 μg/ml of vanadate, a known inhibitor of H^+ -ATPase, almost completely inhibited the liberation of inorganic phosphate by ATP hydrolysis, suggesting an indirect effect of CPM on H^+ -ATPase (Fig 6B) (28).

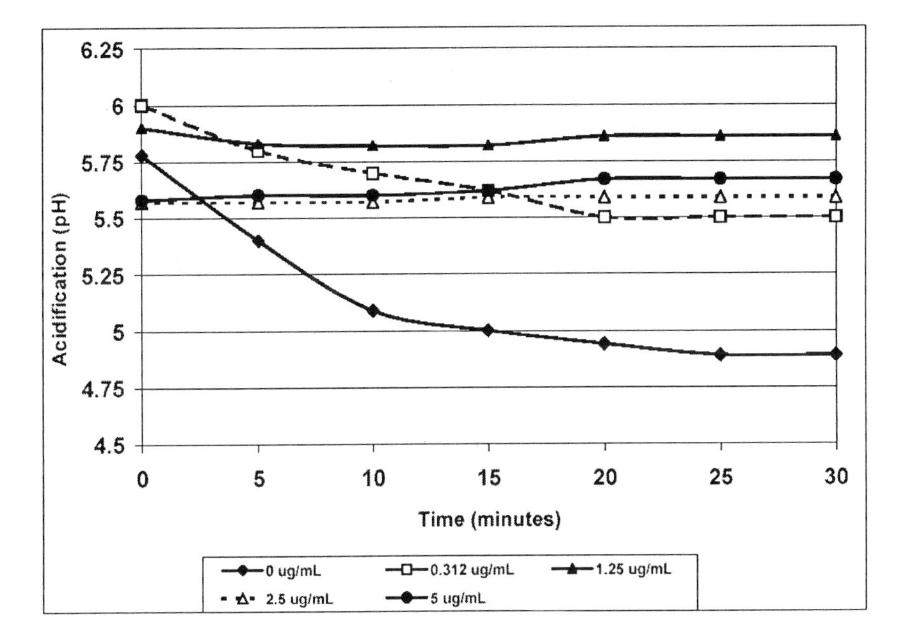

Figure 5. Effect of cationic polysaccharide from Mucor rouxii *(CPM) on the proton pumping (as measured by the acidification of external medium) of* C. Albicans.

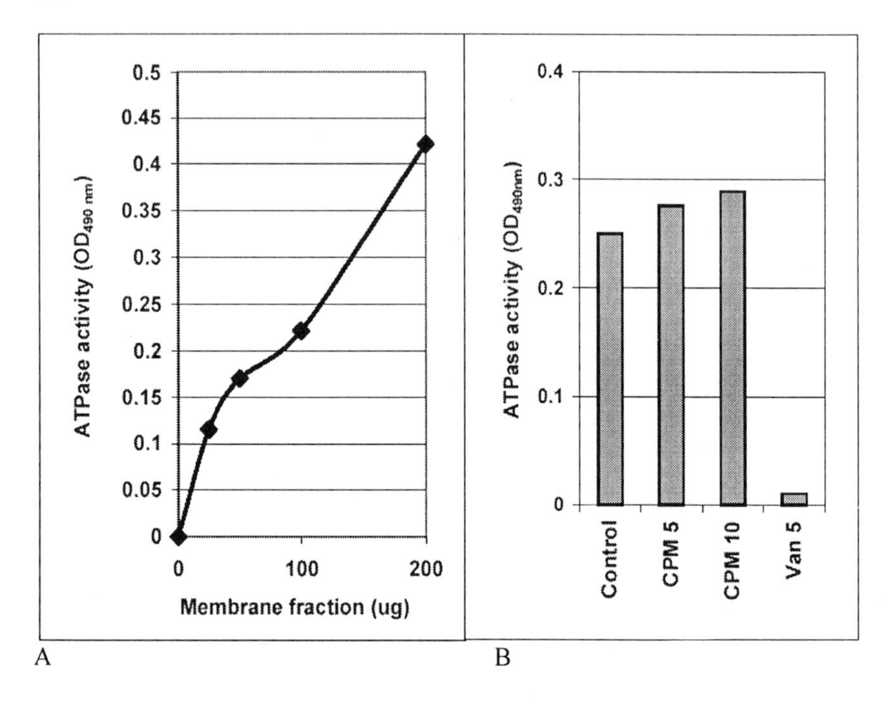

Figure 6. Effect of cationic polysaccharide from Mucor rouxii *(CPM) on ATP hydrolysis by purified membrane fraction from* C. albicans. *(A) Relationship between the amount of membrane fraction and H^+-ATPase activity. (B) Effect of CPM and* vanadate *on ATP hydrolysis by purified fraction.*

In Vivo activity

The therapeutic effect of the combination of chitosan and pyrithione was tested on cutaneous candidiasis caused by *C.albicans* in guinea pigs.

A total of 175 female Duncan-Hartley guinea pigs were used to study the therapeutic efficacy of CPP. Among laboratory animals, guinea pigs were found to be very susceptible to cutaneous candidiasis. The therapeutic efficacy of CPP, as a 4% aqueous solution, was tested in treating cutaneous infection of *C. albicans*, and was compared to commercially available miconazole 2% (Table 2) (33). With short-duration therapy of 1 to 3 days, CPP 4% showed significantly better results than treatment with miconazole 2% (P = 0.0012 and 0.0071, on days 1 and 3 respectively). With long-duration therapy of 5 and 7 days, there were no significant differences between the two drugs.

Table 2. Therapeutic efficiency of topical treatment with sodium pyrithionate (CPP) compared with miconazole 2% in cutaneous candidiasis assessed by the number of culture-positive skin biopsies at completion of therapy.

Number of days treated	**1**	**3**	**5**	**7**
Negative control (not treated)	77/80	73/100	40/60	61/70
Negative control (treated with PBS)	20/20	20/20	20/20	ND
Positive control (miconazole 2%)	6/80***	5/90**	1/60	0/80*
CPP 0.125%	34/40****	0/40	ND	ND
CPP 0.25%	1/40	0/40	ND	ND
CPP 0.5%	5/40	0/40	1/40	3/40
CPP 1%	2/40	0/40	0/40	1/40
CPP 4%	0/130***	0/120**	0/100	0/80

* $P = 0.0145$.
** $P = 0.0071$.
*** $P = 0.0012$.
**** $P = 0.0000$.

The efficacy of CPP was found to be concentration dependent, and application of CPP 0.125% for 1 day proved to be ineffective in clearing the infection. However, when the treatment period was extended to 3 days, even this low concentration of CPP cleared the infection in all tested animals (33).

Conclusion

Further studies of the mode of action of cationic polysaccharide from *Mucor rouxii* and the soluble chitosan and combination with pyrithione are essential for the development of this novel complex carbohydrate as an antifungal agent. This compound and its derivative possess several highly desirable characteristics for the next generation of antifungal agents. First, being novel compounds, unrelated to the existing antifungal drugs, there is no cross resistance between the compounds and the currently used antifungal agents. Second is the importance of the rapidity of action, low MIC values and the lethal effect of this compound against a wide spectrum of pathogenic yeasts. This will help reduce treatment duration and the development of untoward effects. Third, these compounds are heat and light-stable, making them ideal to use for topical applications as was proven by the in vivo experiments with CPP. Fourth, being a large molecule with highly charged residues, these compounds are acting on external targets of the cell membrane. Thus, it is unlikely to develop resistance to CPM by the efflux mechanism that is a major cause of drug resistance in microorganisms including fungi.

In summary, CPM and CPP are potent antifungal agents both in vitro and in vivo. These compounds hold great promise to become clinically useful.

References

1. Bodey, GP. The emergence of fungi as major hospital pathogens. J. hosp Infect. 1988; 11: 411-26.
2. Banerjee, SN, Enori, TG, & Culver, DH. Secular trends in noscomial primary bllodstream infections in the United States, 1980-1989. Am. J. Med. 1991; 91 (Suppl. 3B): 86S-9S.
3. Edwards, JR. Invasive Candida infections: evolution of a fungal pathogen. New Engl. J. Med. 1991; 324: 1060-2.
4. Bech-sague, CM, Jarvis, WR. The national Nosocomial Infections Surveillance System. Secular trends in the epidemiology of mosocomial infections in the United States, 1890-1990. J, Infet. Dis. 1993; 267: 1247-51.
5. Polak, A, Hartman, PG. Antifungal chemotherapy are we winning? Prog. Drug Res. 1991; 37: 31-269.
6. Gallis, HA, Drew, H, Pickard, WW. Amphotericin B: 30 years of experience. Rev. Inect. Dis. 1990; 12: 308-329
7. Baily, GG, Perry, FM, Denning, DW, Mandel, BK. Fluconazole resistance candidosis in an HIV cohort. AIDS. 1994; 8: 787-92.
8. Komshian, SV, Uwaydah, A, Sobel, JD, Crane, LR. Fungemia caused by Candida species and Torulopsis glabrata in hospitalized patients. Frequent charchcteristic, and evaluation of factors influencing outome. Rev. Infec. Dis. 1989; 11: 379-90.
9. Nobre, G, Mences, E, Charrua, M, Cruz, O. Ketoconazole resistance in Torulopsis glabrata. Mycopathologia. 1989; 107: 51-5.
10. Vazquez, JA, Lynch, M., Sobel, JD. In vitro activity of a new pneumocandin antifungal agent, L-733,560 against azole-susceptible and resistant Candida and Torulopsis species. Antimicrob. Agents Chemother. 1995; 39: 2689-91.
11. Warnock, D & al. Fluconazole resistance in Candida glabrata. Lancet. 1988; I: 1310

12. Wingard, JR & al. Increase in Candida krusi infection among patient with bone marrow transplantation and neutropenia treared prophylactically with fluconazole. N. England J. Med. 1991; 18:1274-77.
13. Mazar Ben-Josef, A, Manavnthu, EK, Platt, D, Sobel, JD. In vitro activity of CAN-296: a naturally occurring complex carbohydrate. J. of antibiotics. 1997; 50, 71-7.
14. Drickamer, K, Taylor, ME. Biology of animal lectins. Annual review of Cell Biology. 1993; 9, 237-64.
15. Morell, AG & al. Physical and chemical studies on ceruloplasmin: V. metabolic studies on sialic acid-free ceruloplasmin in vivo. Journal of Biological Chemistry. 1968; 243, 155-9.
16. Stahl, P & al. Receptor-mediated pinocytosis of mannose glyconjugates by macrophages: charhcterization and evidence for receptor recycling. Cell. 1980; 19, 207-15.
17. Fischer, HD, Gonzales-Noriega, A, Sly, WS, Morre, DJ. Phosphomannosyl-enzyme receptors in rat liver. Subcellular distribution and role in intracellular transport of lysosymal enzymes. Journal of Biology Chemistry. 1980; 255: 9608-15.
18. Lee, YC. Biochemistry of carbohydrate-protein interaction. FASEB J. 1992; 6:3193-200.
19. Drickamer, K. Ca++ - dependent sugar recognition by animal lectins. Biochemical Society Transactions. 1996; 24: 146-50.
20. Lee, RT. & al. Ligand-binding characteristic of rat serum-type mannose-binding protein (MBP-A). Homology of binding site architecture with mammalian and chicken hepatic lectins. Journal of biological Chemistry. 1991; 266: 4810-15.
21. Drikamer, K. Recognition of complex carbohydrate by Ca^{2+} -dependent animal lectins. Biochemical Society Transactions. 1993; 21: 456-9.
22. Sawada, Y & al. Calcium-dependent anticandidal action of pradimicin A. J. of Antibiotics. 1990; 43: 715-21.
23. Humes, HD, Sastrasinh, M, Weinberg, JM. Calcium is a competitive inhibitor of gentamicin-renal membranebinding interactions and dietary calcium supplementation protects against gentamicin nephrotoxicity. J. Of Clinical investigation. 1984; 73: 134-47.
24. Weinberg, JM, Humes, HD. Mechanisms of Gentamicin-induced dysfunction of renal mitochondria. 1. Effects on mitochondrial respiration. Archives of Biochemistry and Biophysics. 1980; 205: 222-31.
25. Sastrasinh, M, Knauss, TC, Weinberg, JM, Humes, HD. Identification of the aminoglycoside binding site in rat renal brush border membrane. J. of Pharmacology and Experimental Therapeutics. 1982; 222:350-8.
26. Schacht, J. Isolation of an aminoglycoside receptor from guinea pig inner ear tissues and kidney. Archives of Oto-Rhino-Laryngology. 1979; 224: 129-34.

27. Mazar Ben-Josef, A, Manavnthu, EK, Platt, D, Sobel, JD. Involvement of calcium inhibitable binding to the cell wall in the fungicidal activity of CPM. J. of Antimicrobial Chemotherapy. 1999; 44: 217-222.
28. Mazar Ben-Josef, A, Manavnthu, EK, Platt, D, Sobel, JD. Proton translocating ATPase mediated fungicidal activity of a novel complex carbohydrate: CAN-296. International J. of Antimicrobial Agents. 2000; 13: 287-95.
29. Moller, JV, Juul, B, Le Maire, M. Structural organization, ion transport, and energy transduction of P-type ATPase. Biochem. Biophys. Acta. 1996; 1286: 1-51.
30. Axelsen, KB, Palmgren, MG. Evolution of substrate specificities in the p-type ATPase superfamily. J. Mol. Evol. 1998; 46: 84-101.
31. Serrano, RM. Structure and function of proton translocating ATPase in plasma membranes of plantes and fungi. Biochem. Biophys. Acta. 1988; 947: 1-28.
32. Perlin, DS, Harber, JE. Genetic approaches to structure-function analysis in the yeast plasma membrane H^+ -ATPase. Adv. Mol. Cell Biol. 1998; 23A: 143-66.
33. Mazar Ben-Josef, A, Cutright, JL, Manavnthu, EK, Sobel, JD. CAN-296P is effective against cutaneous candidiasis in guinea pigs. International J. of Antimicrobial Agents. 2003; 22: 168-171.

New Approached in Synthesis and Computational Studies

Chapter 9

Synthetic Methods to Incorporate α-Linked 2-Amino-2-Deoxy-D-Glucopyranoside and 2-Amino-2-Deoxy-D-Galactopyranoside Residues into Glycoconjugate Structures

Robert J. Kerns and Peng Wei

Division of Medicinal and Natural Products Chemistry, University of Iowa, Iowa City, IA 52242

Numerous bioactive glycoconjugates contain D-glucosamine and D-galactosamine residues within their core structures. Diverse protecting group strategies for the 2-amino group of these sugars that impart neighboring group participation in the glycosylation process afford many strategies for the stereoselective synthesis of beta-linked (1,2-*trans*) glycosides. In contrast, stereoselective synthesis of alpha-linked (1,2-*cis*) glycosides of D-glucosamine and D-galactosamine has historically been more challenging because of a more limited repertoire of effective glycosylation methodologies, including a limited number of protecting group or latentiation strategies for the 2-amine group and poor consistency of stereocontrol and yield during glycosylation reactions. Synthetic methods to form alpha-linked glycosides of D-glucosamine and D-galactosamine are presented here, with a focus on the application and utility of both new methods and old strategies toward the synthesis of bioactive natural glycoconjugates and their analogs.

Overview of the 2-amino-2-deoxy-D-hexopyranosides D-glucosamine and D-galactosamine

Numerous bioactive saccharides and glycoconjugates contain 2-amino-2-deoxy-D-hexopyranoside residues as an integral component of their molecular structure.[1-3] Many of these glycoconjugates contain the beta-linked isomer of the 2-amino-2-deoxy-D-hexopyranoside sugar residue. In particular, *O*-substituted derivatives of the 2-acetamido-2-deoxy-ß-D-glucopyranoside (*N*-acetyl-ß-D-glucosamine) and 2-acetamido-2-deoxy-ß-D-galactopyranoside (*N*-acetyl-ß-D-galactosamine) core residues are found as constituents in a diverse array of prokaryotic and eukaryotic glycoconjugate structures (Figure 1). In contrast to the diverse types of glycoconjugates that contain variably substituted beta-linked D-glucosamine and D-galactosamine residues, alpha-linked D-glucosamine and D-galactosamine residues are found in a more limited number of structural classes, or structural types, of bioactive glycoconjugates (Figure 1).

The more limited distribution in the number of unique structures containing alpha-linked D-glucosamine and D-galactosamine residues is offset, however, by the high level of molecular diversity found within many of the classes of bioactive structures containing these alpha-linked 2-amino sugars. For example, *N*-acetyl and *N*-sulfo substituted α-D-glucosamine residues are a core component of heparin and heparan sulfate, while *N*-acetyl-α-D-galactosamine is an important constituent of *O*-linked glycopeptide and glycoprotein structures. Within each of these types or classes of bioactive glycoconjugates, there are many structural variations through either the substituents on the 2-amino sugar or in the structure, sequence, substitution pattern, and/or linking position of additional sugar residues attached to the amino sugar. Thus, efficient and versatile methods to introduce variably substituted and differentially modified α-D-glucosamine and α-D-galactosamine residues into glycoconjugate structures are vital to the synthesis of many important bioactive glycoconjugates and their structural analogs.

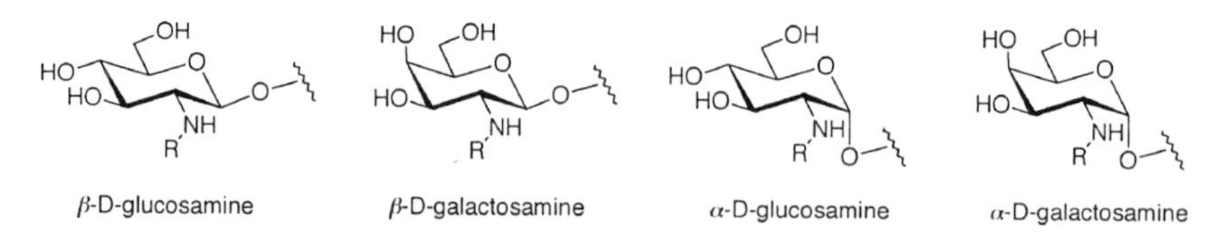

Figure 1. Structures of α- and β-glycosides of D-glucosamine and D-galactosamine. These 2-amino sugars are present in many bioactive natural products glycosidically linked to glycan or aglycone structures. The 2-amino groups are typically R = Ac, H, or SO_3^-, although other N-substituted derivatives are known. The 3, 4, and 6 hydroxyls can be substituted with saccharide residues or other functional groups such as OSO_3^-, NH_2, etc.

Bioactive glycoconjugates that contain alpha-linked D-glucosamine or alpha-linked D-galactosamine residues within their core structure

Alpha-linked glycosides of D-glucosamine and D-galactosamine and their diversely substituted derivatives are important structural constituents in a significant number of bioactive natural products. Examples representative of the different classes of bioactive glycoconjugates containing these alpha-linked 2-amino sugars are shown (Figure 2).

Heparin and heparan sulfate are structurally similar members of the glycosaminoglycan class of acidic polysaccharides.[4] Heparin is primarily found in the granules of mast cells and has been employed therapeutically for decades as an anticoagulant agent. Heparan sulfate is found on the surface of nearly all mammalian cell types where it plays a profound role in numerous physiological processes.[4] Natural heparin and heparan sulfate sequences as well as analogs and mimics of these structures have been the focus of many syntheses with the goal of developing therapeutic agents that selectively bind heparin/heparan-sulfate-binding proteins.

The tunicamycins[5] and aminoglycosides[6] are structural classes of natural antibiotics that are bacterial in origin (Figure 2). Certain aminoglycosides have found clinical utility. The continued development of efficient synthesis of these antibiotics and their analogs holds promise for the development of new antibiotics.

Mucin-type *O*-glycans are representative of a major class of *O*-linked glycopeptide/glycoprotein structures, where alpha-linked D-galactosamine affords a key linkage to serine or threonine amino acid residues (Figure 2).[7,8] Certain core structures of the mucin-type *O*-glycans also contain alpha-linked D-galactosamine residues attached to the 3-position or the 6-position of the amino acid-linked D-galactosamine residue. Efficient formation of these alpha-linked D-galactosamine conjugates has been extensively studied in efforts to synthesize and evaluate the biological function of numerous cell-surface glycopeptide structures and as vaccine candidates.[9]

A wide variety of different oligosaccharide structures are found on the surface of bacteria in the form of capsular polysaccharides, lipooligosaccharides and lipopolysaccharides, of which the alpha-linked D-glucosamine containing *O*-antigen structure found on the surface of *Shigella dysenteriae* is shown (Figure 2).[10] Mycothiol is a low molecular weight thiol found intracellularly in most actinomycetes including mycobacteria, where it serves as an antioxidant by maintaining a reducing environment (Figure 2).[11] The biosynthesis and metabolism of mycothiol is a putative new target for antituberculosis drug development. The glycosylphosphatidylinositols (GPIs) are a diverse family of glycolipids expressed by eukaryotic cell types, where they serve to anchor extracellular protein and other glycoconjugate structures to the cell membrane.[12] The core structure of a representative membrane protein GPIs anchor is shown (Figure 2).

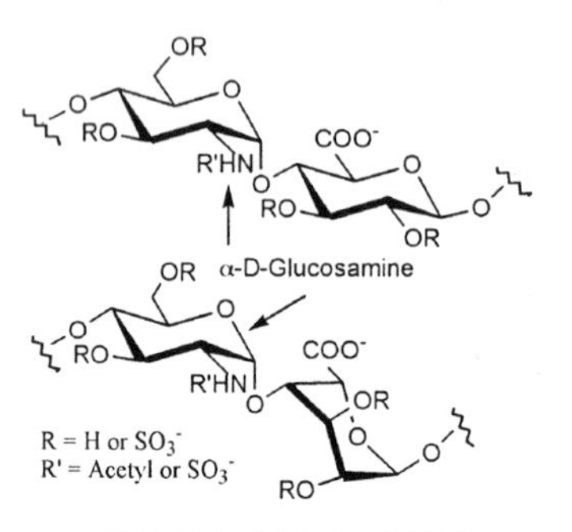

(Heparin and Heparan Sulfate)
(major repeating disaccharide units shown)

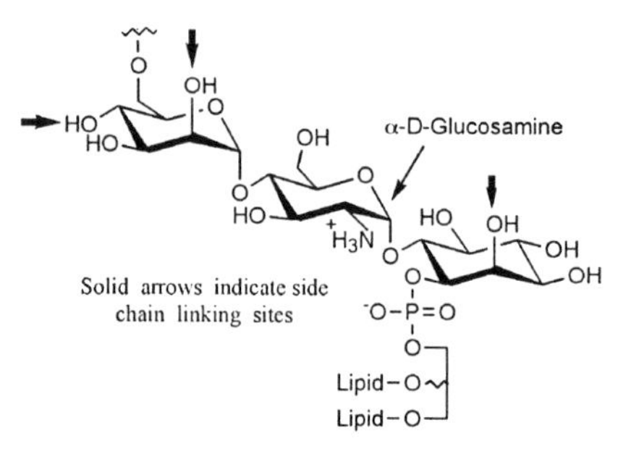

(Structural core of extracellular GPI anchors)

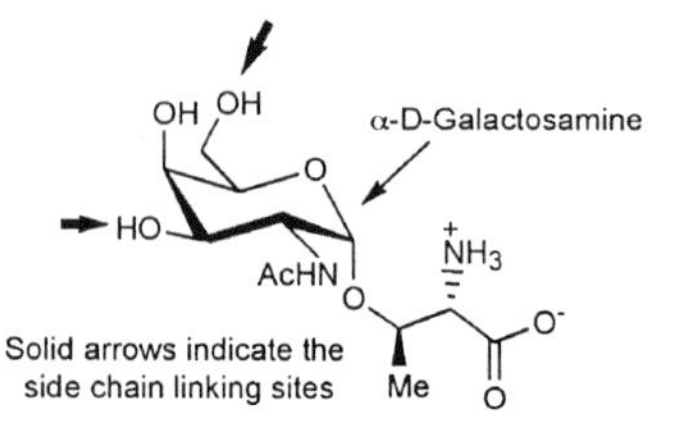

(Core structure of certain mucin-type *O*-glycans)

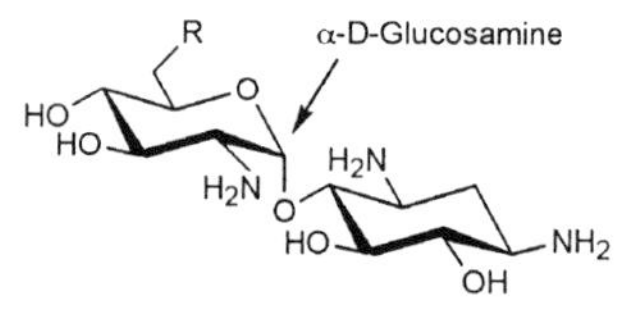

(Core component of aminoglycoside antibiotics)

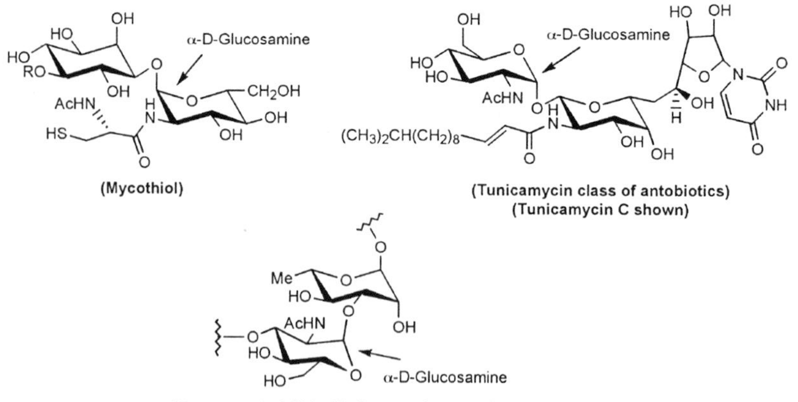

(Mycothiol)

(Tunicamycin class of antobiotics)
(Tunicamycin C shown)

(Component of *Shigella Dysenteriae* type 1 *O*-antigen)

Figure 2. Representative bioactive natural glycoconjugates that contain alpha-linked D-glucosamine or alpha-linked D-galactosamine residues as a core component of their structure.

General considerations for stereochemical control in the glycosidation of D-glucosamine and D-galactosamine using traditional glycosyl donor approaches

Formation of the beta glycosidic linkage of various D-glucosamine and D-galactosamine glycosyl donors during the introduction of these residues into glycoconjugate structures typically relies on the use of protecting groups for the 2-amino group that afford neighboring group participation in the glycosylation reaction (Figure 3).[13] A C-2 participating group provides anchimeric assistance and thus stereochemical control to afford 1,2-trans glycoside products. Many amine protecting groups that participate in glycosylation of 2-amino-2-deoxy-D-hexopyranoside derivatives to afford the 1,2-trans product are known, which affords numerous orthogonal protection/deprotection strategies. An occasional exception to the above-described role of C-2 participating groups arises with select donor activation systems employed during glycosylation of certain D-galactosamine derivatives. These derivatives have large or bulky protecting groups on the C-4 hydroxyl moiety that may impede formation of the beta glycoside by sterically hindering attack of the acceptor molecule from the same face of the amino sugar. In these unique cases the steric effect of the C-4 substituent appears to override the C-2 participating effect to afford high yields of the alpha-linked (1,2-*cis*) glycoside.

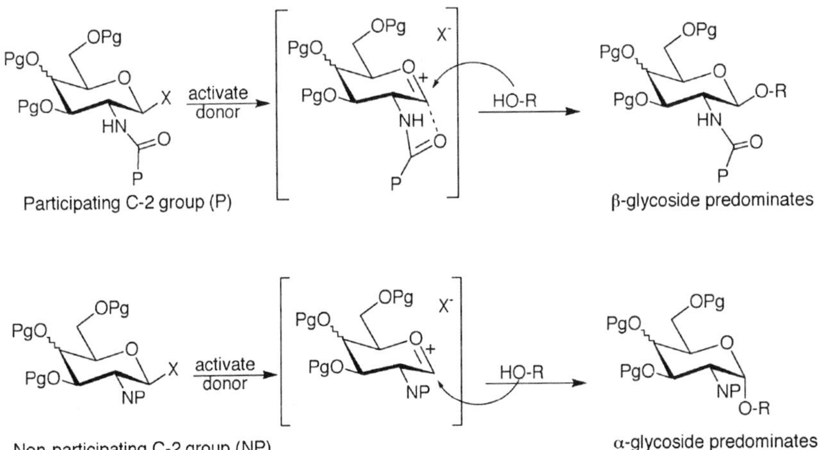

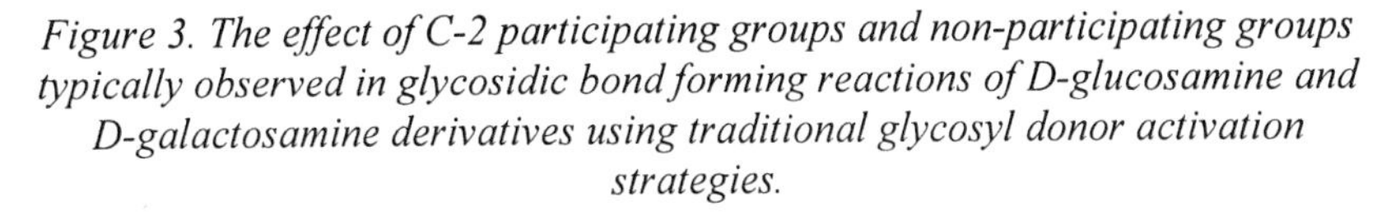

Figure 3. The effect of C-2 participating groups and non-participating groups typically observed in glycosidic bond forming reactions of D-glucosamine and D-galactosamine derivatives using traditional glycosyl donor activation strategies.

In contrast to the stereochemical control typically afforded by a neighboring group during the formation of 1,2-*trans* glycosides, introduction of D-glucosamine and D-galactosamine residues as their 1,2-cis (alpha-linked) glycosides into glycoconjugate structures using tradition glycosylation strategies cannot, in principal, possess a participating substituent at the C-2 position. Therefore, in order to preferentially obtain the 1,2-*cis* glycoside product an amine protecting group or a latent form of amine that will not participate in the glycosylation reaction is requisite at the C-2 position of the glycosyl donor (Figure 3). The fundamental principles of stereocontrol during the synthesis of 1,2-*cis* glycosides, with a focus on 2-*O*-sugars, have been recently reviewed.[14]

Discussed throughout the various sections below are a number of putative non-participating C-2 groups that have been employed over the years with varied success in the synthesis of the alpha-linked (1,2-*cis*) glycosides of D-glucosamine and D-galactosamine. Furthermore, a variety of unique glycosyl donor strategies and novel glycoside bond forming methodologies that do not fit the tradition donor activation approaches shown in Figure 3 have also been developed and exploited toward the synthesis glycoconjugates containing α-D-glucosamine and/or α-D-galactosamine residues.

Overview of 2-azido sugars in the synthesis of α-Linked glycosides of D-glucosamine and D-galactosamine

The most common strategies employed over the years for incorporating α-D-glucosamine and α-D-galactosamine residues into glycoconjugate structures utilize an appropriate glycosyl donor or other functionalized saccharide having a protected or latent amine-precursor at the C-2 position, where formation of the glycosidic bond affords the desired alpha-linked glycoside. Subsequent chemical transformation of the C-2 substituent affords the desired 2-amino substituent. For roughly 30 years the azide group has served as a non-participating C-2 substituent and has provided the most versatile method in the synthesis of alpha-linked glycosides of D-glucosamine and D-galactosamine (Figure 3, NP = N_3). Indeed, appropriately substituted 2-azido sugars have been employed in at least one, and in many cases all reports describing total synthesis of individual glycoconjugates containing α-D-glucosamine or α-D-galactosamine residues (see Figures 2 and 4).

The chemical compatibility of a wide range of glycosyl donors and anomeric activation chemistries employed for glycosylation reactions using 2-azido-2-deoxy-D-hexopyranoside donors affords broad applicability of the 2-azido group. Furthermore, the azido group can be efficiently and selectively converted to amine using a variety of chemical methods, affording a range of orthogonal protecting group strategies with numerous other protecting groups employed in glycoconjugate synthesis.

While 2-azido derivatives of D-glucosamine and D-galactosamine glycosyl donors have served well for many years, synthesis of variably *O*-protected 2-azido glycosyl donors can be lengthy, and costly. Furthermore, careful control of solvent and temperature while comparing varied anomeric leaving groups will often yield a final synthesis of the alpha-linked conjugate with good to high stereoselectivity but rarely stereospecificity. Indeed, as shown in Figures 4-10, the 2-azido sugars are versatile and consistent in affording desired alpha-linked glycoside, but overall stereoselectivity and yield are often limited.

Because of the wide application of 2-azido sugars in the synthesis of bioactive glycoconjugates containing alpha-linked D-glucosamine and D-galactosamine residues, these donors and various donor strategies have been covered in many glycosylation reviews. Outlined below in Figures 4-10 are representative examples showing application of the 2-azide substituted D-glucosamine and D-galactosamine glycosyl donors in the formation of alpha-linked glycosides in the synthesis of the bioactive glycoconjugates presented in Figure 2. Examples shown in Figures 4-10 were chosen from literally thousands of examples to display a sample of various anomeric groups and activation strategies employed with 2-azido sugars to form alpha-linked glycosides. These examples also exemplify the need for new synthetic methods that afford more consistent stereochemical control for the introduction of alpha-linked D-glucosamine and D-galactosamine into the structure of bioactive glycoconjugates and their analogs.

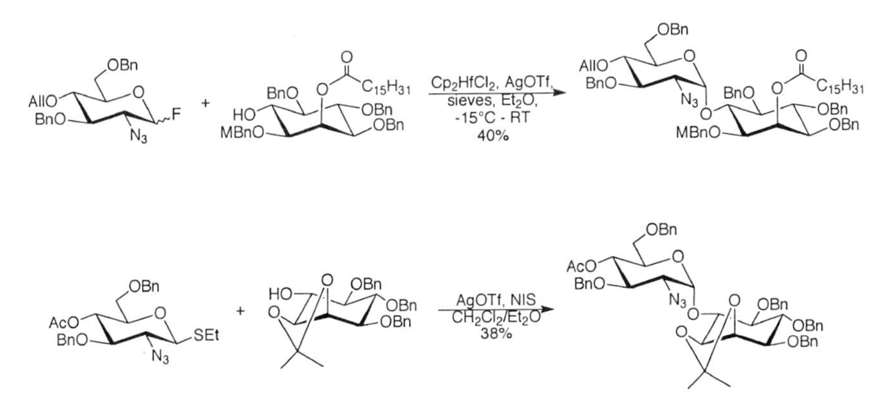

Figure 4. Glycosylation reactions published during 2002 that demonstrate the coupling of 2-azido-2-deoxy-D-glucopyranose donors to inositol structures toward the synthesis of GPI core structures.[15,16]

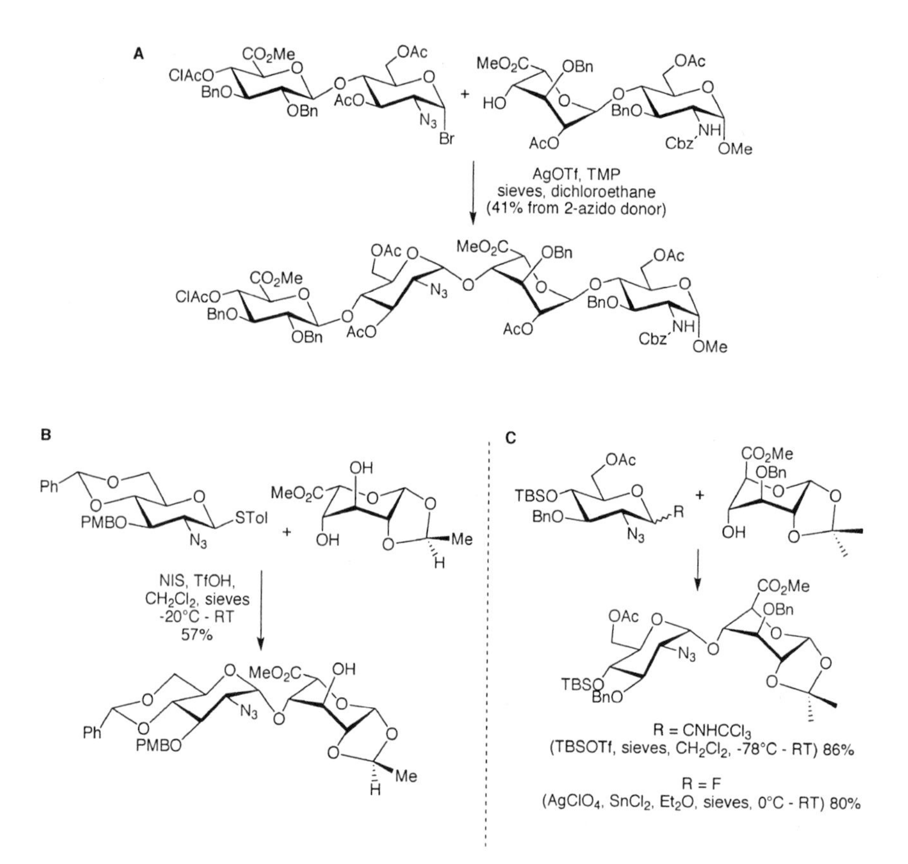

Figure 5. Select examples showing the application of 2-azido-2-deoxy-D-glucopyranosyl donors toward the synthesis of heparin and heparan sulfate oligosaccharide sequences. Panel A) Original coupling methodology reported by Petitou et al toward the synthesis of the first synthetic heparin-based anticoagulant (Fondaparinux) to be approved for clinical use.[17] Panel B) The regioselective and stereoselective glycosylation of a conformationally locked iduronic acid diol reported in 2004.[18] Panel C) Comparison of two anomeric activation strategies employed in the glycosylation of a conformationally locked glucuronic acid residue reported in 2002.[19]

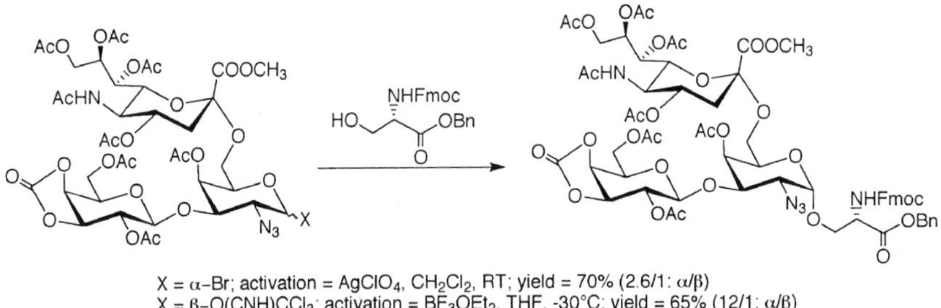

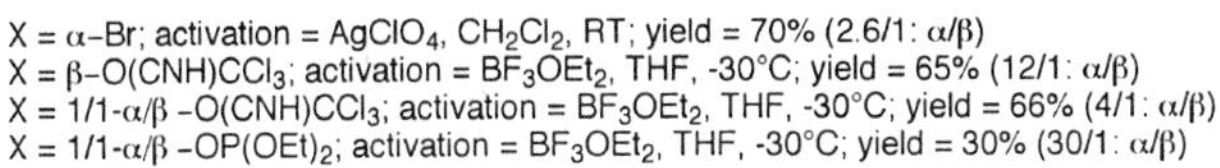

Figure 6. Example from a 1997 report showing different donor strategies compared during the synthesis of a tumor-associated mucin motif.[20] In this example a fully elaborated 2-azido-2-deoxy-D-galactose glycan was employed to form the alpha-linkage to serine. There are hundreds of reports for the coupling of variably protected monosaccharides of 2-azido-2-deoxy-D-glactopyranosyl donors to serine and/or threonine followed by subsequent elaboration of the glycan on the core glycopeptide structure.

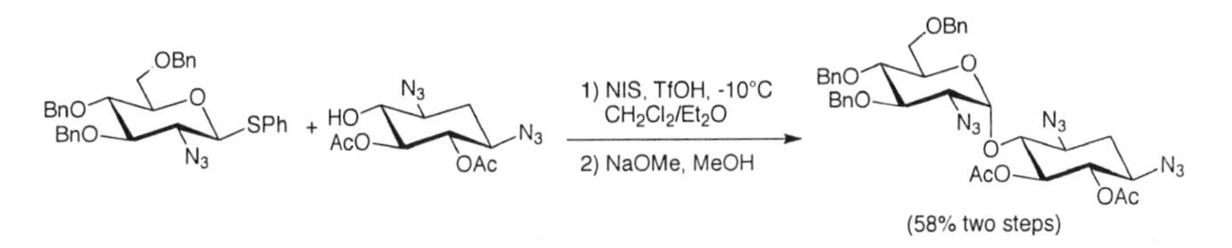

Figure 7. Glycosylation reaction reported in 1999 demonstrating the use of a non-participating 2-azide group in forming the alpha-linked glucosamine pseudodisaccharide core of aminoglycoside antibiotics, neamine.[21]

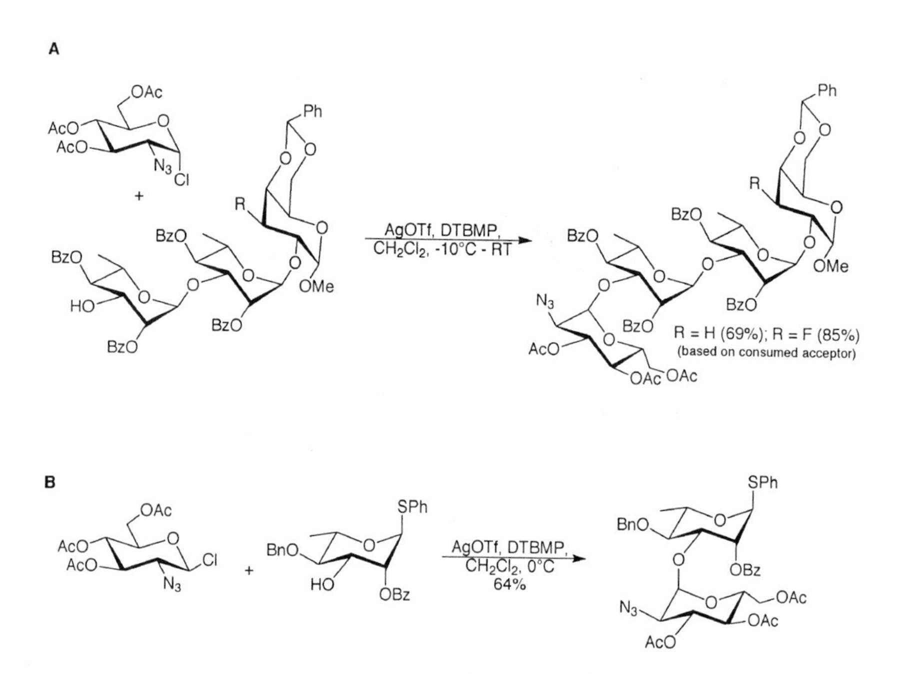

Figure 8. Two coupling reactions reported in 1998 that utilize the non-participating 2-azide group to obtain the alpha-linkage of D-glucosamine in the synthesis of Shigella dysenteriae type 1 O-specific polysaccharide structures toward the potential development of vaccines against Shigella dysenteriae type 1. Panel A) Penultimate glycosidic coupling toward tetrasaccharide fragments, deoxygenated and fluorinated at a terminal galactose residue.[22] Panel B) Initial coupling to afford a core disaccharide intermediate for further chain elongation.[23]

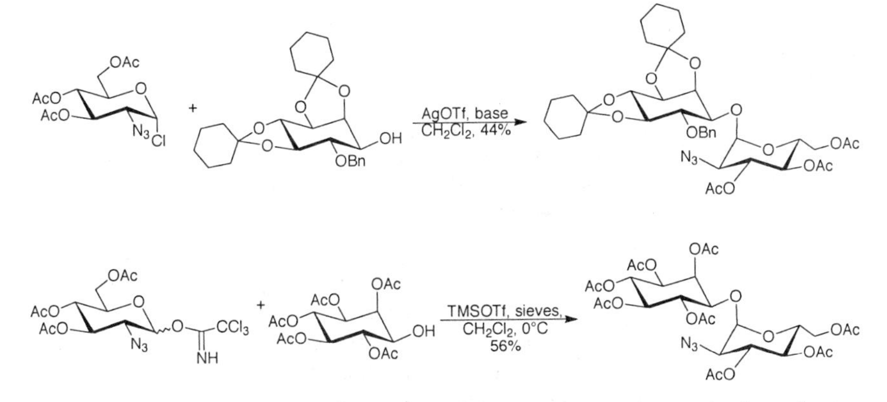

Figure 9. Synthesis of key intermediates in the synthesis of mycothiol employing 2-azido-2-deoxy-D-glucopyranosyl donors. Top) From a 2003 report describing the synthesis of structures used to evaluate the substrate specificity of AcGI deacetylase found in M. tuberculosis.[24] *Bottom) Key synthetic step in the first total synthesis of mycothiol and mycothiol disulfide reported in 2002.*[25]

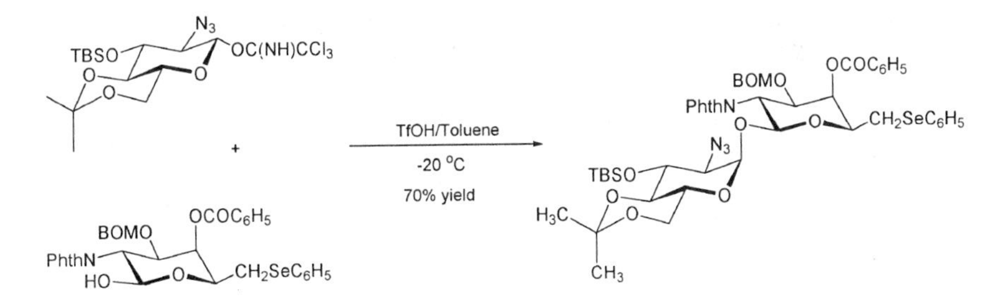

Figure 10. Example from a 1993 report showing synthesis of the 1,1-linked core disaccharide of tunicamycin employing a 2-azido donor to afford the alpha-linked D-glucosamine residue.[26]

Why are old and new alternatives to the 2-azido strategy for forming alpha-linked glycosides of D-glucosamine and D-galactosamine important?

The use of 2-azido-2-deoxy glycosyl donors of D-glucose and D-galactose in the formation of 1,2-*cis* glycosides revolutionized the synthesis of glycoconjugates containing α-D-glucosamine and α-D-galactosamine residues. The versatility of the 2-azido group as a non-participating C-2 substituent has dramatically facilitated laboratory-scale synthesis of bioactive glycoconjugates and their analogs toward understanding biological function and exploring structure-activity-relationships. However, the cost, number of synthetic steps, and often low efficiency in preparing various 2-azido glycosyl donors is considered a major drawback to application of this methodology to the large scale synthesis of single target structures. In addition, while extremely versatile and reliable, alpha/beta ratios of the resulting glycosides generated using 2-azido glycosyl donors can range from exceptional to modest depending on the donor activation system and the acceptor alcohol. Because of these limitations, efforts continue toward the development of new methods for the efficient and stereoselective introduction of alpha-linked glycosides of D-glucosamine and D-galactosamine into glycoconjugate structures.

Over the past 40-plus years a fairly large number of methods have been intentionally developed or peripherally observed to afford alpha-linked glycosides of D-glucosamine and D-galactosamine, albeit often with mixed results from a versatility and consistency standpoint. As the pharmaceutical potential and therapeutic reality of synthetic glycoconjugates continues to grow, the cost and efficiency of synthesizing specific clinical candidates on large scale will necessitate evaluation of the most appropriate approach to prepare specific glycoconjugates on an industrial scale. Here, versatility of a method becomes secondary to cost and industrial scale efficiently in preparing the target structure. In this context, it is expected that old, new and even aberrant strategies that have been shown to afford alpha-linked glycosides of D-glucosamine and D-galactosamine derivatives may ultimately yield the most direct and cost effective syntheses of specific target structures.

In the sections below we discuss our work as well as the work of many other groups over many years that have yielded a somewhat under appreciated number and diverse strategies to introduce alpha-linked glycosides of D-glucosamine and D-galactosamine into glycoconjugate structures. It is possible that no current or future method will achieve the versatility of 2-azido sugars in the synthetic formation of alpha-linked glycosides of D-glucosamine and D-galactosamine derivatives. However, each of the unique methods presented below has its own advantages and disadvantages; and may ultimately afford the simplicity, cost effectiveness and chemical compatibility required for the large-scale synthesis of individual pharmaceutical glycoconjugates.

Alternatives to 2-azido sugars: Methods for forming alpha-linked glycosides of D-glucosamine and D-galactosamine and application of these methods in the synthesis of bioactive glycoconjugates

Fused-Ring 2,3-Oxazolidinone protection of phenyl 2-amino-2-deoxy-1-thio-glucopyranosides: a C-2 non-participating group strategy

In 2001 we reported the utility of ring-fused 2,3-oxazolidinone derivatives of phenyl 2-amino-2-deoxy-1-thio-glucopyranosides in the synthesis of alpha-linked glucosamine derivatives (Figure 11).[27] A notable feature of this class of alpha-selective D-glucosaminyl donors is a high yield and versatile synthesis; where the oxazolidinone group affords simultaneous protection of the 2-*N* and 3-*O* positions (Figure 11).

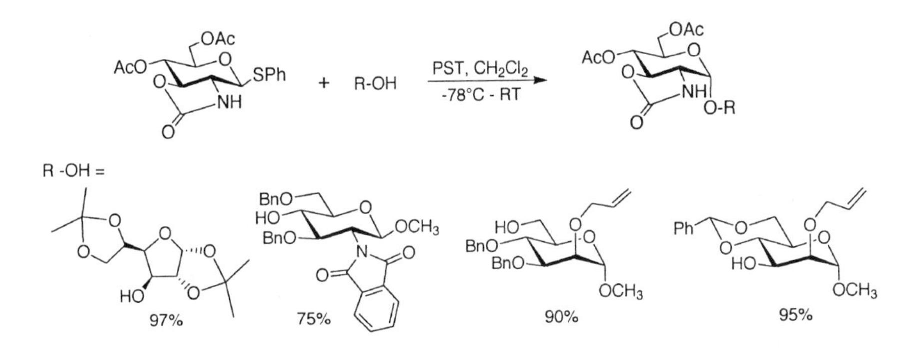

Figure 11. Ring-fused 2,3-oxazolidinone derivatives of phenyl 2-amino-2-deoxy-1-thio-glucopyranosides afford alpha-linked glycosides of D-glucosamine.

The ring-fused 2,3-oxazolidinones have also been shown to serve as versatile intermediates for more complex protecting group manipulations. Selective opening of the oxazolidinone ring with alcohols affords selective deprotection of the 3-hydroxyl group with concomitant protection of the 2-amine group as the carbamate, affording entry into thioglycoside donors bearing *N*-protection capable of neighboring group participation in the formation of beta-linked glycosides. In addition, entry into the D-galactosamine series of thioglycoside donors from 2,3-oxazolidinone protected glucosamine derivatives is readily achieved through inversion of stereochemistry at the C-4 position (Figure 12).[27]

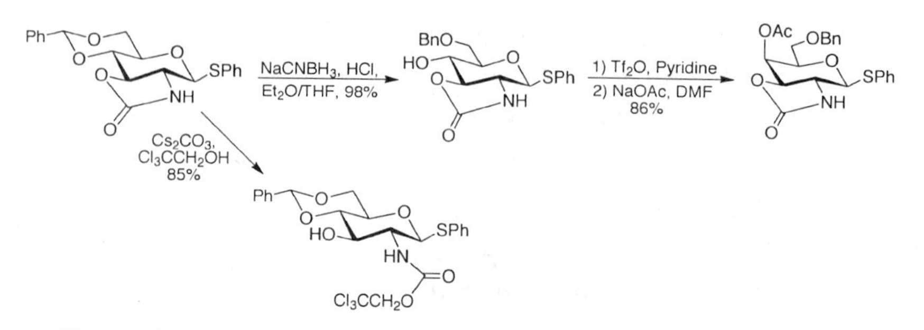

Figure 12. Ring fused 2,3-oxazolidinone derivatives of D-glucosamine are readily converted into D-galactosamine derivatives and C-2-carbamate derivatives, the later of which are putative participating groups for the formation of beta-glycosides.

Glycosylation of a variety of sugar acceptors using a ring-fused 2,3-oxazolidinone protected thioglycoside donor of glucosamine upon activation with phenylsulfenyl triflate was shown to afford high yield, stereoselective, formation of the alpha-linked glycosides (Figure 11). This methodology has been applied to the formation of the repeating disaccharide unit of heparan sulfate and synthesis of alpha-linked serine and threonine derivatives (Figure 13).[27,28]

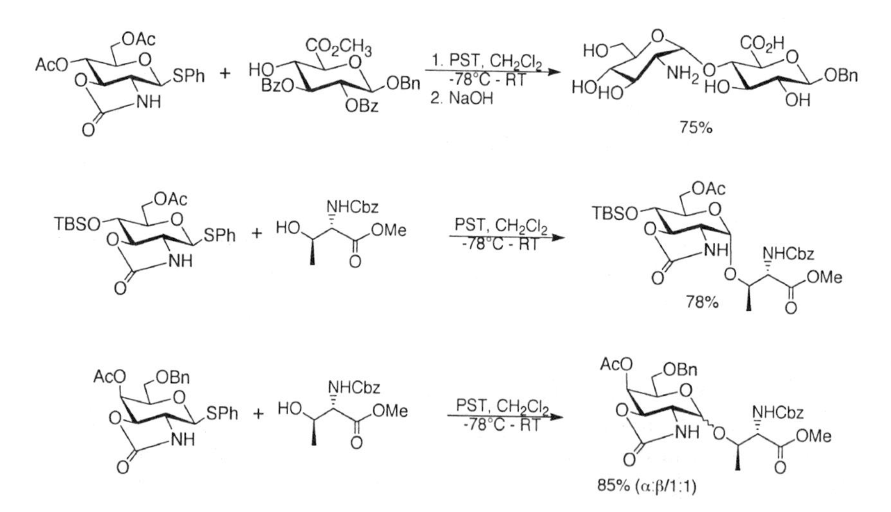

Figure 13. Examples of the application of ring-fused 2,3-oxazolidinone protected thioglycoside donors in the synthesis of a heparan sulfate disaccharide and alpha-linked conjugates with amino acid residues.

The utility of a ring-fused 2,3-oxazolidinone protected phenyl 2-amino-2-deoxy-1-thio-glucopyranoside as glycosyl acceptor for a variety of glucuronic acid glycosyl donors in high yield has also been shown, demonstrating stability of the oxazolidinone protecting group to a number of different glycosyl donor activation methods (Figure 14).

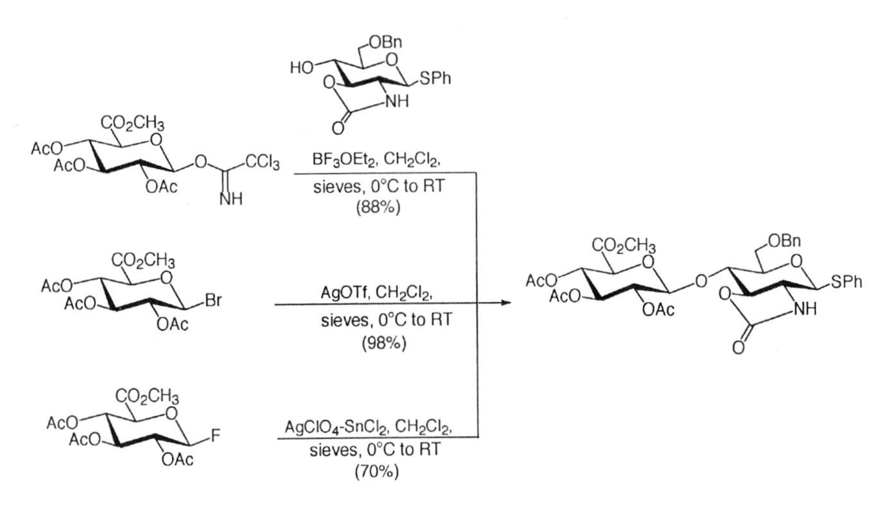

Figure 14. Chemical compatibility of ring-fused 2,3-oxazolidinone protection to a variety of glycosylation procedures.

Limitations of 2,3-oxazolidinone protected sugars in the synthesis of alpha-linked glycosides of D-glucosamine and D-galactosamine are evident. First, the 2,3-oxazolidinone protected thioglycoside donors are fairly low in reactivity, requiring phenylsulfenyl triflate activation, which is not a readily obtainable and employable reagent. Furthermore, deactivation of these donors through certain substituents at the 4 and 6 positions has been shown to render some donors virtually unreactive.[28] To date the stereoselectivity of 2,3-oxazolidinone protected donors in the D-galactosamine series has been modest to poor. Finally, application of these donors to stepwise glycosylation methodologies as required for oligosaccharide synthesis may be complicated by varied levels of *N*-glycosylation of the oxazolidinone ring.[28] Continued efforts to overcome these limitations are now focused on evaluating the effect of *N*-substituted 2,3-oxazolidone protection on the stereoselective formation of glycosides.

Nitro-D-galactal based strategies for the synthesis of glycoconjugates containing alpha-linked 2-amino-2-deoxy-D-galactopyranosides

Building on early work by Lemieux and others in the field of glycal and galactal chemistry, Schmidt and coworkers have recently expanded chemical methodology for the Michael-type addition of hydroxy nucleophiles to 2-nitro-D-galactal derivatives to afford alpha-linked *O*-glycosides of D-galactosamine.[29-33] Fully *O*-benzyl protected 2-nitro-D-galactal has been reported to afford excellent yields of alpha-linked galactosides upon treatment of monosaccharides, serine, or threonine derivatives with strong base (Figure 15).[30,31,33] Subsequent reduction of the nitro group followed by acetylation of the resulting amine affords efficient formation of the alpha-linked 2-acetamido-2-deoxy-D-galactopyranosides in high overall yields.

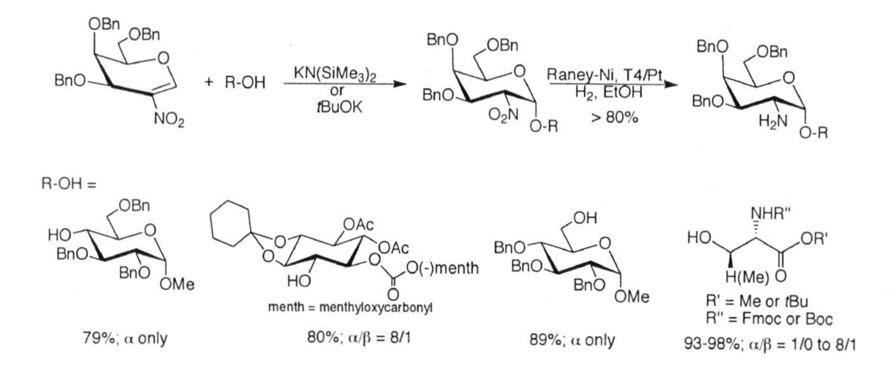

Figure 15. Michael-type addition of alcohols to 2-nitro-D-galactal derivatives followed by reduction of the nitro group affords alpha-linked glycosides of D-galactosamine derivatives.

The 2-nitro-D-galactal methodology has been elaborated using a C-6 *O*-differentiated 2-nitro-D-galactal intermediate to form an alpha-linked 2-acetamido-2-deoxy-D-galactopyranosyl linked threonine conjugate in the synthesis of a sialylated mucin-type *O*-glycan as shown (Figure 16).[32,33] Based on these early, albeit limited reports, this methodology is anticipated to be a highly useful addition to the synthetic arsenal for synthesizing bioactive glycoconjugates containing the alpha-linked galactosamine residues. Efficient formation of the requisite nitro galactals and the strong base employed to form the glycosidic bond will present potential limitations of this methodology in certain glycoconjugate systems.

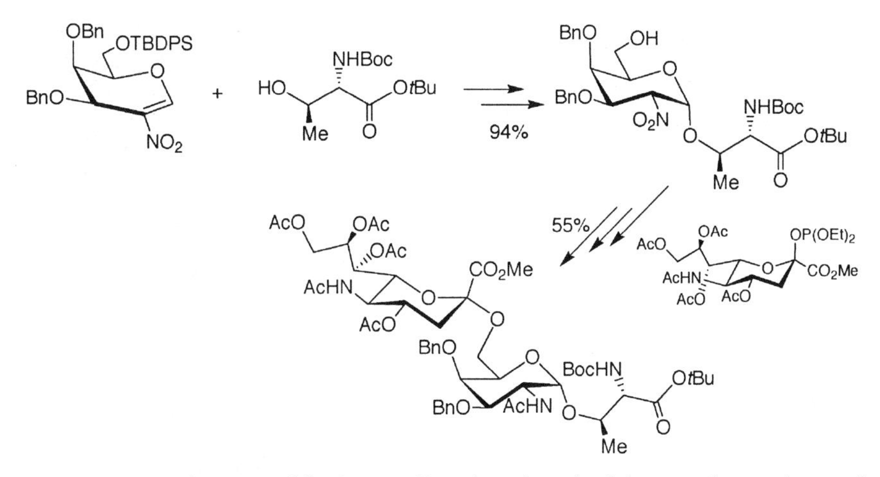

Figure 16. Application of the 2-nitro-D-galactal methodology in the synthesis of a sialylated mucin-type O-glycan.[32,33]

The 2,4-dintrophenyl moiety as a putative non-participating group for the 2-amino group in D-glucosamine and D-galactosamine glycosyl donors

The 2,4-dinitrophenyl group was investigated over 40 years ago as a non-participating protecting group for the 2-amine functionality in formation of alpha-linked (1,2-*cis*) glycosides of D-glucosamine (Figure 17).[34] Application of 2-(2,4-dinitrophenyl amino donors in the synthesis of alpha-linked disaccharides of 2-amino-2-deoxy-D-glucopyranoside has afforded mixed results, where stereocontrol is highly dependent on the activation system employed to activate glycosyl halide donor.[13]

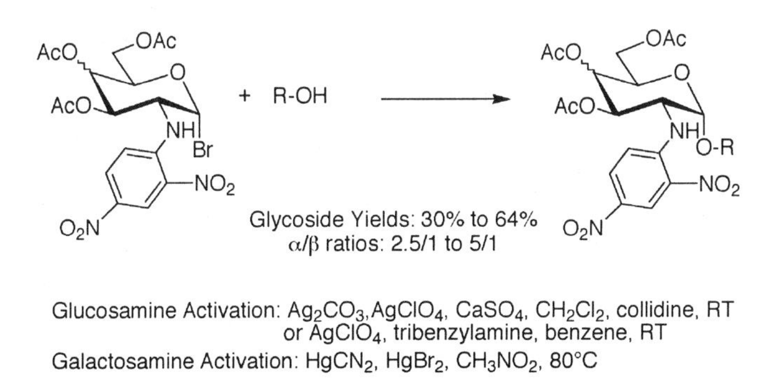

Glucosamine Activation: Ag_2CO_3, $AgClO_4$, $CaSO_4$, CH_2Cl_2, collidine, RT
or $AgClO_4$, tribenzylamine, benzene, RT
Galactosamine Activation: $HgCN_2$, $HgBr_2$, CH_3NO_2, 80°C

Figure 17. Overview of 2-(2,4-dinitrophenyl)amino protected D-glucosamine and D-galactosamine derivatives as potential alpha-selective glycosyl donors.

Syntheses employing 2,4-dinitrophenyl protection of the 2-amino group in D-glucosamine and D-galactosamine donor residues have been reported for a number of bioactive natural product structures including trehalosamine analogs,[35] aminoglycosides,[36] glycosylated amino acids,[37] and glycosylated inositols.[38,39] The coupling reactions employed in these syntheses are exemplified by those shown (Figure 18). Future applications of this approach to the synthesis of alpha-linked glycosides are likely to be limited because relatively low yields and modest enantioselectivity are observed compared to other methods.

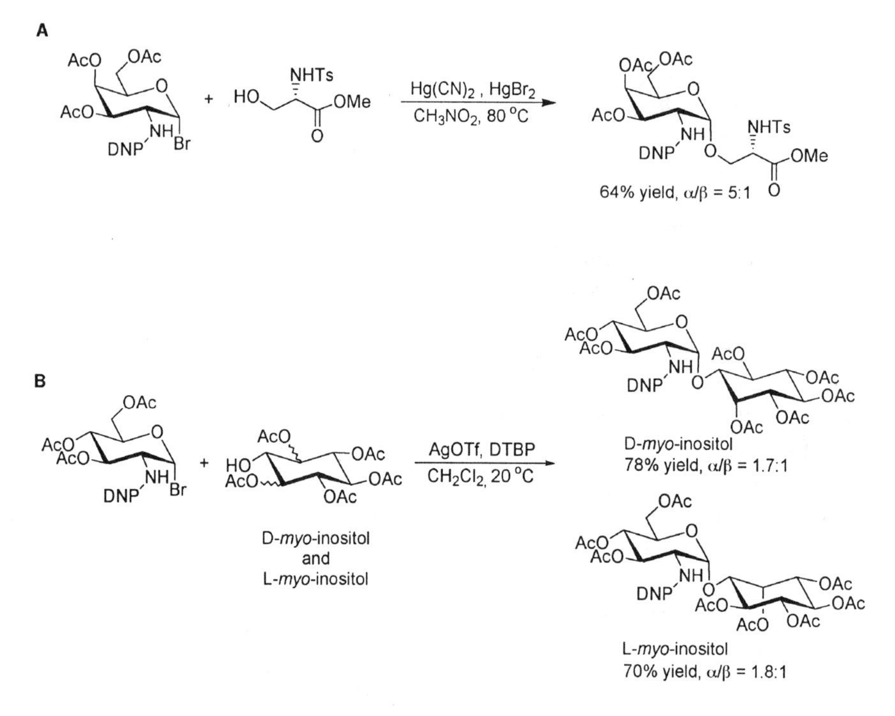

Figure 18. Select coupling reactions employing 2-(2,4-dinitrophenyl)amino glycosyl donors investigated toward the synthesis of natural bioactive glycoconjugates. Panel A) Formation of the α-O-glycopeptide linkage with a serine residue employing DNP protected 2-amino-2-deoxy-D-galactopyranosyl bromide donor.[37] Panel B) Application of DNP protected donors reported in 2002 for studies toward the synthesis of mycothiol derivatives.[38]

Methoxybenzylidene (arylimino) protection for the 2-amino group of D-glucosamine donors as a putative non-participating moiety

Protection of the 2-amine group of 2-amino sugar donors through formation of arylimino moieties was initially employed under the premise that the arylimine group represented a potential non-participating group for the formation of 1,2-*cis* glycosides. Early application of this strategy toward the early synthesis of bioactive natural products such as aminoglycosides,[40-42] heparin,[43] and teichoic acid degradation products[44] employed *O*-protected 2-(4-methoxybenzylideneamino) glycosyl bromides as donor residues (Figure 19). Importantly, the stereochemical outcome of these glycosylations using the 2-(4-methoxybenzylideneamino) protected donors appears to be strongly dependent on the system employed in donor activation, with mercury(II)cyanide affording modest yields of the desired alpha-linked glycoconjugates.[13]

Further evaluation of this methodology in the 1990s during application of this coupling strategy toward synthesis of heparin-like oligosaccharides revealed a possible alternative pathway for formation of the alpha-linked glycosides as a result of mercuric cyanide activation, indicating the 2-imino group does display neighboring group participation during glycosylation reactions.[45]

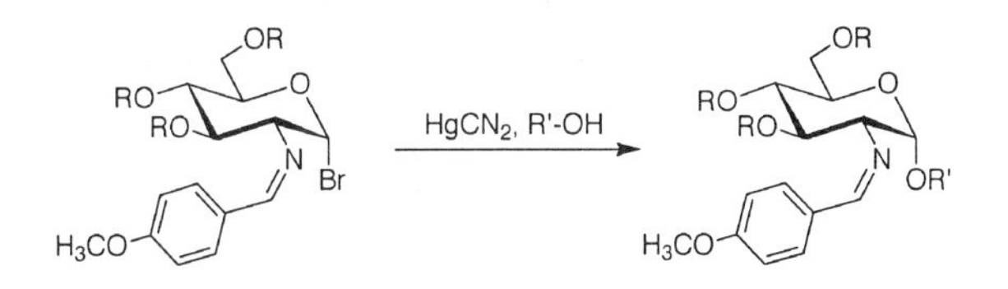

Figure 19. Arylimino protection of the 2-amine group as a putative non-participating C-2 moiety for the synthesis of alpha-linked 2-amino sugars.

In 1989 the stereoselective glycosidation of a 2-(4-methoxybenzylideneamino) protected pentenyl glycoside was reported using a series of monosaccharide acceptors (Figure 20, Panel A).[46] Here, alpha-linked disaccharides were obtained in good yields (60-64%) with the exception of a 4-hydroxy-galactose derivative that afforded 68% yield of 1:1 α/β mixture. Subsequently, a 4,6-benzylidene protected pentenyl glycoside bearing a C-2 4-methoxybenzylideneamino group was employed in the synthesis of an alpha-linked 2-amino-2-deoxy-D-glucosylinositol intermediate, which was employed in synthesis of the pentasaccharide core of the protein membrane anchor found in *Trypanosoma brucei* (Figure 20 Panel B).[47]

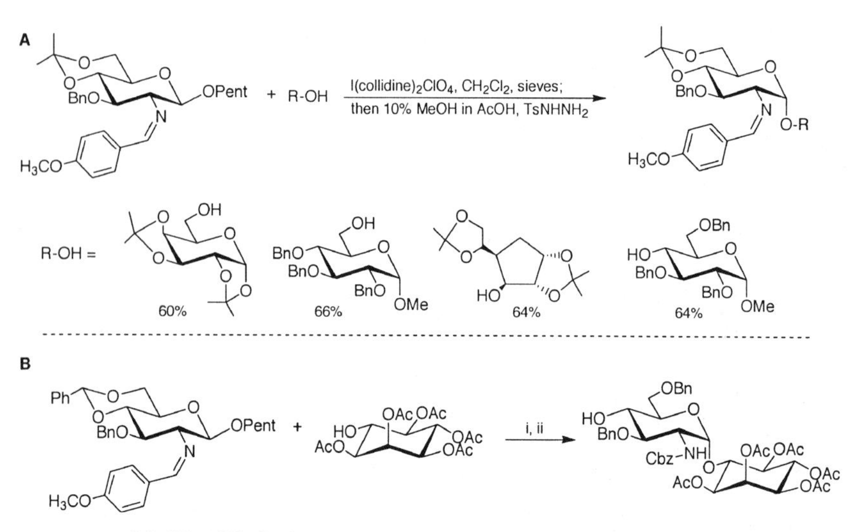

(i) I(collidine)$_2$ClO$_4$, CH$_2$Cl$_2$, sieves; then 10% MeOH in AcOH, TsNHNH$_2$; then Et$_3$N, Cbz-Cl; 65%
(ii) NaBH$_3$(CN), THF-Et$_2$O, HCl, sieves; 75%

Figure 20. Application of 1-O-pentenyl glycosyl donors bearing a C-2 arylimino moiety as a latent amine group. (Panel A) Application to the synthesis of alpha-linked disaccharides.[46] *(Panel B) Application to the synthesis of a α-D-glucosaminyl-inositol intermediate that was further elaborated to prepare the pentasaccharide core of a GPI anchor of Trypanosoma brucei.*[47]

C-2-Oximino protection as a putative non-participating moiety for the 2-amino group of D-glucosamine and D-galactosamine glycosyl donors

The utility of glycal-derived oximino derivatives of hexopyranosides in preparing glycosides of 2-amino-2-deoxy-D-hexopyranosides was first introduced in the 1960s.[48] Formation of a nitroso dimer by addition of nitrosyl chloride to glycal followed by condensation with monosaccharides as well as simple alcohols selectively afforded good yields of alpha-linked 2-oximino glycosides (Figure 21). Reduction of the resulting oximino functionality ultimate determines stereochemistry of the C-2-amine group. Utility of this methodology in the synthesis of 1,2-cis and 1,2-trans glycosides has been reviewed.[13] This methodology was employed in early syntheses of aminoglycoside antibiotics[49-52] and proposed antigenic teichoic acid structures of *Staphylococcus aureus.*[53] Alpha-linked glycosides were typically obtained in 30% to 50% yields during these syntheses.

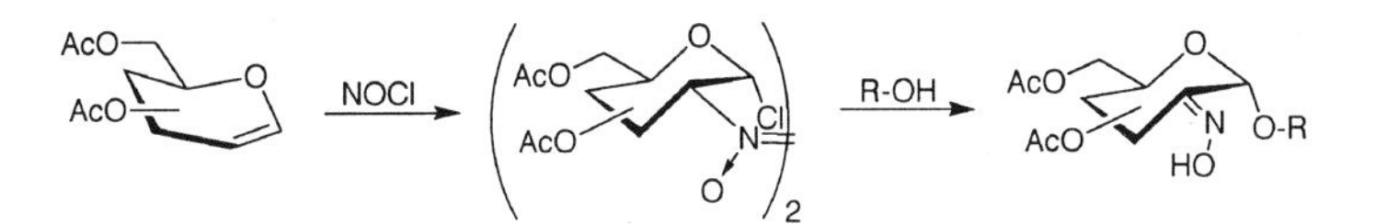

Figure 21. Early methodology employing glycal-derived 2-oximino dimers for the formation of alpha-linked glycosides of 2-amino sugars.

A variation of this method has also been employed, where a C-2 oximino ester is prepared as a putative non-participating C-2 protecting group for traditional promoter-mediated glycosidation of a glycosyl donor. For example, in 1985 Lichtenthaler *et al* reported the coupling of an oximino ester containing disaccharide donor with a monosaccharide acceptor to afford the alpha-linked trisaccharide product in 77% yield (Figure 22, panel A).[54] In 1997 Karpiesiuk and Banaszek employed a 2-oximino derivatized glucosyl bromide toward synthesis of the 1,1-linked disaccharide component in tunicamycin antibiotics (Figure 22, Panel B).[55,56] It must be noted that alpha selectivity in this coupling strategy is sensitive to the donor activation system (activating agent and solvent). Beta-glycosides of the 2-oximino sugars often prevailed until optimal conditions were found to form the alpha-linked glycoside,[54-56] and could even be optimized to preferentially afford high yields of beta glycosides.[57]

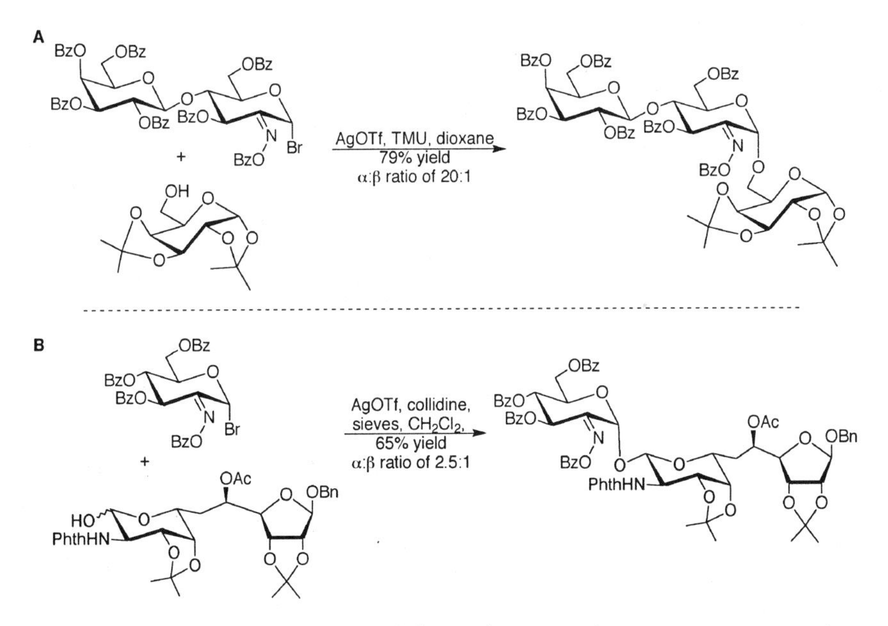

Figure 22. Application of glycosyl donors bearing a 2-oximino ester group in the synthesis of a trisaccharide (panel A), and synthesis of the 1,1-disaccharide of tunicamycin (Panel B).

N-Acetyl 1-diazirine glycosyl donors

Vasella and coworkers have reported the formation of alpha-linked glycosides of *N*-acetyl-D-glucosamine and *N*-acetyl-D-allosamine *via* photolysis of C-1-diazirine derivatives of 2-acetamido-2-deoxy-D-glucopyranose and 2-acetamido-2-deoxy-D-allopyranose (Figure 23).[58,59] This glycosidation strategy is based on photolytic or thermolytic generation of glycosylidene carbenes in the presence of alcohol to afford the *O*-linked glycoside. Employing simple alcohols of different Pka at varied concentrations and temperatures of photolysis provided fair to excellent yields of the alpha-linked 2-acetamido products.[59] The only alpha-linked disaccharide of a 2-acetamido-2-deoxy-D-hexopyranoside derivative prepared using this method employed 1,2:5,6-di-*O*-isopropylidene-α-D-glucofuranose as the acceptor alcohol, affording a modest 30% yield of the alpha-linked product and significant recovery of the unreacted glucofuranose.[59] This method does not require the use of a chemical promoter and accommodates use of the *N*-acetylated monosaccharide aziridine, which eliminates the need to modify the 2-amino group after glycoside bond formation to obtain the 1,2-cis glycosides of *N*-acetyl products. However, modest yields appear to be a limiting factor. In addition, synthesis of the diazirine sugars is a non-trivial multi-step process that requires preparation of appropriate hydroxyl-protected ring-opened oximes in order to obtain the requisite glyconohydroximo-lactone intermediates.

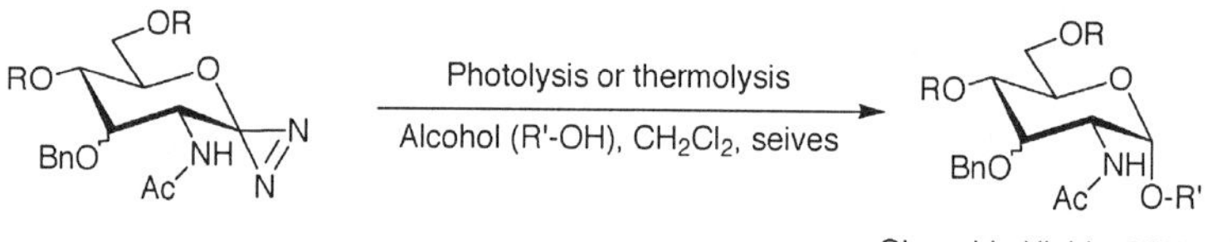

Figure 23. Pictorial representation of Vasella's methodology to utilize glycosylidene carbene-based glycosyl donors in the synthesis of glycosides of N-acetylated D-glucosamine.

Formation of alpha-linked glycosides from D-glucosamine and D-galactosamine glycosyl donors bearing C-2 *N*-alkoxycarbonyl moieties that traditionally afford beta-linked glycosides

Glycosylation reactions using 2-amino-2-deoxy-D-glucopyranosyl and D-galactopyranosyl donors bearing *N*-alkoxycarbonyl (carbamate) protection of the 2-amine group typically afford beta-linked (1,2-*trans*) glycosides, owing to participation of the alkoxycarbonyl group during the glycosylation reaction. However, alpha-selective glycosylation has been reported for certain 2-alkoxycarbonyl protected glycosyl donors under specific conditions. For example in 1990 Higashi *et al* reported good yields and excellent stereoselectivity in obtaining alpha glycosides upon activating an *N*-Troc protected 2-amino-2-deoxy-α-D-glucopyranosyl bromide with zinc halides or silver perchlorate (Figure 24).[60] As would be anticipated, good yields and excellent stereoselectivity in forming the beta-linked (1,2-*trans*) glycosides of 2-*N*-Troc substituted glycosyl donors is achieved under alternative activation conditions.[60,61] Based on this report it is somewhat surprising that activation of *N*-Troc protected 2-amino-2-deoxy-α-D-glucopyranosyl bromides with zinc halides has not been employed in more syntheses of alpha-linked 2-amino-2-deoxy-α-D-glycopyranosides. Although versatility of the 2-azido methodologies and extensive use of 2-*N*-alkoxycarbonyl protection of 2-amino-D-glucopyranosides and 2-amino-D-galactopyranosides to afford the 1,2-*trans* (beta) glycosidic linkage likely overshadow this potentially useful unique system.

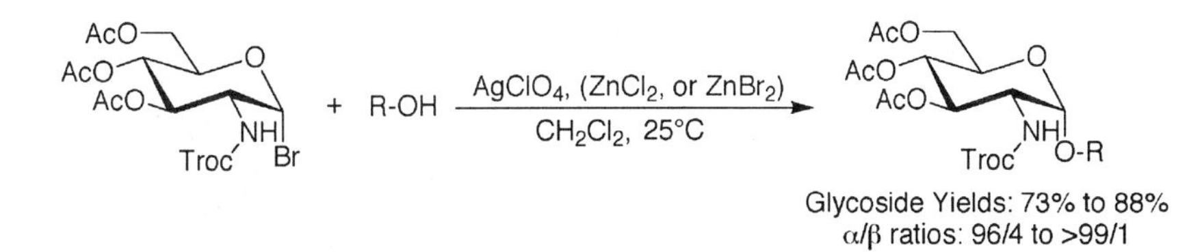

Figure 24. Activation of N-Troc protected 2-amino-2-deoxy-D-glucopyranosyl bromides under specific conditions has been shown to afford unexpectedly high stereoselectivity and yields of the alpha-linked glycoside products.[60]

A 1990 report by Kolar *et al* also showed good yields and stereoselectivity in formation of non-natural etoposide derivatives, where activation of 2-*N*-benzyloxycarbonyl 1-α-lactols with boron trifluoride etherate in dichloromethane afforded the alpha-linked glycosides (Figure 25).[62] The corresponding beta glycosides were preferentially formed when ethyl acetate or ethyl acetate-dichloromethane were employed as reaction solvent. Mutarotation studies demonstrated the stereoselectivity observed in this coupling reaction was controlled by solvent-dependent α/β equilibrium of the 2-*N*-Cbz protected lactol under glycosylation conditions.

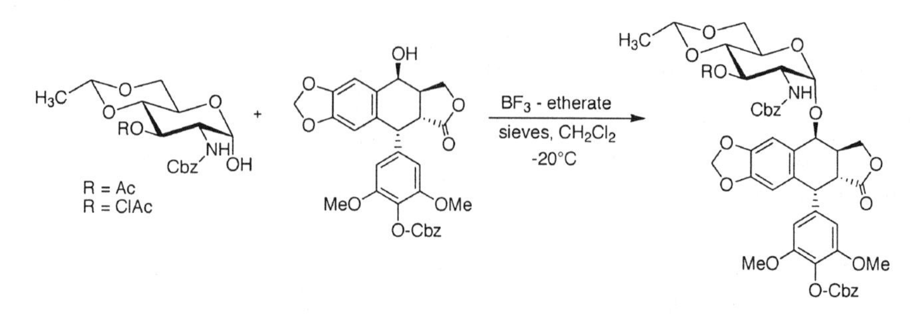

Figure 25. Reported alpha-selective glycosylation where the C-2 amine substituent bears a benzyloxycarbonyl protecting group.[62]

In 2003 Imamura *et.al* reported a potentially more general approach to achieving stereoselective formation of 2-amino-2-deoxy-α-D-galactopyranosides using D-galactosamine donors possessing 2-*N*-alkoxycarbonyl and other 2-*N*-acyl protecting groups.[63] In this approach a 4,6-*O*-di-*tert*-butylsilylene protected thiophenyl glycoside bearing *N*-Troc protection of the C-2 amine group was shown to yield alpha-selective glycosylation of a trisaccharide, two monosaccharides, and two serine derivatives (Figure 26).[63] Alpha-selective glycosylation of alcohols in modest to high yields was also imparted by the 4,6-*O*-di-*tert*-butylsilylene protecting group for derivatives having *N*-Phthaloyl and *N*-Acetyl protection of the 2-amino group. It is notable that the corresponding 4,6-*O*-benzylidene derivatives did not show similar alpha-directing properties.

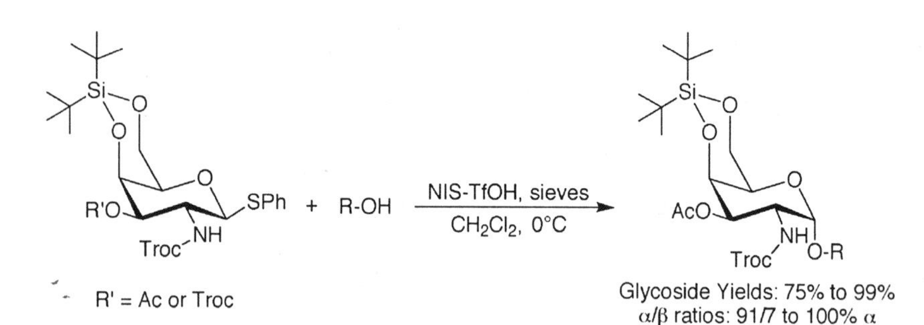

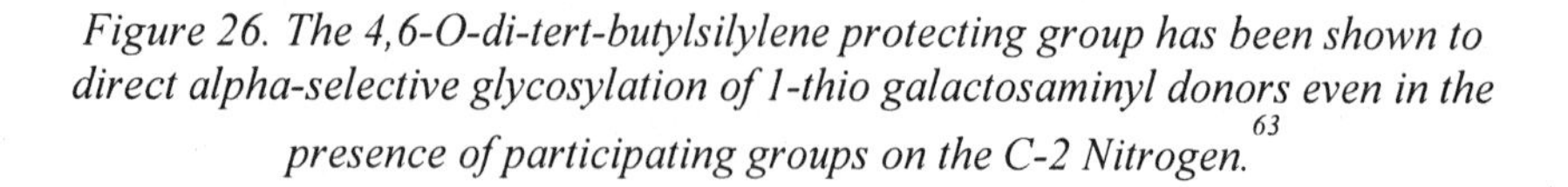

Figure 26. The 4,6-O-di-tert-butylsilylene protecting group has been shown to direct alpha-selective glycosylation of 1-thio galactosaminyl donors even in the presence of participating groups on the C-2 Nitrogen.[63]

Certain 2-acetamido-2-deoxy-D-glucopyranosyl donors have been shown to afford alpha-linked glycosides under thermal glycosylation conditions.[64,65] Thermal glycosylation of alcohols employing 2-acetamido-2-deoxy-3,4,6-tri-*O*-acetyl-α-D-glucopyranosyl chloride at high temperatures (160°C-180°C) in the presence of acid scavengers including α-methylstyrene or N,N,N',N'-tetramethylurea preferentially afforded alpha-linked products in modest yield (Figure 27, Panel A).[64] A complication with this reaction system is acylation of the acceptor alcohol, which presumably forms as a result of intermolecular acetate transfer at high temperatures. At lower temperatures beta-linked products predominate and loss of acceptor to acetylation is minimized.

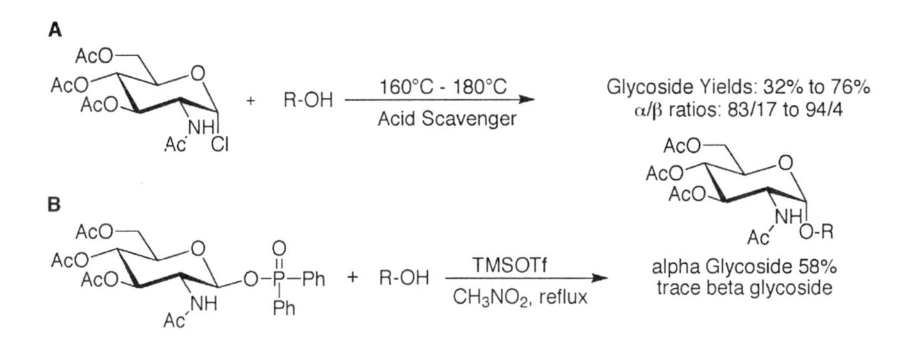

Figure 27. Thermal glycosylation strategies that have been shown to afford alpha-linked glycosides of N-acetyl-D-glucosamine derivatives.

In 2000 an alpha-selective thermal glycosylation was reported using acetyl-protected 2-acetamido-2-deoxy-ß-D-glucopyranosyl diphenylphosphinate donor activated with TMSOTf in refluxing nitromethane (Figure 27, Panel B).[65] Although stereoselectivity for the alpha-linked product appears high, only modest overall yield of the glycoside is obtained. In theory, an advantage of the thermal glycosylation using 2-acetamido glycosyl donors is the glycosylation product contains the intact *N*-acetyl substituent at the C-2 position, thus avoiding manipulation of a latent C-2 amine substituent after glycoside bond formation to obtain *N*-acetyl derivatized alpha-linked products. In practice, it is anticipated that the high temperatures, modest yields, and sometimes modest stereoselectivity of these current thermal glycosylation strategies, when compared to the alternative glycosylation strategies, will limit application of these methods to the synthesis of more complex bioactive glycoconjugates containing alpha-linked 2-acetamido-2-deoxy-α-D-glycopyranoside residues.

Utility of miscellaneous 1-OH glycosyl donors

Glycosylation by way of direct activation of 1-OH sugar derivatives is of interest because such approaches eliminate the need to prepare reactive glycosyl donors. One example of this strategy was discussed above (Figure 25). Koto *et al* reported the synthesis of α-linked glycosides of 2-azido-3,4,6-tri-*O*-benzyl-2-deoxy-D-glucose and using 2-azido-3,4,6-tri-*O*-benzyl-2-deoxy-D-galactose through *in situ* activation of the lactol employing *p*-nitrobenzenesulfonyl chloride (NsCl)-silver triflate-triethylamine as the activation system (Figure 28).[66] Modest yields and stereoselectivity were achieved.

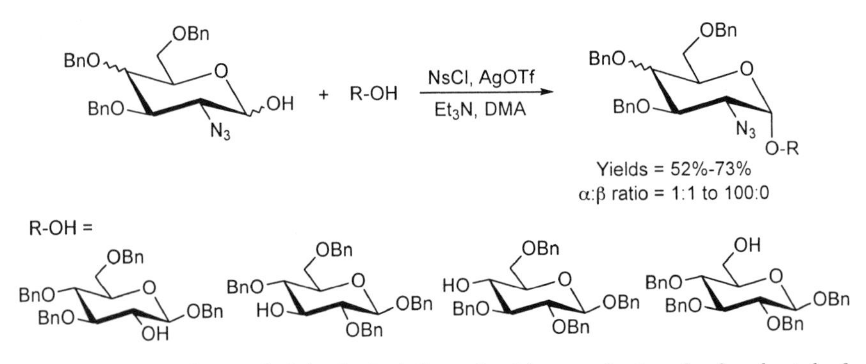

Figure 28. Synthesis of alpha-linked disaccharides employing the free lactol of 2-azido-2-deoxy-D-glucopyranose and 2-azido-2-deoxy-D-galactopyranose reported in 1999.[66]

In general, the existence of an *N*-acetyl group at the C-2 position of hexopyranosyl glycosyl donors preferentially affords β-glycosides upon glycosylation due to anchimeric participation of the *N*-acetyl group. However, Fisher-type glycosidation of *N*-acetyl-D-glucosamine using *n*-octanol or allyl alcohol with boron trifluoride etherate activation and acetonitrile as solvent has been shown to afford the corresponding α-glycosides with modest yields and good stereoselectivity.[67,68]

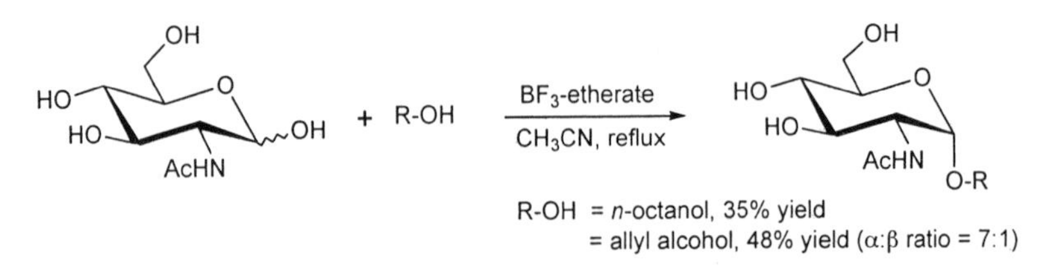

Figure 29. Examples of Fisher-type glycosidation of N-acetyl-D-glucosamine using simple alcohols.

A suitable α-linked spacer at the reducing end of D-galactosamine derivatives has been employed for conjugation of this saccharide residue to macromolecular carriers. For example, the introduction of a α-linked spacer into *N*-acetyl-D-galactosamine has been achieved using Fisher glycosidation as shown (Figure 30). Here, heating *N*-acetyl-D-galactosamine in 2-bromoethanol or 3-bromo-1-propanol at 60-70°C in the presence of catalytic hydrogen chloride afforded the alpha glycosides in good yield and stereoselectivity.[69,70]

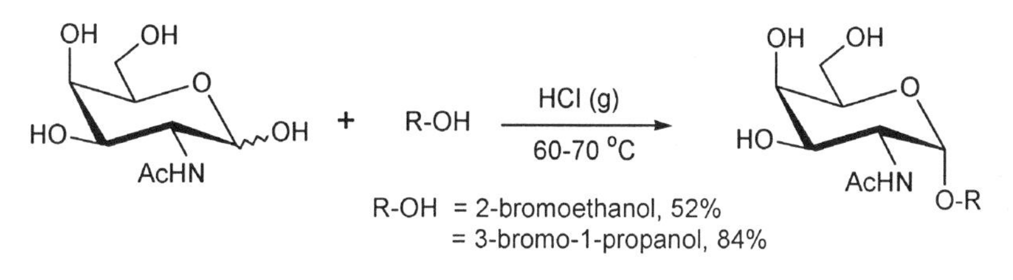

Figure 30. Formation of simple alpha-linked glycoside of N-acetyl-D-galactosamine using Fisher glycosidation.

The later two methods described above are applicable to the formation of glycosides using simple alcohols during the early steps of a synthesis. These methods are typically used to cap the anomeric position rendering a stable glycoside or to introduce a stable glycosidic moiety bearing requisite functional groups for additional, non-glycosidic, conjugation chemistry.

Methods to obtain alpha-linked aryl *O*-glycosides of D-glucosamine and D-galactosamine

The occurrence of aryl *O*-glycosides in nature and the biological activities of these glycoconjugates makes them attractive targets in synthetic organic chemistry. However, due to reduced nucleophilicity of a phenolic hydroxyl group compared to other alcohols, formation of aryl *O*-glycosides typically requires modification of traditional glycosylation strategies. The synthesis of aryl-*O* 2-amino-2-deoxy-β-D-glycosides have been extensively reported, with the corresponding α-glycosides generally presented as side products obtained during these syntheses. However, aryl-*O* glycosides of 2-amino-2-deoxy-α-D-glucopyranose and 2-amino-2-deoxy-α-D-galactopyranose have been synthesized in excellent yields under specific conditions. For example, in 2004 Nishiyama *et al* reported good yields and stereoselectivity in formation of phenyl 2-amino-2-deoxy-α-D-glucopyranoside, the core structure of glycocinnasperimicin D.[71] Glycosidation of peracetylated D-glucosamine with phenol using catalytic tin (IV) chloride in CH_2Cl_2 proceeded smoothly to afford a 4:1 (α:β) anomeric mixture of phenyl glycosides in 87% yield. The stereoselectivity in this glycosylation is reversed using TMSOTf in ether, affording a 1:9 (α:β) anomeric mixture of phenyl glycoside in 50% yield.

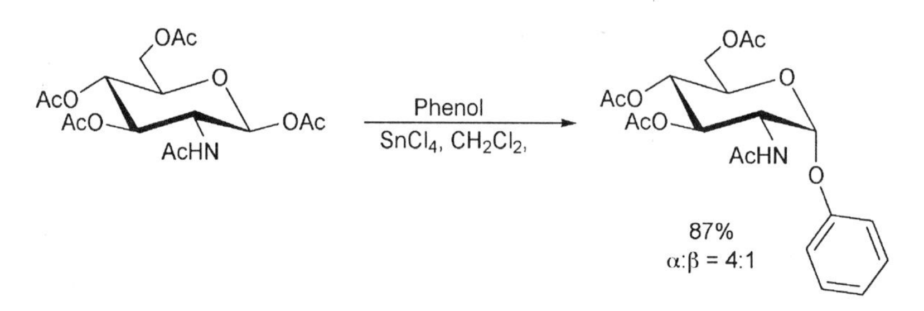

Figure 31.Direct conversion of peracetylated D-glucosamine to alpha-linked aryl-O-glycosides reported in 2004.[71]

In 2003 Schmidt and co-workers reported good yields and excellent stereoselectivity in obtaining aryl-*O* glycosides of 2-acetamido-2-deoxy-α-D-galactopyranosides employing the Michael-type addition of phenols to nitro-D-galactal.[29] Here, *O*-benzyl protected 2-nitro-D-galactal was treated with commercially available *N*-Boc protected tyrosine methyl ester in toluene in the presence of a catalytic amount of potassium *tert*-butoxide, affording stereoselective formation of the α-anomer (α/β = 40/1) in 82% yield (Figure 32). No racemization of the tyrosine moiety was observed under the reaction conditions. Selective reduction of the nitro group with platinised Raney nickel followed by acetylation provided the final 2-acetamido-2-deoxy-α-D-galactopyranoside product.[29]

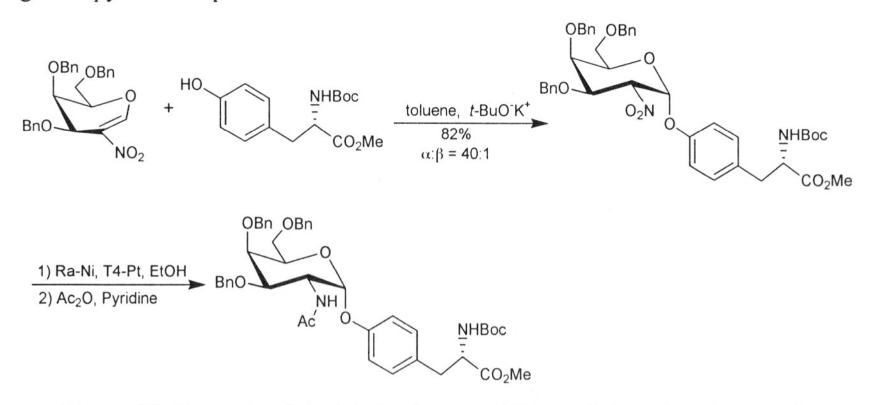

Figure 32. Example of the Michael-type addition of phenols to 2-nitro-D-galactal derivatives reported in 2003 as a method to obtain alpha-linked aryl-O-glycosides of D-galactosamine derivatives.[29]

Araki and coworkers reported a straightforward approach to obtaining aryl *O*-glycosides of D-glucosamine and D-galactosamine *via* use of the non-participating C-2 azide group.[72] The 1-*O*-acetyl and 1-chloro glycosyl donors were coupled with ethyl *p*-hydroxycinnamate using a number of different activation strategies. Modest yields were obtained while preferentially affording the α-glycosides.

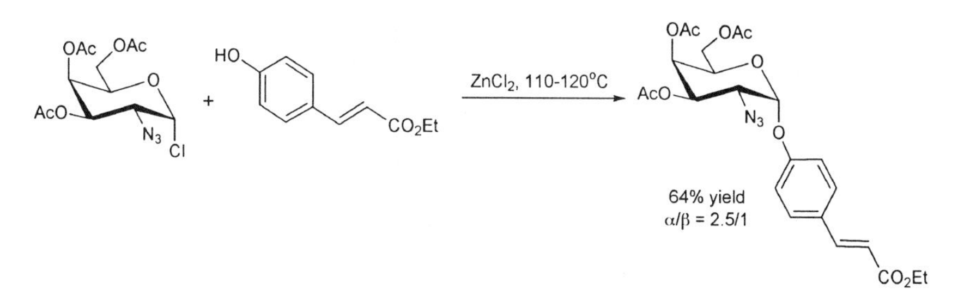

Figure 33. One example of conditions reported for aryl O-glycoside synthesis employing the non-participating 2-azide moiety.[72]

An alternative route for the stereoselective introduction of alpha-linked 2-amino-2-deoxy-D-hexopyranosides into glycoconjugate structures.

The use of fully *O*-protected D-glucosyl and D-galactosyl donors in alpha-selective glycoside bond formation followed by the introduction of an amine or amine precursor has also found utility in glycoconjugate synthesis. Introduction of the amine group or latent amine group is typically achieved through nucleophilic displacement of a C-2 leaving group such as triflate. One example of this indirect strategy for obtaining alpha-linked glycosides of D-glucosamine or D-galactosamine employed by Fraser-Reid in the synthesis of a prototype GPI of *P. falciparum* is shown (Figure 34).[73,74]

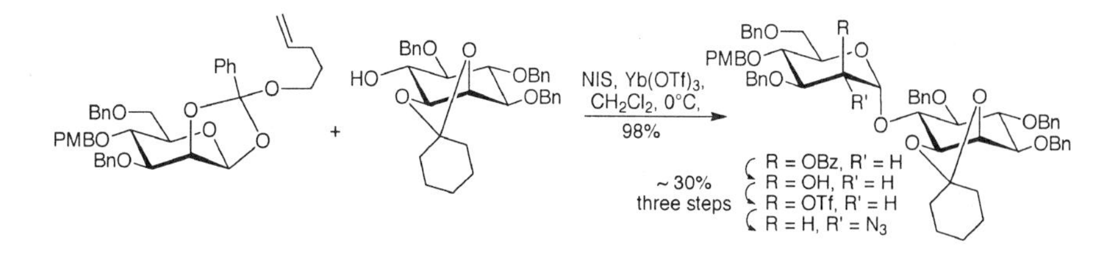

Figure 34. Incorporation of a 2-azido-2-deoxy-α-D-glucopyranoside residue into a GPI precursor structure where the 2-azide group (latent amine group) is introduced after formation of the requisite alpha glycosidic linkage.

References

(1) Dwek, R. A. *Chem. Rev.* **1996**, *96*, 683-720.

(2) Casu, B.; Lindahl, U. *Adv. Carbohydr. Chem. Biochem.* **2001**, *57*, 159-206.

(3) Davis, B. G. *Chem. Rev.* **2002**, *102*, 579-601.

(4) Rabenstein, D. L. *Nat. Prod. Rep.* **2002**, *19*, 312-331.

(5) *Tunicamycin*; Tamura, G., Ed.; Scientific Press: Tokyo, Japan, **1982**.

(6) *Aminoglycoside Antibiotics*; Umezawa, H.; Hooper, I. R., Eds.; Springer-Verlag: New York, Heidelberg, **1982**.

(7) Hart, G. W.; Haltiwanger, R. S.; Holt, G. D.; Kelly, W. G. *Ann. Rev. Biochem.* **1989**, *58*, 841-874.

(8) Montreuil, J. *Adv. Carbohydr. Chem. Biochem.* **1980**, *37*, 157-223.

(9) Taylor, C. M. *Tetrahedron* **1998**, *54*, 11317-11362.

(10) Sturm, S.; Jann, B.; Fortnagel, P.; Timmis, K. N. *Microbial Pathogenesis* **1986**, *1*, 307-324.

(11) Newton, G. L.; Arnold, K.; Price, M. S.; Sherrill, C.; Delcardayre, S. B.; Aharonowitz, Y.; Cohen, G.; Davies, J.; Fahey, R. C.; Davis, C. *J. Bacteriol.* **1996**, 1990.

(12) Ferguson, M. A. J.; Williams, A. F. *Ann. Rev. Biochem.* **1988**, *57*, 285-320.

(13) Banoub, J.; Boullanger, P.; Lafont, D. *Chem. Rev.* **1992**, *92*, 1167-1195.

(14) Demchenko, A. V. *Curr. Org. Chem.* **2003**, *7*, 35-79.

(15) Schofield, L.; Hewitt, M. C.; Evans, C.; Siomas, M.-A.; Seeberger, P. H. *Nature* **2002**, *418*, 785-789.

(16) Xue, J.; Guo, Z. *Bioorg. Med. Chem. Letters* **2002**, *12*, 2015-2018.

(17) Petitou, M.; Duchaussoy, P.; Lederman, I.; Choay, J.; Jacquinet, J.-C.; Sinay, P.; Torri, G. *Carbohydr. Res.* **1987**, *167*, 67-75.

(18) Yu, H. N.; Furukawa, J.-I.; Ikeda, T.; Wong, C. H. *Org. Letters* **2004**, *6*, 723-726.

(19) Orgueira, H. A.; Bartolozzi, A.; Schell, P.; Seeberger, P. H. *Angew. Chem. Int. Ed.* **2002**, *41*, 2129-2131.

(20) Sames, D.; Chen, X.-T.; Danishefsky, S. J. *Nature* **1997**, *389*, 587-591.

(21) Greenberg, W. A.; Priestley, E. S.; Sears, P. S.; Alper, P. B.; Rosenbohm, C.; Hendrix, M.; Hung, S.-C.; Wong, C.-H. *J. Am. Chem. Soc.* **1999**, *121*, 6527-6541.

(22) Mulard, L. A.; Glaudemans, C. P. J. *Carbohydr. Res.* **1998**, *311*, 121-133.

(23) Pozsgay, V. *J. Org. Chem.* **1998**, *63*, 5983-5999.

(24) Nicholas, G. M.; Eckman, L. L.; Kovac, P.; Otero-Quintero, S.; Bewley, C. A. *Bioorg. Med. Chem.* **2003**, *11*, 2641-2647.

(25) Lee, S.; Rosazza, J. P. N. . *Org. Lett.* **2004**, *6*, 365-368.

(26) Myers, A. G.; Gin, D. Y.; Rogers, D. H. *J. Am. Chem. Soc.* **1993**, *115*, 2036-2038.

(27) Benakli, K.; Zha, C.; Kerns, R. J. *J. Am. Chem. Soc.* **2001**, *123*, 9461-9462.

(28) Kerns, R. J.; Zha, C.; Benakli, K.; Liang, Y.-Z. *Tetrahedron Lett.* **2003**, *44*, 8069-8072.
(29) Khodair, A. I.; Winterfeld, G. A.; Schmidt, R. R. *Eur. J. Org. Chem.* **2003**, 1847-1852.
(30) Das, J.; Schmidt, R. R. *Eur. J. Org. Chem.* **1998**, 1609-1613.
(31) Winterfeld, G. A.; Ito, Y.; Ogawa, T.; Schmidt, R. R. *Eur. J. Org. Chem.* **1999**, 1167-1171.
(32) Winterfeld, G. A.; Schmidt, R. R. *Angew. Chem. Int. Ed.* **2001**, *40*, 2654-2657.
(33) Winterfeld, G. A.; Khodair, A. I.; Schmidt, R. R. *Eur. J. Org. Chem.* **2003**, 1009-1021.
(34) Lloyd, P. F.; Stacey, M. *Tetrahedron* **1960**, *9*, 116-124.
(35) Ogawa, S.; Shibata, Y. *Carbohydr. Res.* **1988**, *176*, 309-315.
(36) Umezawa, S.; Koto, S. *Bull. Chem. Soc. Jpn.* **1966**, *39*, 2014-2017.
(37) Kaifu, R.; Osawa, T. *Carbohydr. Res.* **1977**, *58*, 235-239.
(38) Jardine, M. A.; Spies, H. S. C.; Nkambule, C. M.; Gammon, D. W.; Steenkamp, D. J. *Bioorg. Med. Chem.* **2002**, *10*, 875-881.
(39) Plourde, R.; d'Alarcao, M. *Tetrahedron Lett.* **1990**, *31*, 2693-2696.
(40) Umezawa, S.; Sana, H.; Tsuchiya, T. *Bull. Chem. Soc. Jpn.* **1975**, *48*, 556-559.
(41) Watanabe, I.; Tsuchiya, T.; Takase, T.; Umezawa, S.; Umezawa, H. *Bull. Chem. Soc. Jpn.* **1977**, *50*, 2369-2374.
(42) Harayama, A.; Tsuchiya, T.; Umezawa, S. *Bull. Chem. Soc. Jpn.* **1979**, *52*, 3626-3628.
(43) Wolfrom, M. L.; Tomomatsu, H.; Szarek, W. A. *J. Org. Chem.* **1966**, *31*, 1173-1178.
(44) Hardy, F. E.; Buchanan, J. G.; Baddiley, J. *J. Chem. Soc.* **1963**, 3360-3366.
(45) Marra, A.; Sinay, P. *Carbohydr. Res.* **1990**, *200*, 319-337.
(46) Mootoo, D. R.; Fraser-Reid, B. *Tetrahedron Lett.* **1989**, *30*, 2363-2366.
(47) Mootoo, D. R.; Konradsson, P.; Fraser-Reid, B. *J. Am. Chem. Soc.* **1989**, *111*, 8540-8542.
(48) Lemieux, R. U.; Nagabhushan, T. L.; O'Neill, I. K. *Tetrahedron Lett.* **1964**, *29*, 1909-1916.
(49) Paulsen, H.; Stadler, P.; Todter, F. *Chem. Ber.* **1977**, *110*, 1925-1930.
(50) Kugelman, M.; Mallams, A. K.; Vernay, H. F.; Crowe, D. F.; Detre, G.; Tanabe, M.; Yasuda, D. M. *J. Chem. Soc. Perk. I* **1976**, 1097-1113.
(51) Kugelman, M.; Mallams, A. K.; Vernay, H. F. *J. Chem. Soc. Perk. I* **1976**, 1113-1126.
(52) Kugelman, M.; Mallams, A. K.; Vernay, H. F. *J. Chem. Soc. Perk. I* **1976**, 1126-1134.
(53) Loureau, J.-M.; Boullanger, P.; Descotes, G. *Eur. J. Med. Chem.* **1985**, *20*, 455-458.
(54) Lichtenthaler, F. W.; Kaji, E. *Liebigs Ann. Chem.* **1985**, 1659-1668.
(55) Karpiesiuk, W.; Banaszek, A. *Carbohydr. Res.* **1994**, *261*, 243-253.

(56) Karpiesiuk, W.; Banaszek, A. *Carbohydr. Res.* **1997**, *299*, 245-252.
(57) Kaji, E.; Lichtenthaler, F. W.; Nishino, T.; Yamane, A.; Zen, S. *Bull. Chem. Soc. Jpn.* **1988**, *61*, 1291-1297.
(58) Vasella, A. *Pure & Appl. Chem.* **1993**, *65*, 731-752.
(59) Vasella, A.; Witzig, C. *Helv. Chim. Acta* **1995**, *78*, 1971-1982.
(60) Higashi, K.; Nakayama, K.; Soga, T.; Shioya, E.; Uoto, K.; Kusama, T. *Chem. Pharm. Bull.* **1990**, *38*, 3280-3282.
(61) Ellervik, U.; Magnusson, G. *Carbohydr. Res.* **1996**, *280*, 251-260.
(62) Kolar, C.; Dehmel, K.; Wolf, H. *Carbohydr. Res.* **1990**, *206*, 219-231.
(63) Imamura, A.; Ando, H.; Korogi, S.; Tanabe, G.; Muraoka, O.; Ishida, H.; Kiso, M. *Tetrahedron Lett.* **2003**, *44*, 6725-6728.
(64) Nishizawa, M.; Shimomoto, W.; Momil, F.; Yamada, H. *Tetrahedron Lett.* **1992**, *33*, 1907-1908.
(65) Kadokawa, J.-I.; Nagaoka, T.; Ebana, J.; Tagaya, H.; Chiba, K. *Carbohydr. Res.* **2000**, *327*, 341-344.
(66) Koto, S.; Asami, K.; Hirooka, M.; Nagura, K.; Takiazawa, M.; Yamamoto, S.; Okamoto, N.; Sato, M.; Tajima, H.; Yoshida, T.; Nonaka, N.; Sato, T.; Zen, S.; Yago, K.; Tomonaga, F. *Bull. Chem. Soc. Jpn.* **1999**, *72*, 765-777.
(67) Wong, C.-H.; Hendrix, M.; Manning, D. D.; Rosenbohm, C.; Greenberg, W. A. *J. Am. Chem. Soc.* **1998**, *120*, 8319-8327.
(68) Aguilera, B.; Romero-Ramirez, L.; Abad-Rodriguez, J.; Corrales, G.; Nieto-Sampedro, M.; Fernandez-Mayoralas, A. *J. Med. Chem.* **1998**, *41*, 4599-4606.
(69) George, S. K.; Holm, B.; Reis, C. A.; Schwientek, T.; Clausen, H.; Kihlberg, J. *J. Chem. Soc. Perk. I* **2001**, 880-885.
(70) Spijker, N. M.; Keuning, C. A.; Hooglugt, M.; Veeneman, G. H.; van Boeckel, C. A. A. *Tetrahedron* **1996**, *52*, 5945-5960.
(71) Nishiyama, T.; Ichikawa, Y.; Isobe, M. *Synlett* **2004**, 89-92.
(72) Araki, K.; Hashimoto, H.; Yoshimura, J. *Carbohydr. Res.* **1982**, *109*, 143-160.
(73) Lu, J.; Jayaprakash, K. N.; Fraser-Reid, B. *Tetrahedron Lett.* **2004**, *45*, 879-882.
(74) Lu, J.; Jayaprakash, K. N.; Schlueter, U.; Fraser-Reid, B. *J. Am. Chem. Soc.* **2004**, *126*, 7540-7547.

Chapter 10

Systematic Synthesis of Aminosugars and Their Stereoselective Glycosylation

Jinhua Wang and Cheng-Wei Tom Chang*

Department of Chemistry and Biochemistry, Utah State University, 0300 Old Main Hill, Logan, UT 84322–0300

1. Introduction

Synthesis of oligosaccharides and glycoconjugates has become increasingly important for the elucidation of their biological functions. Aminosugar, as one of the main components of these diverse glycoconjugates, has attracted burgeoning interest (1-8). Aminosugars are a group of structurally diverse unusual sugars bearing amino substitution on a normal sugar scaffold. Aminosugars can be found in bacteria, plant, and some mammalian cells, exerting unique but essential biological functions (2-8). Therefore, if there is a library of aminosugars available, a modular approach can be readily employed to construct a library of novel aminosugar-containing glycoconjugates for exploring important activities of interest in areas like biology, chemistry, and medicine. This strategy is termed glycodiversification or glycorandomization, a concept with fruitful applications (Figure 1).

In order to materialize the applications of aminosugar-containing glycoconjugates, numerous efforts have been devoted to the synthesis of aminosugars (9-11). Normal sugars, such as glucose, galactose, and mannose, are commonly used as the starting material for the chemical synthesis of aminosugars due to their intrinsic chirality, availability in large quantity, and lower cost. For the synthesis and application of aminosugars, two major issues are typically needed to be resolved properly: protective group manipulation including both hydroxyl and amino groups, and the stereoselective glycosylation.

Protective group manipulations are necessary to achieve regiospecific reactions for the aminosugar synthesis. For example, the hydroxyl groups which are not involved in the glycosylation must be masked. Protective groups, especially at the C-2 position, will affect not only the reactivity of the glycosyl donor during glycosylation, but also the stereoselectivity. Amino groups often require different reagents for protection and deprotection as compared to hydroxyl groups, creating more synthetic complications. Meanwhile, the protection of the anomeric position must be treated separately from the protection at other positions since it has to be stable enough to endure the reagents and conditions needed for the incorporation of amino groups. Nevertheless, the anomeric group is preferred to be labile so it can be activated at mild conditions for glycosylation. These two requirements regarding the anomeric groups work against each other (Figure 2). All these subtle demands make the preparation and utilization of aminosugars an extremely challenging and time-consuming task.

In addition, there are many reported methods for the preparation of specific aminosugars (9-11). However, the general protocols or methods that will allow the synthesis of other aminosugars for different applications are not available. As a result, when initiating a synthesis for a desired aminosugar, one may still

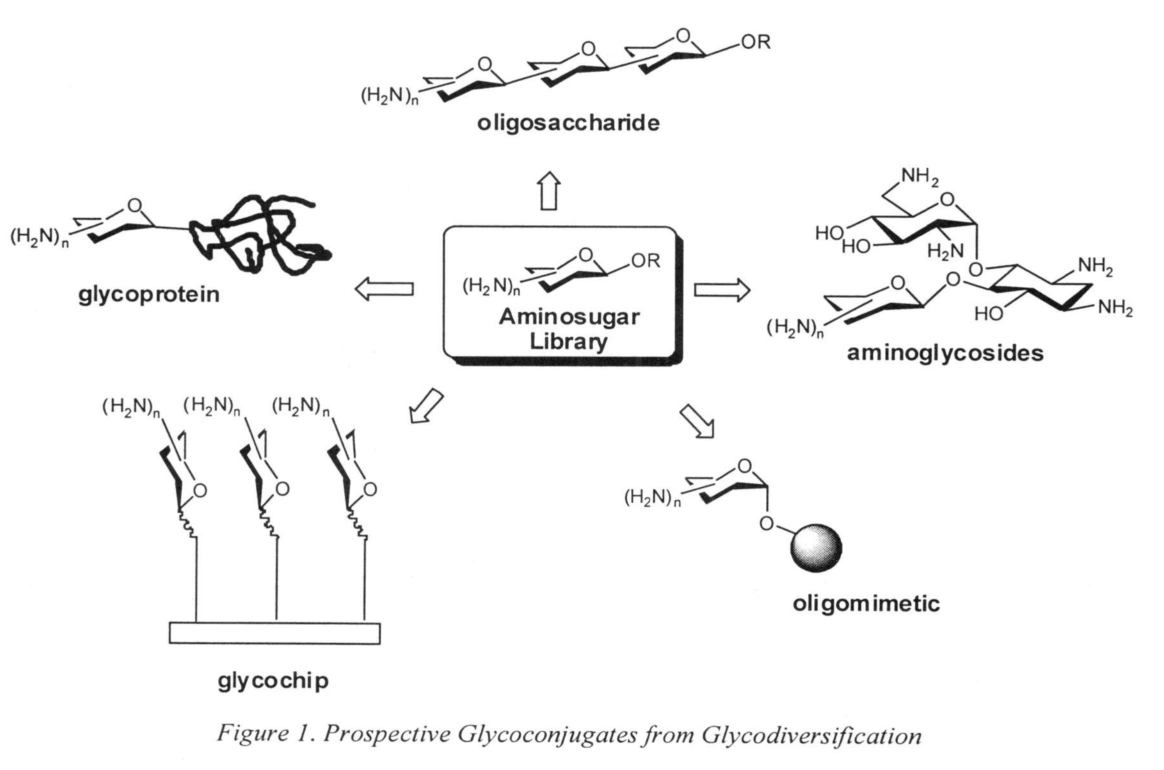

Figure 1. Prospective Glycoconjugates from Glycodiversification

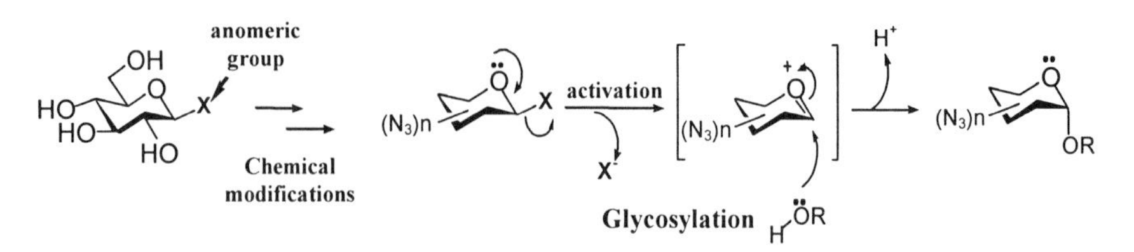

x group needs to be **stable** enough to endure the conditions for chemical modifications while being **labile** enough to be activated for glycosylation.

Figure 2. Challenges in Making the Aminosugar Donors

need to contribute a significant amount of effort in searching and optimizing the best approach among the vast number of documented prior reports. There may not be an ideal answer for this issue. Nevertheless, we wish to provide somesolution to alleviate the synthetic burden through our efforts in this area. In this article, we will focus on the identification of general protocols that can be used for the systematic construction of two libraries of aminopyranoses, and provide a brief coverage of their use in the preparation of novel aminoglycosides, pyranmycins and kanamycin analogues.

2. Synthesis of Aminosugars

2.1. Choice of Starting Sugars

Based on a survey of current reviews, glycosyl donors can be generally classified into thirteen types based on the anomeric functional groups and their activating methods (Table I) (12-16). The predominantly used methods among them include glycosyl halide (17), thioglycoside (18-19), trichloroimidate (15-20), and 1-*O*-acyl sugar (21).

However, when considering the systematic synthesis of aminosugars, and their glycosylation, not all these donors are suitable. As shown in Table I, most of the commonly employed glycosyl donors, such as glycosyl halides, glycosyl acetates, and glycosyl trichloroacetamidates, are not stable enough for the procedures of aminosugar synthesis, especially the incorporation of amino group and deoxygenation. 4-Pentenyl glycoside and glycal could be ideal options. Nevertheless, it is not very cost effective to use these molecules for the preparation of corresponding materials in large quantity.

In addition, the "armed" and "disarmed" effects of protecting groups and azido groups on the reactivity of pyranose further limit the options for suitable donors (22). The reactivity of glycosyl donors can be enhanced with an electron-

Table I. Commonly Used Anomeric Functional Groups Employed in Glycosyl Donors

Name	Structures	Stability in epimerization conditions	Stability in hydride or radical-mediated deoxygenation	Stability in azido group substitution
Glycosyl Halide	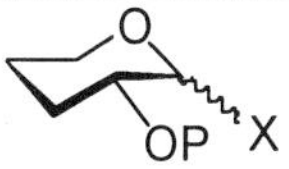 OP X X = F, Cl or Br	Not stable	Not stable	Not stable
Thioglycoside	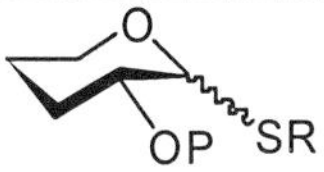 OP SR	Stable	Not stable in radical-mediated deoxygenation	Stable
1-*O*-Acyl Sugar	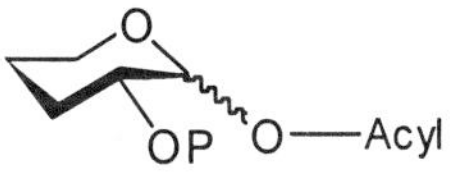 OP O—Acyl	Could be stable	Not stable in hydride-mediated deoxygenation	Could be stable
Ortho Ester	 O O R R'	Stable	Stable	Stable
1-*O*- and *S*-Carbonate	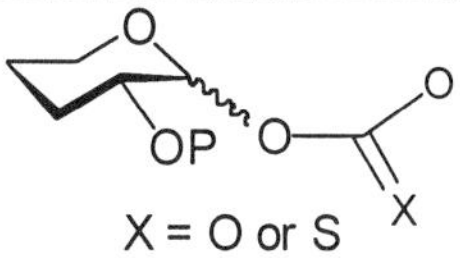 OP O X X = O or S	Not stable	Not stable	Not stable
Trichloro-acetimidate	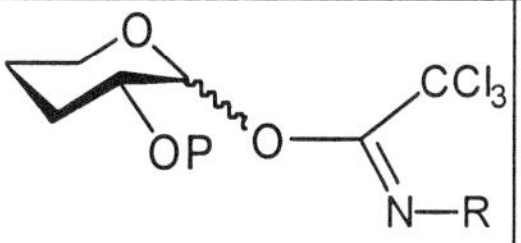 OP O CCl_3 N—R	Not stable	Not stable	Not stable
4-Pentenyl Glycoside	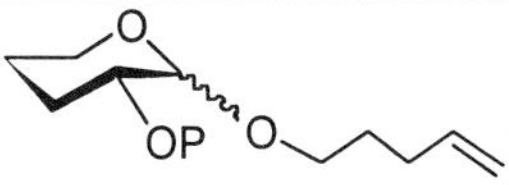 OP O	Stable	Not stable in radical-mediated deoxygenation	Stable

Continued on next page.

Table I. *Continued.*

Name	Structures	Stability in epimerization conditions	Stability in hydride or radical-mediated deoxygenation	Stability in azido group substitution
Phosphate Derivatives	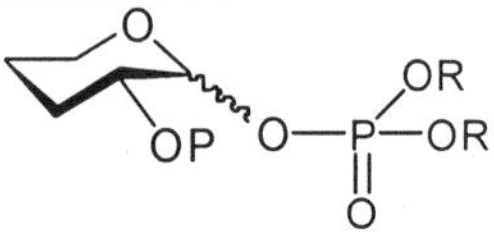 OP, O—P(=O)(OR)—OR	Not stable	Not stable	Not stable
Sulfoxide	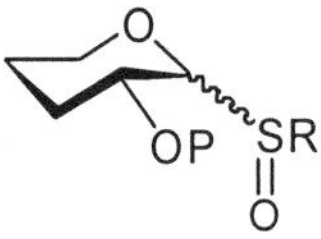 OP, S(=O)R	Not stable in the presence of Tf_2O	Not Stable	Could be stable
1-*O*-Silylated Glycoside	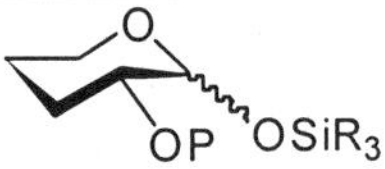 OP, $OSiR_3$	Could be stable	Could be stable	Could be stable
1,2-Anhydro Sugar	 O, O	Not stable	Not stable	Not stable
1-Hydroxyl Sugar	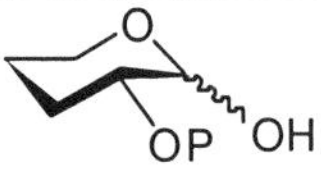 OP, OH	Not stable	Not stable	Not stable
Glycal	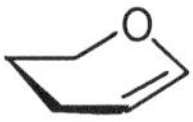 O	Stable	Not stable	Stable

donating protecting group, such as Bn, leading to the term: armed glycosyl donor. On the other hand, having an electron-withdrawing protecting group, such as Ac, Bz, or azido group, will decrease the reactivity of the glycosyl donor, which is termed as a disarmed donor.

One solution toward this problem is to have a rigid anomeric group that will enable the incorporation of an amino group and other modifications. After which, the anomeric group can be transformed into those functional groups that can serve as glycosyl donors. Following this strategy, methyl glucopyranosides that have a relatively stable anomeric methoxy group, and are available at lower cost, are commonly employed for synthesis of aminosugars. Combining with the concept of divergent synthesis, our group has examined a panel of reagents and completed the synthesis of 4- and/or 6-aminopyranoses. A general protocol with modest to excellent yields for converting diverse modified methyl glycosides into acetyl glycosides was also developed. Acetyl glycopyranosides can be transformed into two different glycosyl donors, glycosyl trichloroacetimidate and phenylthioglycosides, for "disarmed" and "armed" pyranoses following the reported procedures (Scheme 1).[23]

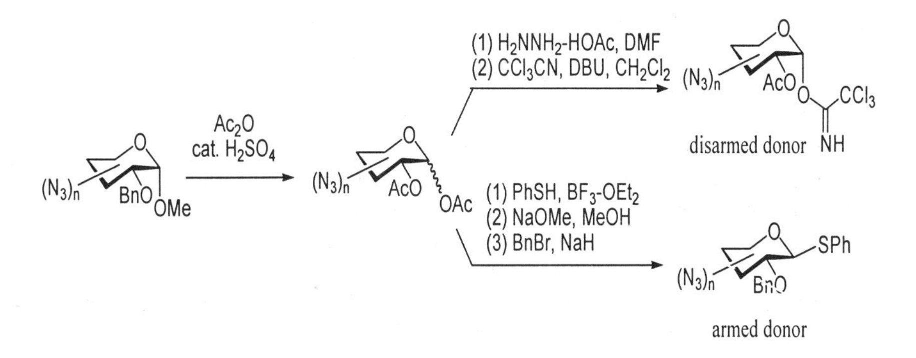

Scheme 1. General Procedure for the Synthesis of Glycosyl Donors

1,2:5,6-di-*O*-isopropylidine-D-glucofuranose is another important starting material since it gives immediate access to the modifications at the 3-OH group. Also, because it can be selectively hydrolyzed to 1,2-*O*-isopropylidine derivatives, further manipulation leading to the synthesis of 3,6-diaminopyranose can be readily achieved (Scheme 2) (24). Glucosamine is commonly used for the synthesis of 2-aminopyranoses since it has an amino group at the C-2 position and is of lower cost than galactosamine (25).

Arylthio or alkylthio groups are resistant toward many organic operations used in aminosugar synthesis, and can be activated for glycosylation directly. Therefore, arylthio or alkylthioglycosides are often used for the synthesis of

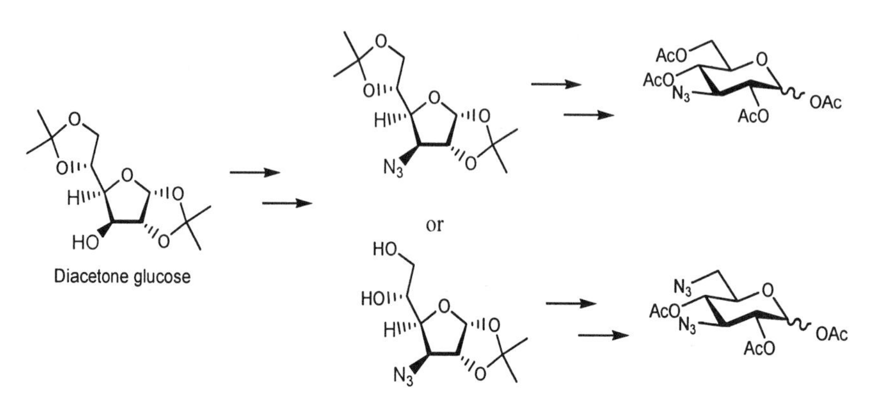

Scheme 2. Examples of Modification of 1,2:5,6-Di-O-isopropylidine-D-glucofuranose

aminosugars and corresponding derivatives (19). Although ethylthioglycopyranoside, tolylthioglycopyranoside or phenylthioglycopyranoside are expensive to purchase, these compounds can be prepared in large quantities from treating glycose pentaacetate with the corresponding ethanethiol, thiophenol, or *p*-thiocresol under the catalysis of Lewis acids (26). Our group favors the use of phenylthioglycopyranoside for two reasons. First, unlike ethylthioglycopyranoside, phenylthioglycopyranoside is easy to crystallize and thereby avoids the formidable task of column chromatography. Second, to ensure the completed conversion of glycosyl acetate to the corresponding thioglycoside, excess thiol is often required (27). The excess thiophenol can be readily removed by co-evaporating with other organic solvents, and subsequently bleached for proper disposal (28). However, *p*-thiocresol is a solid at room temperature, making its removal more challenging.

The overall strategy of constructing aminosugar libraries is summarized in Scheme 3. We prefer the use of disarmed trichloroacetimidate glycosyl donor for the formation of the β-glycosidic bond, and armed phenylthio glycosyl donor for the formation of the α-glycosidic bond. In general, trichloroacetimidate donors are more reactive than phenylthio glycosyl donors. Thus, the electron-withdrawing effect in the disarmed donors can be better countered by the higher reactivity of trichloroacetimidate glycosyl donor, while the electron-donating effect in the phenylthio glycosyl donors is advantageous for the less reactive disarmed donors.

Azido group incorporated glycosyl acetates that can be derived from methyl glucoside are better for the synthesis of disarmed donors bearing an acetyl protecting group at C-2 position, which is designed for the formation of β-linked glycosides. If necessary, these glycosyl acetates can also be converted into

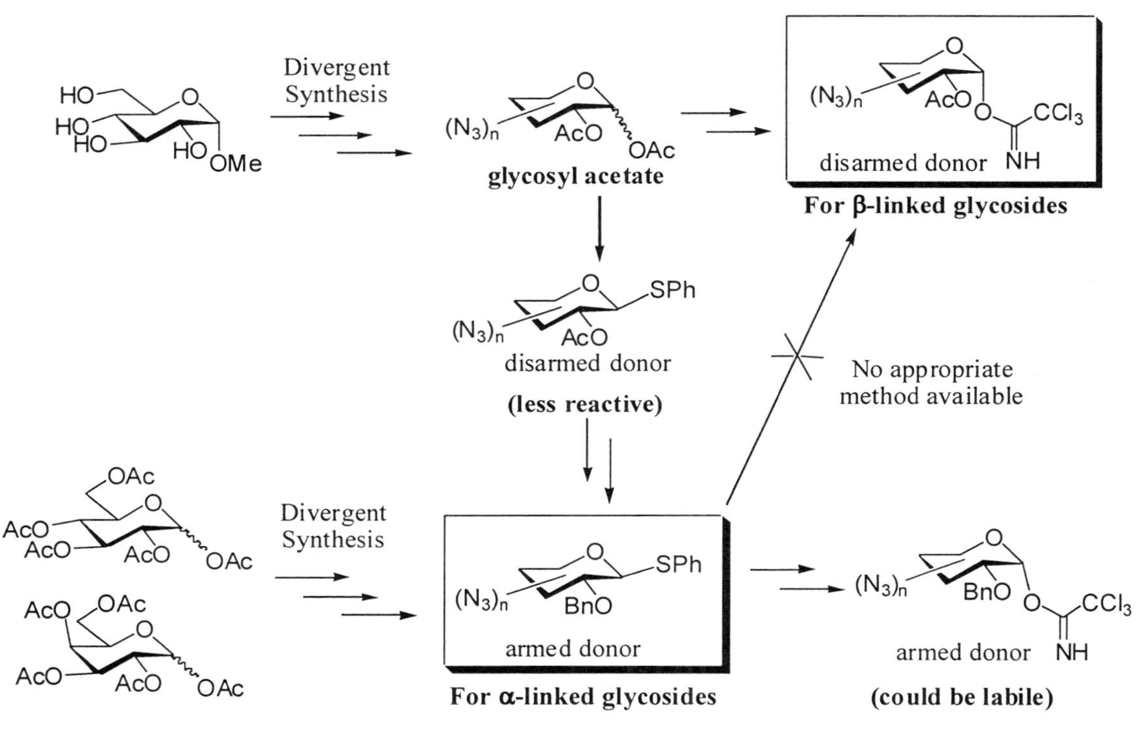

Scheme 3. Overall Concept of Aminosugar Construction

phenylthioglycosides. Although the disarmed donors like phenylthioglycosides can be less reactive and give rise to unsatisfactory yield in glycosylation, they can be converted to more reactive armed donors with Bn protecting groups favoring the formation of α-linked glycosides.

Based on our experience, however, it would be more convenient and cost effective to synthesize azido groups incorporated phenylthioglycosides from glucose pentaacetate or galacto pentaacetate. The disadvantage is that these glycosides can only lead to the synthesis of armed donors. To our knowledge and through our unsuccessful attempts, there is no simple method to convert armed donors into disarmed donors once the Bn groups have been incorporated. Incorporation of Bn groups prior to azido group incorporation is, however, necessary since acetyl groups are not stable enough under the chemical conditions employed. The reactivity of armed phenylthio glycosyl donors can be further enhanced by converting the phenylthio group into a trichloroacetimidate group. Nevertheless, such highly reactive armed donors could be too reactive to be properly purified in some of the aminosugar constructs.

2.2. General Synthetic Protocols

To achieve divergent synthesis of aminosugar, four synthetic transformations for the development of general protocols should be considered in advance: regioselective protection and deprotection of hydroxyl groups (for example: cleavage of *O*-benzylidine acetale), epimerization of hydroxyl group, azido (amino) group incorporation, and regioselective deoxygenation. As long as the general protocols for the above-mentioned reactions can be established, a systematic approach for the synthesis designed aminosugars can be envisioned.

2.2.1. Synthesis and Regioselective Cleavage of *O*-benzylidine Acetales

There are many protecting groups for pyranoses reported in the literature (29, 30). We are especially interested in the benzylidene-type of protecting group because the 4,6-benzylidene pyranoses can lead to several regioselectively deprotected pyranoses, which is valuable in our divergent synthetic approach (Scheme 4). For example, by using appropriate solvent and reducing agent, the 4,6-benzylidene-protected pyranose can lead to the synthesis of derivatives with either 4-OH or 6-OH (31). Alternatively, hydrolysis of benzylidene results in the expose of 4,6-dihydroxyl groups (32). In our opinion, the selective protection methods have been well developed for many pyranoses, such as glucose, galactose, and mannose, and hence will not be discussed in detail in this article.

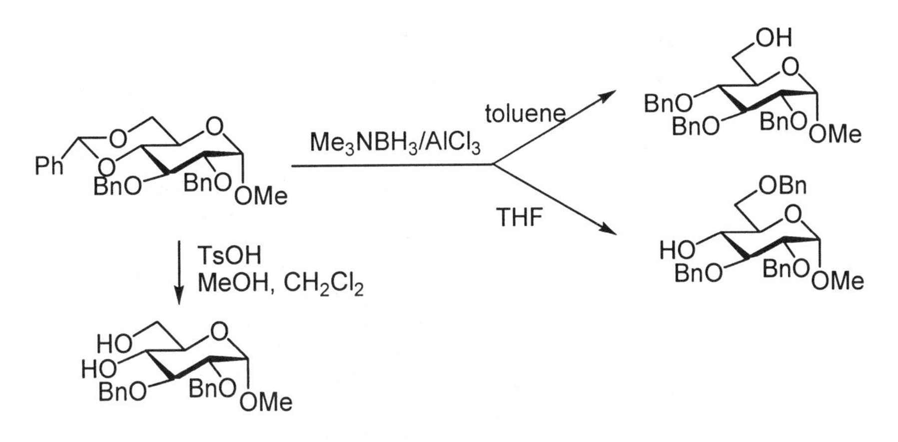

Scheme 4. Sample of Regioselective Opening of 4,6-O-Benzelidene Acetals

2.2.2. Epimerization of Hydroxyl Group

Since the substitution of a hydroxyl group with a nucleophile is often carried out via S_N2 substitution, it is often necessary to stereoselectively epimerize a hydroxyl group on a pyranose scaffold that will allow the installation of the nucleophile with the desired stereochemistry. Four out of five hydroxyl groups in hexopyranose are chiral, therefore stereoselectively epimerizing a hydroxyl group on the glucopyranose and galactopyranose scaffold is essential to expand the possibilities for more sugar manipulation.

There are many well-established methods in the literature, which can be grouped into two types including oxidation-reduction and nucleophilic substitution processes. The former involves an oxidation of a hydroxyl group to a keto group followed by stereoselective hydride reduction. In this method, the vicinal protecting groups may influence the stereoselectivity (33).

It is also worthy to point out that oxidation of a hydroxyl group using Swern oxidation and reduction of ketone using $NaBH_4$ are compatible to the presence of phenylthiol and azido groups, respectively.

The second method involves converting the secondary hydroxyl group into a leaving group using Tf_2O followed by S_N2 substitution using, for example, OAc^- or NO_2^-, as the nucleophiles. While Tf_2O may activate glycosyl sulfoxide, it is compatible to the presence of phenylthio group. Thus, both oxidation-reduction and nucleophilic substitution reactions can be applied in the synthesis of a glycosyl trichloroacetimidate library and phenyl thioglucopyranoside library. Three commonly used protocols in our laboratory are summarized in Table II.

Table II. Common Protocols for Epimerization of Hydroxyl Group

Types of Transformations	Examples of Reagents	Comments	References
Oxidation/reduction (Swern oxidation)	(1) $(COCl)_2$, DMSO DIPEA (2) $NaBH_4$	Vicinal protecting groups (*i.e.* Bn) are essential for the selectivity. However, the selectivity may vary among different sugars. Others (*i.e.* Bz) may offer lower or no selectivity toward epimerization.	33
S_N2 substitution	(1) Tf_2O (2) $(n\text{-Bu})_4N^+NO_2^-$	$(n\text{-Bu})_4N^+NO_2^-$ can be soluble in CH_2Cl_2, providing better results than reagents like $NaNO_2$. In general, this method offers stereospecific epimerization.	34
S_N2 substitution	(1) Tf_2O (2) $(n\text{-Bu})_4N^+OAc^-$ (3) hydrolysis of Oac	$(n\text{-Bu})_4N^+OAc^-$ can be soluble in CH_2Cl_2, providing better result than reagent like KOAc or CsOAc. In general, this method offers stereospecific epimerization.	35

2.2.3. Azido Group Incorporation

The amino group needs to be protected in order to make the synthesis of the glycosyl donor and glycosylation feasible. We favor use of the azido group as an amino group surrogate for the synthesis of aminosugars (or azidosugars). There are several advantages that make the azido group a popular choice. First, the azido group is relatively stable to many reductive and oxidative conditions. Second, unlike the carbamate type protecting group for amines, azido compounds have good solubility in organic media, allowing an expedient chromatographic purification. Third, azido groups can be converted into amino groups conveniently by hydrogenation or the Staudinger reaction. More importantly, an azido group can be easily installed from an activated hydroxyl group via S_N2 substitution and, unlike amino group, no sequential protection is needed. Substitution of a hydroxyl group using the amino or carbamate-protected amino groups as the nucleophiles can be relatively difficult since the protected amino group is not a good nucleophile, while a free amino group can act as a base and cause undesired elimination. Finally, the azido group can be applied to the "click" chemistry creating more diversification opportunities (36).

Despite the advantage of easy manipulation, one-pot method for the incorporation of azido group is not recommended since it often generates a mixture that complicates the purification of the desired product. In the two-step method, selective tosylation of a primary hydroxyl group in the presence of secondary hydroxyl groups is one of the advantages of employing TsCl. The commonly used protocols are summarized in Table III. The azido group can be converted to amino group by methods, such as hydrogenation, Staudinger reaction, hydride-reducing agents (37).

Table III. Common Protocols for Azide Substitution

Types of Transformations	Reagents	Typical Conditions	Types of Hydroxyl Groups	Notes	References
One-pot	DPPA (or HN_3), PPh_3 and DEAD	-40° to 0°C, overnight	1° and 2°	Complex mixture may be obtained, difficult to purify	38
Two-step	(1) TsCl (2) NaN_3	(1) 0°C to R.T., overnight (2) 80°C, overnight	1° (2° tosylate is difficult to be replaced with N_3^-)	Can be used for selective azide substitution	39
Two-step	(1) MsCl (2) NaN_3	(1) 0°C to R.T., couple hours (2) 120°C, overnight	1° and 2°	MsCl is cheaper than Tf_2O	40
Two-step	(1) Tf_2O (2) NaN_3	(1) 0°C, 0.5 hour (2) R.T., overnight	1° and 2°	Most expedient	39

2.2.4. Regioselective Deoxygenation

Many aminosugars contain the features of deoxygenation. Table IV summarizes the common protocols for regioselective deoxygenation in the literature. Since azido groups can be reduced under conditions of hydride reduction (e.g. $LiAlH_4$) or radical-initiated deoxygenation (e.g. Barton reduction or dehalogenation) (41, 44), deoxygenation generally proceeds before the introduction of the azido group.

Table IV. Common Protocols for Regioselective Deoxygenation

Types of Transformations	Reactions	References
6-deoxy	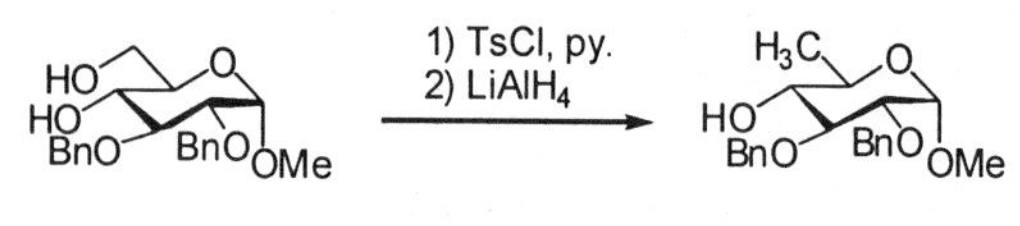 HO, HO, BnO, BnO, O, OMe → 1) TsCl, py. 2) $LiAlH_4$ → H_3C, HO, BnO, BnO, O, OMe	39
4-deoxy	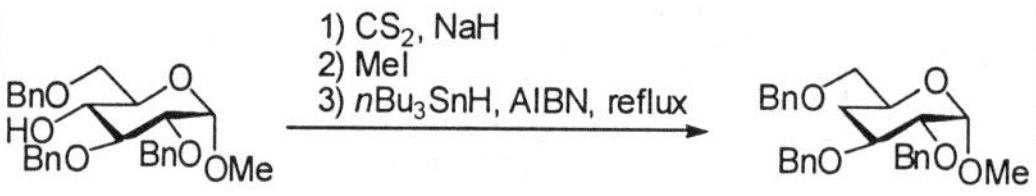 BnO, HO, BnO, BnO, O, OMe → 1) CS_2, NaH 2) MeI 3) $n Bu_3SnH$, AIBN, reflux → BnO, BnO, BnO, O, OMe	41
3-deoxy	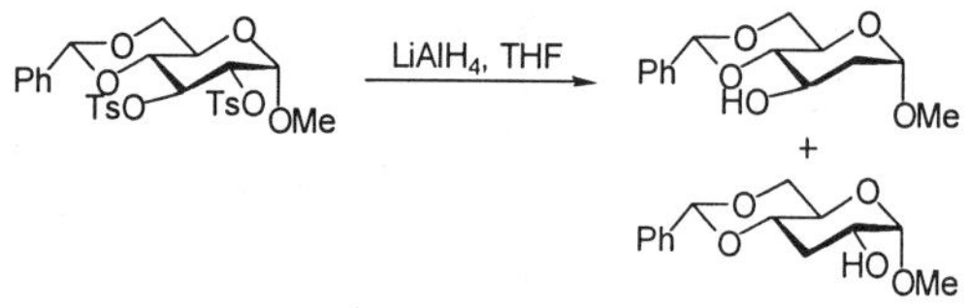 Ph, O, O, O, TsO, TsO, OMe → $LiAlH_4$, THF → Ph, O, O, O, HO, OMe + Ph, O, O, O, HO, OMe	42
3-deoxy	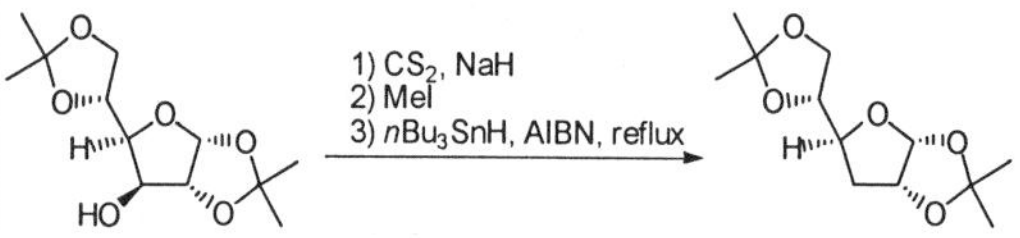 O, O, H, O, O, O, HO → 1) CS_2, NaH 2) MeI 3) $n Bu_3SnH$, AIBN, reflux → O, O, H, O, O, O	41
2-deoxy	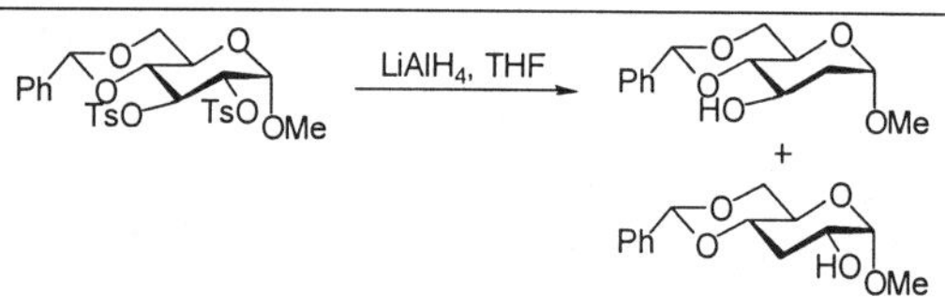 Ph, O, O, O, TsO, TsO, OMe → $LiAlH_4$, THF → Ph, O, O, O, HO, OMe + Ph, O, O, O, HO, OMe	42
2-deoxy	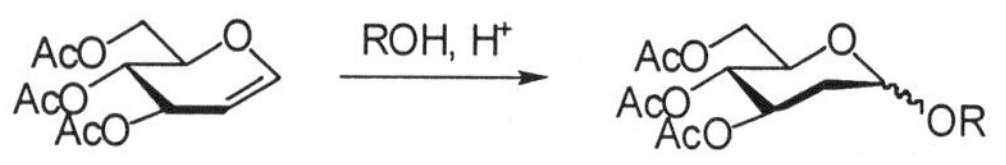 AcO, AcO, AcO, O → ROH, H^+ → AcO, AcO, AcO, O, OR	43
4,6-dideoxy	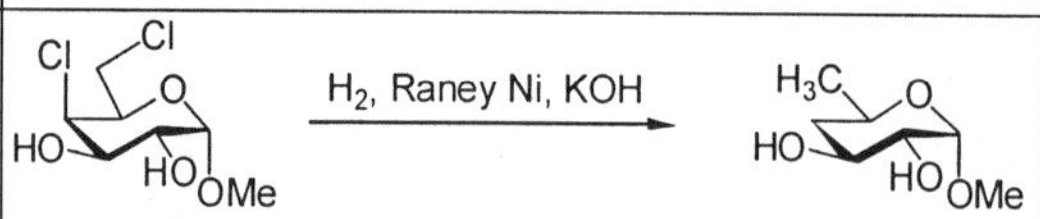 Cl, Cl, HO, HO, O, OMe → H_2, Raney Ni, KOH → H_3C, HO, HO, O, OMe	44

2.3. Divergent Synthesis of Glycosyl Trichloroacetimidate Library

The synthesis of our glycosyl trichloroacetimidate aminosugar library began with the commercially available methyl glucopyranoside (scheme 5). Benzylidine protection, followed by benzylation of C-2 and C-3 hydroxyl groups (31,32), yielded compound **1**. Compound **1** was treated with $NaBH_3CN/HCl$ in THF to selectively deprotect the C-4 hydroxyl group providing compound **2** (31,32). Triflation followed by azide substitution of compound **2** generated azidosugar **3** in the *galacto*-configuration. Epimerization of the 4-OH of **2** was achieved with Swern oxidation and $NaBH_4$ reduction offering compound **4**, which allowed a S_N2 azide substitution that furnished azidosugar, **5** in the desired *gluco*-configuration.

In another route, compound **1** was treated with TsOH to give compound **6** with free hydroxyl groups at the C-4 and C-6 positions, which branched into three distinct routes. In the first route, triflation followed by azide substitution of compound **6** provided azidosugar **7** with C-4 and C-6 diazido substituted. In the second route, compound **6** was selectively deoxygenated at the C-6 position by sequential tosylation and $LiAlH_4$ reduction. Compound **8** was subjected to azide substitution generating azidosugar, **9** with C-6 deoxygenation in the *galacto*-configuration. Alternatively, compound **8** underwent the Swern oxidation/$NaBH_4$ reduction protocol to invert the C-4 hydroxyl group that allowed the synthesis of azidosugar, **11** with C-6 deoxygenation in the *gluco*-configuration. In the last path, compound **6** was treated with TsCl, followed by azide substitution to selectively place an azido group on the C-6 position. The free C-4 hydroxyl group in the equatorial position of compound **12** was converted to the axial position yielding compound **13**, which enabled the synthesis of azidosugar **14** with C-4 and C-6 diazido in the *gluco*-configuration.

As described in the previous section, the anomeric methoxy group and all the benzyl groups of these azidosugars can be converted into acetyl groups using Ac_2O with a catalytic amount of H_2SO_4 (Scheme 1). The resulting acetyl glycosides can then be transformed into the glycosyl trichloroacetimidate as the glycosyl donor (23), which could then be coupled to the acceptor of choice.

2.4. Divergent Synthesis of Phenylthioglucopyranoside Library

By employing the philosophy of divergent synthesis, and phenylthioglucopyranoside as the starting material, the other library, thioglycopyranoside-based aminosugar library can be constructed in a similar fashion (Scheme 6). We modify the literature procedure, which allows the purification of phenylthioglucopyranoside via recrystallization, and hence enables its large scale synthesis (26-28).

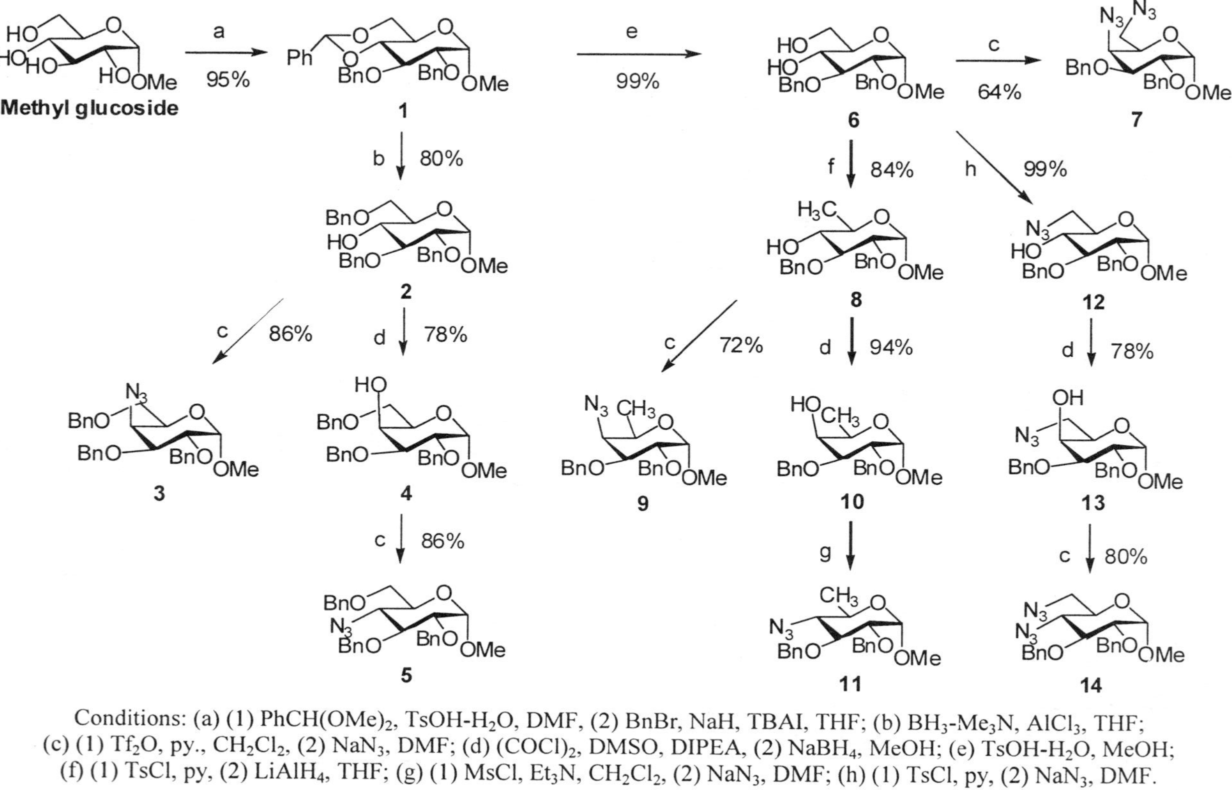

Conditions: (a) (1) $PhCH(OMe)_2$, $TsOH$-H_2O, DMF, (2) BnBr, NaH, TBAI, THF; (b) BH_3-Me_3N, $AlCl_3$, THF; (c) (1) Tf_2O, py., CH_2Cl_2, (2) NaN_3, DMF; (d) $(COCl)_2$, DMSO, DIPEA, (2) $NaBH_4$, MeOH; (e) $TsOH$-H_2O, MeOH; (f) (1) TsCl, py, (2) $LiAlH_4$, THF; (g) (1) MsCl, Et_3N, CH_2Cl_2, (2) NaN_3, DMF; (h) (1) TsCl, py, (2) NaN_3, DMF.

Scheme 5. Divergent Synthesis of Glycosyl Trichloroacetimidate Library

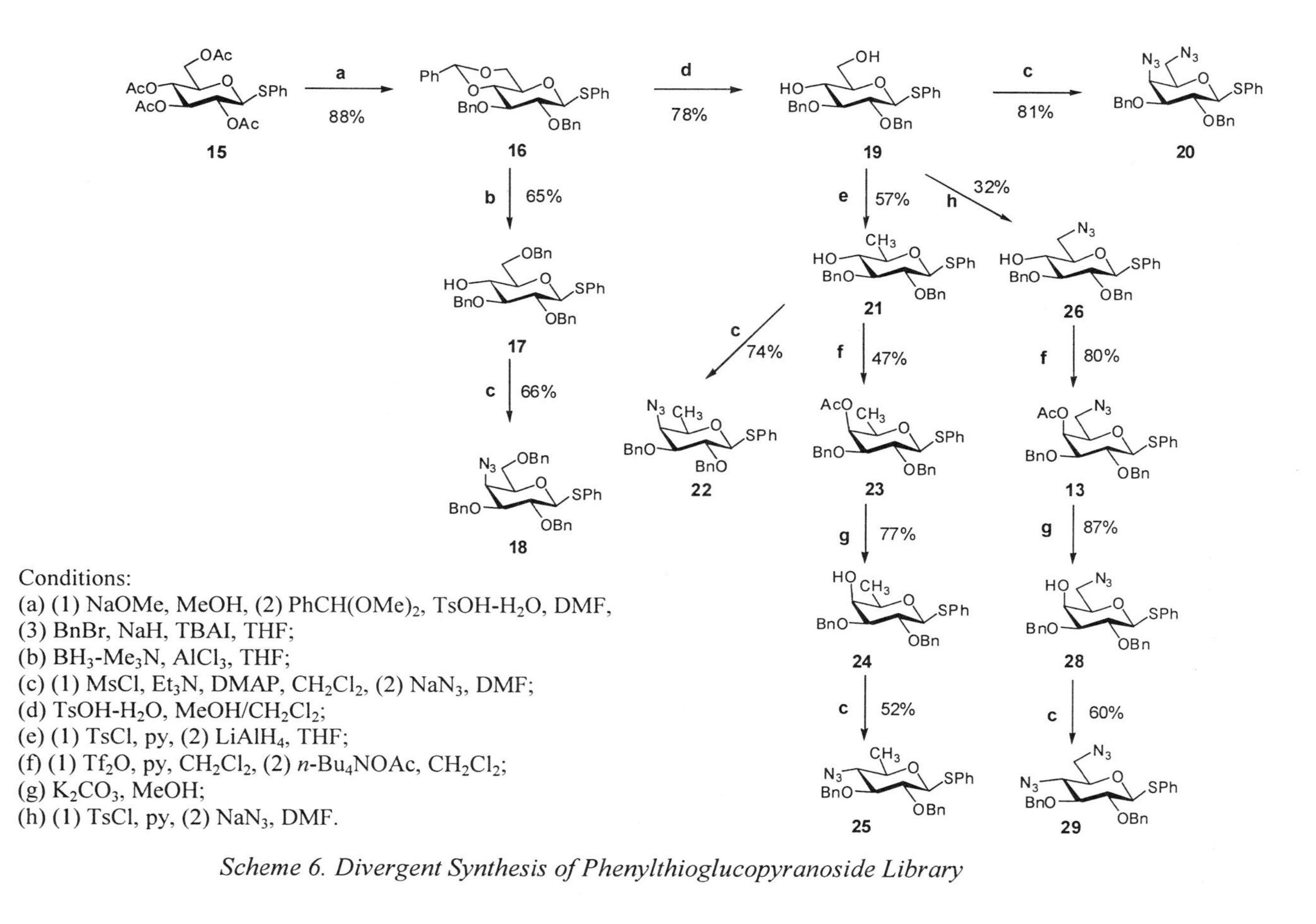

Conditions:
(a) (1) NaOMe, MeOH, (2) $PhCH(OMe)_2$, $TsOH-H_2O$, DMF,
(3) BnBr, NaH, TBAI, THF;
(b) BH_3-Me_3N, $AlCl_3$, THF;
(c) (1) MsCl, Et_3N, DMAP, CH_2Cl_2, (2) NaN_3, DMF;
(d) $TsOH-H_2O$, $MeOH/CH_2Cl_2$;
(e) (1) TsCl, py, (2) $LiAlH_4$, THF;
(f) (1) Tf_2O, py, CH_2Cl_2, (2) n-Bu_4NOAc, CH_2Cl_2;
(g) K_2CO_3, MeOH;
(h) (1) TsCl, py, (2) NaN_3, DMF.

Scheme 6. Divergent Synthesis of Phenylthioglucopyranoside Library

To avoid the cumbersome epimerization of C-4 free hydroxyl, an alternative route is to use the 2,3-di-*O*-benzyl-4,6-*O*-benzylidene-1-phenylthio-β-D-galactopyranoside instead of glucopyranoside as starting material (Scheme 7).

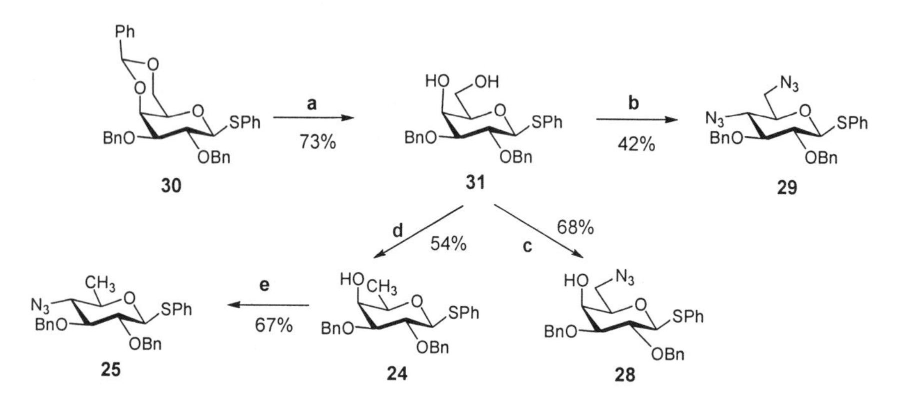

Conditions: (a) TsOH-H_2O, MeOH/CH_2Cl_2; (b) (1) Tf_2O, py, CH_2Cl_2, (2) NaN_3, DMF; (c) (1) TsCl, py, (2) NaN_3, DMF; (d) (1) TsCl, py, (2) $LiAlH_4$, THF; (e) (1) MsCl, Et_3N, DMAP, CH_2Cl_2, (2) NaN_3, DMF.

Scheme 7. Alternative Synthesis Route

As mentioned previously, the C-3 azido group-incorporated sugars can be synthesized from 1,2:5,6-di-*O*-isopropylidine-D-glucofuranose. Further manipulations can be used to synthesize 3,6- or 3,4-diazidopyranoses (Scheme 8).

3. Stereoselective Glycosylation

3.1. Background in Glycosylation

The development of methodologies for efficient glycosylation reaction, especially *O*-glycosylation reaction, has been a major issue for practical synthesis of oligosaccharides and glycoconjugates. As a result, intense research activities have been devoted to the study of stereoselective glycosylation. However, stereochemical problems in glycosidic bond formation have not been solved completely (12, 14, 45-47).

Based on the general structure of pyranoses and the chirality of the anomeric center, four possible products can be formed (Figure 3) (48). The glycosylation, which involves α and β *manno*-type pyranose has been studied extensively by Crich (49,50). Therefore, we will only discuss the α- and β- linkage of *gluco*-type.

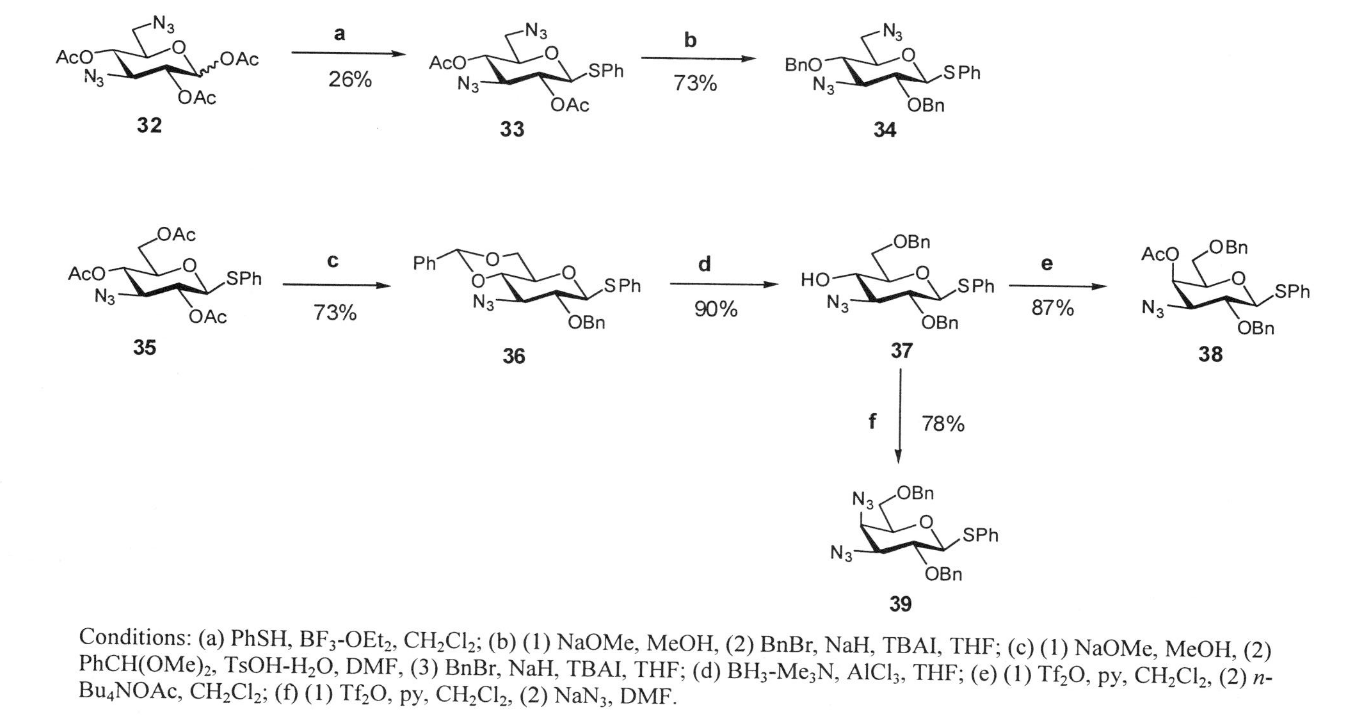

Conditions: (a) PhSH, BF_3-OEt_2, CH_2Cl_2; (b) (1) NaOMe, MeOH, (2) BnBr, NaH, TBAI, THF; (c) (1) NaOMe, MeOH, (2) $PhCH(OMe)_2$, TsOH-H_2O, DMF, (3) BnBr, NaH, TBAI, THF; (d) BH_3-Me_3N, $AlCl_3$, THF; (e) (1) Tf_2O, py, CH_2Cl_2, (2) *n*-Bu_4NOAc, CH_2Cl_2; (f) (1) Tf_2O, py, CH_2Cl_2, (2) NaN_3, DMF.

Scheme 8. Synthesis of C3-azido Phenylthioglucopyranoside Compounds

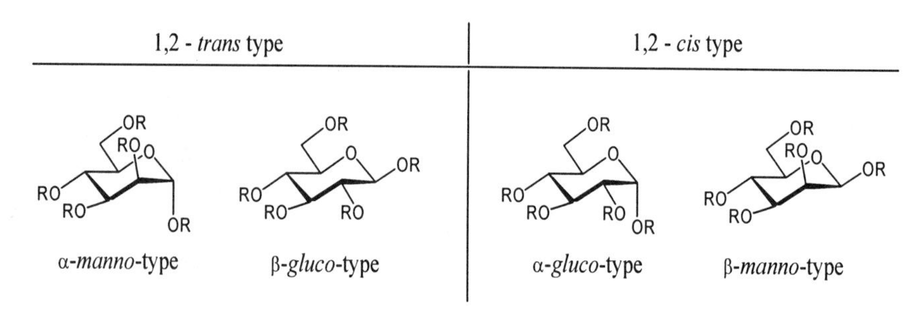

Figure 3. Four Possible Products of Glycosidation

3.1.1. Anomeric effect

In substituted cyclohexanes, the substituents usually prefer the equatorial position due to the steric effects (Figure 4). In rings containing oxygen atoms and with adjacent electronegative substituents there is a preference for this substituent to be *axial.* This is referred as the anomeric effect (51). The anomeric effect favors the formation of kinetic product, the α-anomer, for most glycosides.

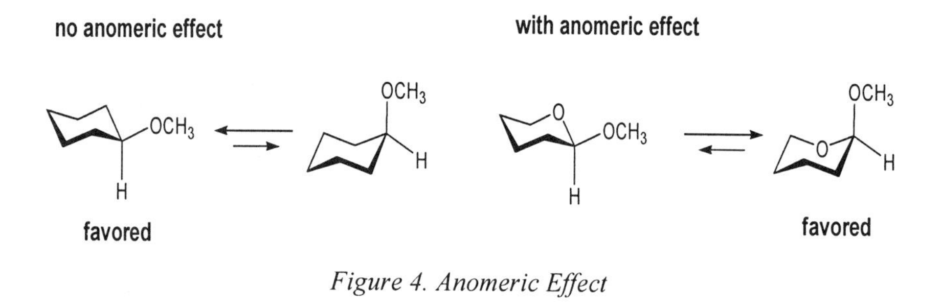

Figure 4. Anomeric Effect

3.1.2. Protecting Group on *O*-2 of the Glycosyl Donor

The stereospecific formation of a β-glycosidic bond can be achieved by the presence of an acyl protecting group at the *O*-2 position via neighboring group participation (Figure 5). The formation of an α-glycosidic bond is, however, more challenging, despite great advances. The stereocontrolled synthesis of α-glycosides can be affected by factors such as electronic effects, steric hindrance, solvent, and conformation. To our knowledge, there is no satisfactory general protocol for stereospecific glycosylation for formation of an α-glycosidic bond despite numerous efforts.

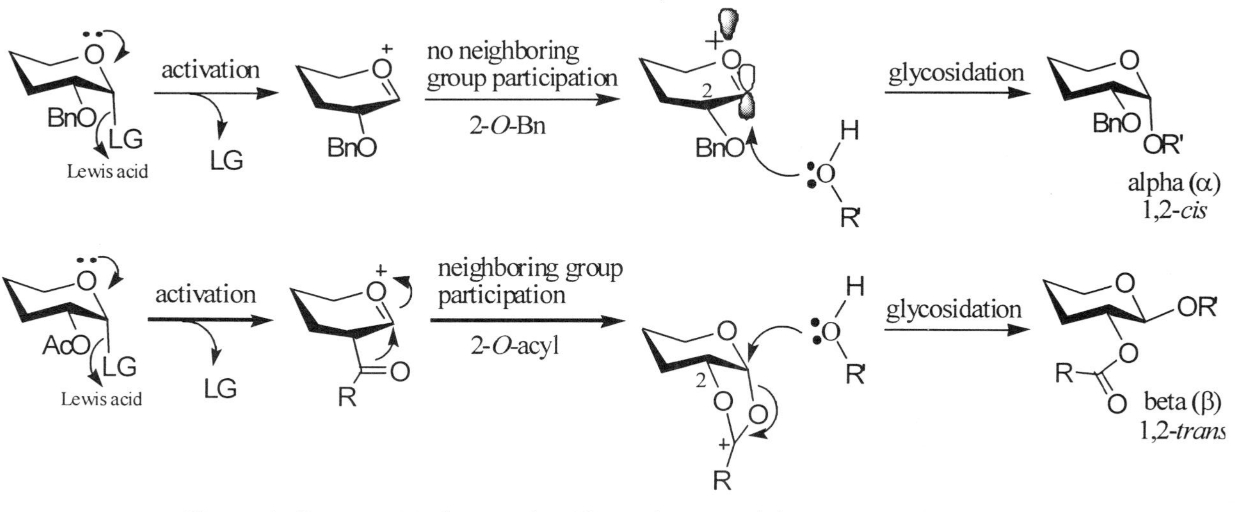

Figure 5. Common Mechanism for Glycosidation and the Associated Selectivity

3.2. Formation of β-Glycosidic Bond: Preparation of Pyranmycin Library

As mentioned previously, our group favors the use of glycosyl trichloroacetimidate library as the glycosyl donor for the formation of β-glycosidic bond leading to the development of the pyranmycin library, a library of neomycin B analogues (Figure 6).

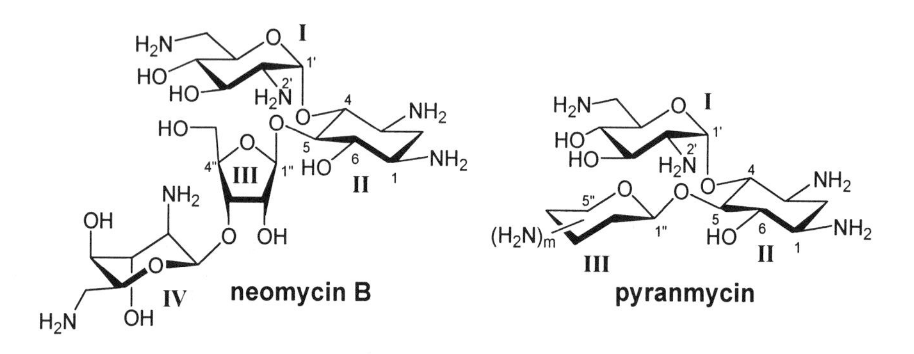

Figure 6. Structures of Neomycin B and Pyranmycin

Neomycin belongs to a group of aminoglycoside antibiotics containing a 4,5-disubstituted 2-deoxystreptamine core (ring II) and has been used against both gram-positive and gram-negative bacteria for more than fifty years (5, 52). The neomycin exerts its antibacterial activity by binding selectively to the A-site decoding region of the 16S ribosomal RNA of bacteria, and thereby disrupts the protein synthesis of these microorganisms.

The design of pyranmycin contains a ring III pyranose that is linked to the *O*-5 of neamine (rings I and II) via a β-glycosidic bond. It is known that neomycin is labile under acidic conditions due to the presence of a glycosidic bond from the ring III furanose. As a result, neomycin degrades readily into less active neamine (rings I and II) and inactive neobiosamine (rings III and IV). Since the corresponding glycosidic bond of pyranmycin is made from a pyranose, this gives pyranmycin better stability to acidic conditions (53). Therefore, replacement of neobiosamine with a pyranose will yield a novel aminoglycoside with improved stability in acidic media.

Reported study has also showed that an intramolecular hydrogen bonding between the 2'-amino group of ring I and the *O*-4" atom of ring III helps to orient ring I for specific binding toward RNA (54). Thus, the attachment of ring III via a β-glycosidic bond is crucial in offering the intramolecular hydrogen bonding similar to that in the neomycin. Therefore, our glycosyl trichloroacetimidate library is ideal for incorporation of the desired β-linked pyranoses. With the appropriate neamine aglycon, we have prepared a library of pyranmycin that show comparable activity against *Escherichia coli* (ATCC 25922) and *Staphylococcus aureus* (ATCC 25923) (Scheme 9) (23, 39, 55).

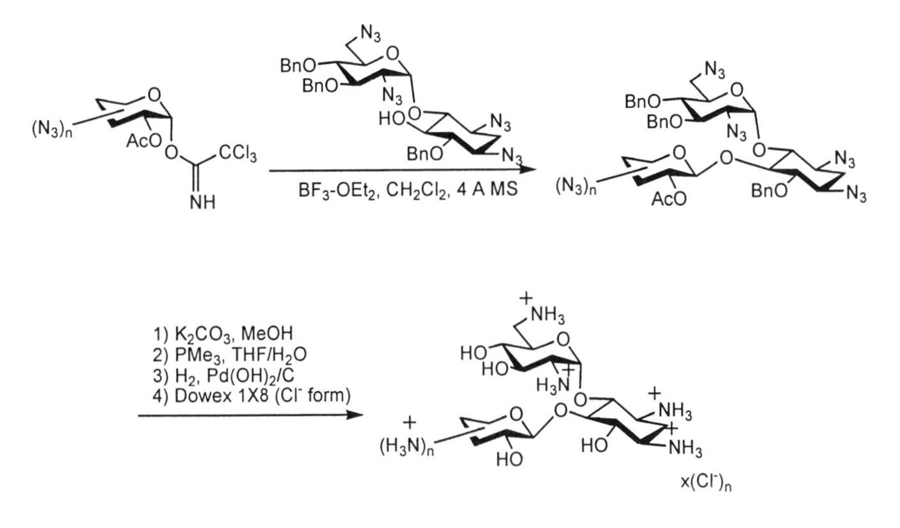

Scheme 9. Synthesis of Pyranmycin Library

3.3. Formation of α-Glycosidic Bond: Preparation of Kanamycin Library

Kanamycin belongs to a group of aminoglycoside antibiotics with 4,6-disubstituted 2-deoxystreptamine (Figure 7) (5, 52). Like neomycin, kanamycin also exerts prominent antibacterial activity against both gram positive and gram negative susceptible strains of bacteria. Nevertheless, kanamycin has become clinically obsolete due to the emergence of aminoglycoside resistant bacteria (56). In order to revive the activity of kanamycin against drug resistant bacteria, numerous attempts have been devoted to the chemical modification of kanamycin (52, 57). Except for a few publications (58), most works use various carbamates as protecting groups for kanamycin resulting in the production of kanamycin with polycarbamate groups. Two drawbacks were often encountered: the poor solubility of polycarbamates, which produce great difficulties in purification and characterization of these compounds, and the limited options for structural modifications imposed by the kanamycin scaffold.

The α-glycosidic bond between rings II and III is important as the kanamycin analogous with a β-glycosidic bond manifests much weaker antibacterial activity (59). Having a 2-*O*-Bn group, our phenylthioglycoside library is preferable for the formation of the needed α-glycosidic bond due to the anomeric effect. The neamine acceptor underwent regiospecific glycosylation at the *O*-6 position

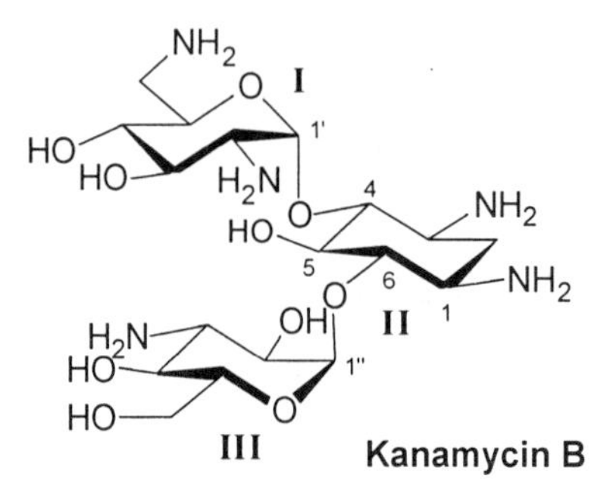

Figure 7. Structures of Kanamycin B

resulting in the desired 4,6-disubstituted 2-deoxystreptamine motif. The optimal stereoselectivity for the formation of the α-glycosidic bond can be accomplished by carrying out the reaction in a solution of Et_2O and CH_2Cl_2 in a 3 : 1 ratio (60). The desired α-glycosylated compounds were often mixed with their inseparable β-epimer. After hydrolysis of the acetyl groups, the triols can be obtained in good purity and improved α/β ratio. The final products (in pure α form or α/β from 10/1 to 7/1) were synthesized as chloride salts using the Staudinger reaction followed by hydrogenation and ion-exchange (Scheme 10) (61). A library of kanamycin B has been prepared via the concept of glycodiversification. These compounds also show comparable activity against *Escherichia coli* (ATCC 25922) and *Staphylococcus aureus* (ATCC 25923).

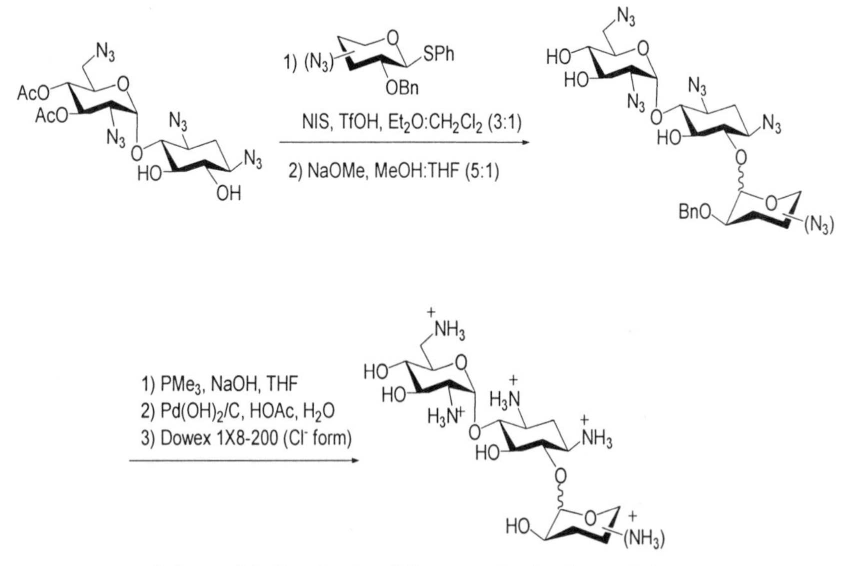

Scheme 10. Synthesis of Kanamycin Analogue Library

4. Conclusion

Carbohydrate synthesis is one of the most formidable tasks in organic synthesis. The synthesis of aminosugar libraries for practical applications represents an even greater challenge. Nevertheless, through the use of standardized protocols and a divergent synthetic approach, systematic procedures have been developed. The advantage of employing two separate aminosugar libraries for stereoselective glycosylation has also been demonstrated in the library construction of pyranmycin and kanamycin B analogues. It is our hope that part of the problem associated with the synthesis and application of aminosugars can be resolved. However, there are several aspects that still require applicable solutions. For example, there is no convenient synthesis of 2-aminopyranose that will favor the stereospecific formation of α-glycosidic bond. The deoxygenation method that is compatible with the presence of an azido group, stereospecific glycosylation for the formation of α-glycosidic bond, and a convenient protocol for the stereoselective glycosylation of 2-deoxyglycopyranoses are several such examples that still need further perfection.

5. References

(1) Dwek, R. A. *Chem. Rev.* **1996**, *96*, 683-720.

(2) Kirschning, A.; Bechthold, A. F.-W.; Rohr, J. *Top. Curr. Chem.* **1997**, *188*, 1-84.

(3) Johnson, D. A.; Liu, H.-w. *Curr. Opin. Chem. Biol.* **1998**, *2*, 642-649.

(4) Weymouth-Wilson, A. C. *Nat. Prod. Rep.* **1997**, 99-110.

(5) Hooper, I. R. *Aminoglycoside Antibiotics*; Springer-Verlag: New York, **1982**.

(6) Priebe, W.; Van, N. T.; Burke, T. G.; Perez-Soler, R. *Anticancer Drugs* **1993**, *4*, 37-48.

(7) Lothstein, L.; Sweatman, T, W.; Priebe, W. *Bioorg. Med. Chem. Lett.* **1995**, *5*, 1807-1812.

(8) Kren, V.; Martinkova, L. *Curr. Med. Chem.* **2001**, *8*, 1313-1338.

(9) Brimacombe, J. S. *Angew. Chem. Int. Ed.* **1971**, *10*, 236-248.

(10) Umezawa, S.; Tsuchiya, T. in *Aminoglycoside Antibiotics*, Springer-Verlag **1982**, New York. p37-110.

(11) Jurczak, J. in *Preparative Carbohydrate Chemistry*, Hanessian, S. Ed., Marcel Dekker, Inc. **1997**, NewYork, p595-614.

(12) Toshima, K.; and Tatsuta, K. *Chem. Rev.* **1993,** *93*, 1503-1531.

(13) Nicolaou, K. C.; Mitchell, H. J. *Angew. Chem. Int. Ed.* **2001**, *40*, 1576-1624.

(14) Paulsen, H. *Angew. Chem. Int. Ed.* **1982**, *21*, 155-173.

(15) Schmidt, R. R. *Angew. Chem. Int. Ed.* **1986**, *25*, 212-235.

(16) Demchenko, A. V. *Synlett.* **2003**, *9*, 1225-1240.

(17) For examples: (a) Lemieux, R. U.; Hendriks, K. S.; Stick, R. V.; James, K. *J. Am. Chem. Soc.* **1975**, *97*, 4056-4062. (b) Paulsen, H. *Angew. Chem. Int. Ed.* **1982**, *21*, 184-201.
(18) For examples: (a) Nicolaou, K. C.; Seitz, S. P.; Papahatjis, D. P. *J. Am. Chem. Soc.* **1983**, *105*, 2430-2434. (b) Konradsaon, P.; Udodong, U. E.; Fraeer-Reid, B. *Tetrahedron Lett.* **1990**, *31*, 4313-4316. (c) Vwneman, G. H.; van Leeuwen, S. H.; van Boom, J. H. *Tetrahedron Lett.* **1990**, *31*, 1331-1334. (d) Dasgupta, F.; Garegg, P. J. *Carbohydr. Res.* **1988**, *177*, c13-c17. (e) Fugedi, P.; Garegg, P. J. *Carbohydr. Res.* **1986**, *149*, c9-c12.
(19) Garegg, P.J. *Adv. Carbohydr. Chem. Biochem.* **1997**, *52*, 179-205.
(20) For examples: (a) Halcomb, R. L.; Boyer, S. H.; Danishefeky, S. J. *Angew. Chem. Int. Ed.* **1992**, *31*, 338-340. (b) Nicolaou, K. C.; Schreiner, E. P.; Iwabuchi, Y.; *Suzuki,* T. *Angew. Chem. Int. Ed.* **1992**, *31*, 340-342.
(21) For examples: (a) Jutten, P.; Scharf, H. D.; Raabe, G.; *J. Org. Chem.* **1991**, *56*, 7144-7149. (b) Gurjar, M. K.; Viswanadham, G. *Tetrahedron Lett.* **1991**, *32*, 6191-6194.
(22) (a) Zhang, Z.; Ollmann, I. R.; Ye, X.-S.; Wischnat, R.; Baasov, T.; Wong, C.-H.; *J. Am. Chem. Soc.* **1999**, *121*, 734-753. (b) Ye, X.-S.; Wong, C.-H.; *J. Org. Chem.* **2000**, *65*, 2410-2431.
(23) Chang, C.-W. T.; Hui, Y.; Elchert, B.; Wang, J.; Li, J.; Rai, R. *Org. Lett.* **2002**, *4*, 4603-4606.
(24) Schmidt, O.T. *Methods Carbohydr. Chem.* **1963,** *2*, 318.
(25) For example: Alper, P. B.; Hung, S.-C.; Wong, C.-H. *Tetrahedron Lett.* **1996**, *37*, 6029-6032.
(26) For examples: (a) Tai, C.-A.; Kulkarni, S. S.; Hung, S.-C.; *J. Org. Chem.* **2003**, *68*, 8719-8722. (b) Benakli, K.; Zha, C.; Kerns, R. J.; *J. Am. Chem. Soc.* **2001**, *123*, 9461-9462.
(27) A recent publication has used 60 mol% MeSSMe for the synthesis of methylthioglycoside (Mukhopadhyay, B.; Kartha, K. P. R.; Russell, D. A.; Field, R. A. *J. Org. Chem.* **2004**, *69*, 7758-7760). However, the maximum scale for the reported synthesis is 20g. In order to facilitate the library construction of phenylthioglycosides, it is preferable to start from 100 – 200g scale.
(28) We have developed method for large-scale synthesis of phenylthioglycosides (100 – 200g scale), and the products can be purified via recrystallization (patenting).
(29) Greene, T. W.; Wuts, P. G. M. *Protective groups in Organic Synthesis,* 3rd ed., John Wiley & Sons, New York, **1998**.
(30) Kocienski, P. J. *Protective groups,* Georg Thieme Verlag, Stuttgart, New York, **1994**.
(31) Garegg, P. J. in *Preparative Carbohydrate Chemistry*, Hanessian, S. Ed., Marcel Dekker, Inc. **1997**, New York, p53-67.
(32) Garegg, P.J. *Pure Appl. Chem.* **1984**, *56*, 845-858, and references cited therein.

(33) Chang, C.-W. T.; Hui, Y.; Elchert, Y. *Tetrahedron Letter*, **2001**, *42*, 7019-7023.

(34) For examples: (a) Poirot, E.; Chang, A. H. C.; Horton, D.; Kovac, P. *Carbohydr. Res.* **2001**, *334*, 195-205. (b) Marcaurelle, L. A.; Bertozzi, C. R. *J. Am. Chem. Soc.* **2001**, *123*, 1587-1595.

(35) For example: Dohi, H.; Nishida, Y.; Furuta, Y.; Uzawa, H.; Yokoyama, S.-I.; Ito, S.; Mori, H.; Kobayashi, K.; *Org. Lett.* **2002**, *4*, 355-357.

(36) For examples: (a) Tornoe, C. W.; Christensen, C.; Meldal, M.; *J. Org. Chem.* **2002**, *67*, 3057-3064. (b) Kuijpers, B. H. M.; Groothuys, S.; Keereweer, A. R.; Quaedflieg, P. J. L. M.; Blaauw, R. H.; Delft, F. L.; Rutjes, F. P. J. T. *Org. Lett.* **2004**, *6*, 3123-3126. (c) Yang, J.; Hoffmeister, D.; Liu, L., Fu, X.; Thorson, J. S. *Bioorg. Med. Chem.* **2004**, *12*, 1577-1584. (d) Lin, H.; Walsh, C. T. *J. Am. Chem. Soc.* **2004**, *126*, 13998-14003. (e) Mocharla, V. P.; Colasson, C.; Lee, L. V.; Röper, S.; Sharpless, K. B.; Wong, C.-H.; Kolb, H. C. *Angew. Chem. Int. Ed.* **2005**, *44*, 116-120.

(37) For reviewing: Scriven, E.F.N.; Turnbull, K. *Chem. Rev.* **1988**, *88*, 297-368.

(38) For example: Tanaka, K. S. E.; Winters, G. C.; Batchelor, R. J.; Einstein, F. W. B.; Bennet, A. J. *J. Am. Chem. Soc.* **2001**, *123*, 998-999.

(39) Elchert, B.; Li, J.; Wang, J.; Hui, Y.; Rai, R.; Ptak, R.; Ward, P.; Takemoto, J. Y.; Bensaci, M.; Chang, C.-W. T. *J. Org. Chem.* **2004**, *69*, 1513-1523.

(40) For example: Takayanagi, M.; Flessner, T.; Wong, C.-H. *J. Org. Chem.* **2000**, *65*, 3811-3815.

(41) (a) Barton, D.H.R.; McCombie, S.W. *J. Chem. Soc., Perkin Trans. I,* ***1975***, 1574-1585. (b) Robins, M.J.; Wilson, J.S.; Hansske, F. *J. Am. Chem. Soc.,* ***1983***, *105*, 4059-4065. (c) Barton, D. H. R.; Ferreira, J. A.; Jaszberenyi, J. C. in *Preparative Carbohydrate Chemistry*, Hanessian, S. Ed., Marcel Dekker, Inc. **1997**, New York, p151-172, and references cited therein.

(42) Chang, C.-W. T.; Clark, T.; Ngaara, M. *Tetrahedron Letter*, **2001**, *42*, 6797-6801.

(43) Sabesan, S.; Neira, S.; *J. Org. Chem.* **1991**, *56*, 5468-5472.

(44) Szarek, W. A.; Kong, X. in *Preparative Carbohydrate Chemistry*, Hanessian, S. Ed., Marcel Dekker, Inc. **1997**, New York, p105-125, and references cited therein.

(45) Boons, G.J. *Tetrahedron.* **1996**, *52*, 1095-1121.

(46) Jung, K.-H.; Muller, M.; Schmidt, R. R. *Chem. Rev.* **2000**, 100, 4423-4442.

(47) Hanessian, S.; Lou, B. *Chem. Rev.* **2000**, 100, 4443-4463.

(48) Schmidt, R. R. Jung, K.-H.; in *Preparative Carbohydrate Chemistry*, Hanessian, S. Ed., Marcel Dekker, Inc. **1997**, New York, p283-313, and references cited therein.

(49) For examples: (a) Crich, D.; Sun, S. *J. Am. Chem. Soc.* **1998**, *120*, 435-436. (b) Crich, D.; Sun, S. *J. Org. Chem.* **1997**, *62*, 1198-1199. (c) Crich,

D.; Smith, M. *Org. Lett.* **2000**, *2*, 4067-4069. (d) Crich, D.; Smith, M. *J. Am. Chem. Soc.* **2001**, *123*, 9015-9020.

(50) Crich, D. in *Glycochemistry Principles, Synthesis, and Applications,* Wang, P.G. and Bertozzi, C. R. Ed. Marcel Dekker, Inc. **2001**; p53-75.

(51) Lemieux, R.U. *Pure Appl. Chem.* **1971**, *25*, 527-548.

(52) Haddad, J; Kotra, L. P.; Mobashery, S. in *Glycochemistry Principles, Synthesis, and Applications,* Wang, P.G. and Bertozzi, C. R. Ed. Marcel Dekker, Inc. **2001**; p307-351.

(53) Bochkov, A. F.; Zaikov, G. E., *Chemistry of the O-Glycosidic Bond: Formation and Cleavage*, Pergamon Press, Elmsford, New York, **1979**.

(54) Fourmy, D.; Recht, M. I.; Blanchard, S. C.; Puglisi, J. D. *Science*, **1996**, *274*, 1367-1371.

(55) (a) Wang, J.; Li, J.; Tuttle, D.; Takemoto, J.; Chang, C.-W. T.; *Org. Lett.* **2002**, *4*, 3997-4000. (b) Li, J.; Wang, J.; Hui, Y.; Chang, C.-W. T.; *Org. Lett.* **2003**, *5*, 431-434. (c) Wang, J.; Li, J.; Czyryca, P. G.; Chang, H.; Kao, J.; Chang, C.-W. T.; *Bioorg. Med. Chem. Lett.* **2004**, *4*, 4389-4393.

(56) For examples: (a) Mingeot-Leclercq, M.-P.; Glupczynski, Y.; Tulkens, P. M. *Antimicrob. Agents Chemother.* **1997**, *43*, 727-737. (b) Kotra, L. P.; Haddad, J.; Mobashery, S. *Antimicrob. Agents Chemother.* **2000**, *44*, 3249-3256. (c) Cohen, M. L. *Science* **2002**, *257*, 1050-1055. (d) Neu, H. C. *Science* **2002**, *257*, 1064-1072.

(57) For examples: (a) Tanaka, H.; Nishida, Y.; Furuta, Y.; Kobayashi, K. *Bioorg. Med. Chem. Lett.* **2002**, *12*, 1723-1726, (b) Hanessian, S.; Tremblay, M.; Swayze, E. E. *Tetrahedron*, **2003**, *59*, 983-993, (c) Hanessian, S.; Kornienko, A.; Swayze, E. E. *Tetrahedron*, **2003**, *59*, 995-1007.

(58) (a) Seeberger, P. H.; Baumann, M.; Zhang, G.; Kanemitsu, T.; Swayze, E. E.; Hofstadler, S. A.; Griffey, R. H. *Synlett.* **2003**, 1323-1326. (b) Chou, C.-H.; Wu, C.-S.; Chen, C.-H.; Lu, L.-D.; Kulkarni, S. S.; Wong, C.-H.; Hung, S.-C. *Org, Lett.* **2004**, *6*, 585-588.

(59) Suami, T.; Nashiyama, S.; Ishikawa, Y.; Katsura, S. *Carbohydr. Res.* **1976**, *52*, 187-196.

(60) Similar solvent effect in the selectivity of glycosylation can be found in: Greenberg, W. A,; Priestley, E. S.; Sears, P. S.; Alper, P. B.; Rosenbohm, C.; Hendrix, M.; Hung, S.-C.; Wong, C.-H. *J. Am. Chem. Soc.* **1999**, *121*, 6527-6541.

(61) Li, J.; Wang, J.; Czyryca, P. G.; Chang, H.; Orsak, T.W.; Evanson, R.; Chang, C.-W. T. *Org. Lett.* **2004**, *6*, 1381-1384.

Chapter 11

Practical Applications of Computational Studies of the Glycosylation Reaction

Dennis M. Whitfield[1] and Tomoo Nukada[2]

[1]Institute for Biological Sciences, National Research Council of Canada, 100 Sussex Drive, Ottawa, Ontario K1A 0R6, Canada
[2]The Institute for Physical and Chemical Research (RIKEN), Wako-shi, 351–01 Saitama, Japan

A qualitative summary of our canonical vector representation of the conformations of six membered rings and its quantitative application to sugar pyranose rings is presented first. This representation is illustrated for the results of Density Functional Theory (DFT) optimized structures of per *O*-methylated D-*lyxo*, D-*manno* and D-*gluco*-configured oxacarbenium ions where two conformers are consistently found. Nucleophilic attack modelled with methanol is found to lead to intramolecular hydrogen bonding from the hydroxylic proton to electronegative oxygens of the oxacarbenium ion. As well, the ring conformation is found to change with the C-5-O-5--C-1-C-2 torsion angle moving from nearly planar in the isolated cations to more negative values in the case of α-approach and to more positive values for β-attack. This can be favorable or unfavorable depending on the protecting groups present with for example 4,6-benzylidene protection favoring β-attack with D-*manno*-configured ions whereas for D-*gluco*-configured ions α-approach is favorable. Extending the DFT studies to 2-*O*-acetyl analogues finds similar pairs of cations and raises the important issue of the order of ring inversion or O-2 bond rotation during the formation of ring inverted dioxolenium ions which in some cases are calculated to be very stable. Finally, an example using complex constraints is described which led to the design and development of 2,6-dimethylbenzoyl and 2,6-dimethoxybenzoyl as *O*-2 protecting groups which greatly minimize the acyl transfer side reaction.

Organic chemistry has a long established principle that by understanding the mechanism of a reaction it should be possible to optimize that reaction. What we wish to present in this chapter is a synopsis of our attempts to understand the mechanism of glycosylation reactions with the goal of optimizing this important reaction. Our technique is the use of modern computational methods, see experimental. Since it is not possible to investigate all aspects of the glycosylation reaction we have focussed on the role of ring conformations. It is hoped that our results will assist experimentalists like us and our collaborators to develop better glycosylation technologies.

This chapter is divided into four sections. The first outlines the new algebra that we have developed to describe the conformations of 6-membered rings that dominate the chemistry of carbohydrates (1). The second section shows our current understanding of the conformations of pyranose sugars with all ether protecting groups. The third section considers the implications of section 2 when C-2 is substituted with an acyl group capable of neighboring group participation. The fourth section considers a side reaction, namely acyl transfer to the acceptor alcohol, that is a problematic side reaction associated with neighboring group participation. This is not the chronological order in which these studies were made. It is, however, a more logical choice and should provide the reader with a logical progression of our ideas.

Section 1. Quantitative Description of Pyranose Ring Conformation Following the IUPAC Nomenclature.

In this communication we characterize each pyranose conformer which is derived from the output of a density functional theory (DFT) calculation by a system that uses the familiar IUPAC nomenclature. This definition is not deduced from a 3-dimensional view of a graphical representation of a molecule. In practice the differentiation of a skew conformer from a boat conformer is difficult by visual examination of graphical models. Instead, our definition is based on a numerical basis. We first introduce the concepts of our new definition method. A detailed description is given in Ref. (1), here we attempt to give a sufficiently simple explanation that all chemists can use.

Why do we need a new definition? The reason follows from the ambiguities of visual examination which partly results from the fact that traditional pyranose ring conformation characterizations are qualitative (2). These traditional methods describe visual conformational changes based on symmetry elements, which are intuitively easy to understand, but it is not easy to describe conformational interconversions qualitatively with these methods.

On the other hand, quantitative characterizations have been also developed (3). These quantitative methods are mathematically precise and afford a set of numbers which is unique to one pyranose conformation. The basic idea of all of these methods is a projection of pyranose endocyclic torsion angles onto a trigonometric orthonormal system. Such projections use a kind of Fourier transformation and this mathematical rigor insures their preciseness. These mathematical methods are advantageous for a quantitative recognition. However, they lack the visual significances that qualitative methods have. Our new definition aims to have both the visual significance familiar to an organic chemists and the quantitative preciseness for computational purposes.

In particular we aim to elucidate the mechanisms of both enzymatic and experimental glycosylations with computational chemistry. These processes involve conformational changes. of pyranose rings and so we need a robust method to calculate these changes. We also should provide conformational figures which researchers in many fields can easily understand. Therefore, we come to need a new definition having advantages both of qualitative and quantitative methods.

Quantitative description of pyranose ring conformation following the IUPAC conformational nomenclature.

Since internal ring dihedral angles give the best descriptions of ring distortions, we choose to use the set of six endocyclic dihedral angles as a geometrical parameter. We use the following definitions τ_1(C-1C-2--C-3C-4), τ_2(C-2C-3--C-4C-5), τ_3(C-3C-4--C-5O-5), τ_4(C-4C-5--O-5C-1), τ_5(C-5O-5--C-1C-2) and τ_6(O-5C-1--C-2C-3). A six dimensional space containing only these sets is a closed vector space which includes torsion angle sets of all typical conformations, namely, chairs, boats, skew-boats, half-chairs, and envelopes.

A basic concept of Fourier transformation is a projection of XYZ-Cartesian coordinates onto another orthonormal system. We reasoned that we could use a direct projection of any set onto the vector space defined by sets of typical conformer's torsion angles. As a first step, we should make this vector space defined by typical conformers conforming to the IUPAC conformational nomenclature. This procedure amounts to making quantitative representations of visual conformational changes which chemists can intuitively understand.

Canonical vector space and redundancy vector space

We define the canonical vector space (a vector space defined by typical conformers) on the basis of a 3-dimensional sphere as qualitatively analyzed by Hendrickson (4). The interrelation diagram among typical conformers can be described as in Figure 1. The two chair conformers occupy the poles and the

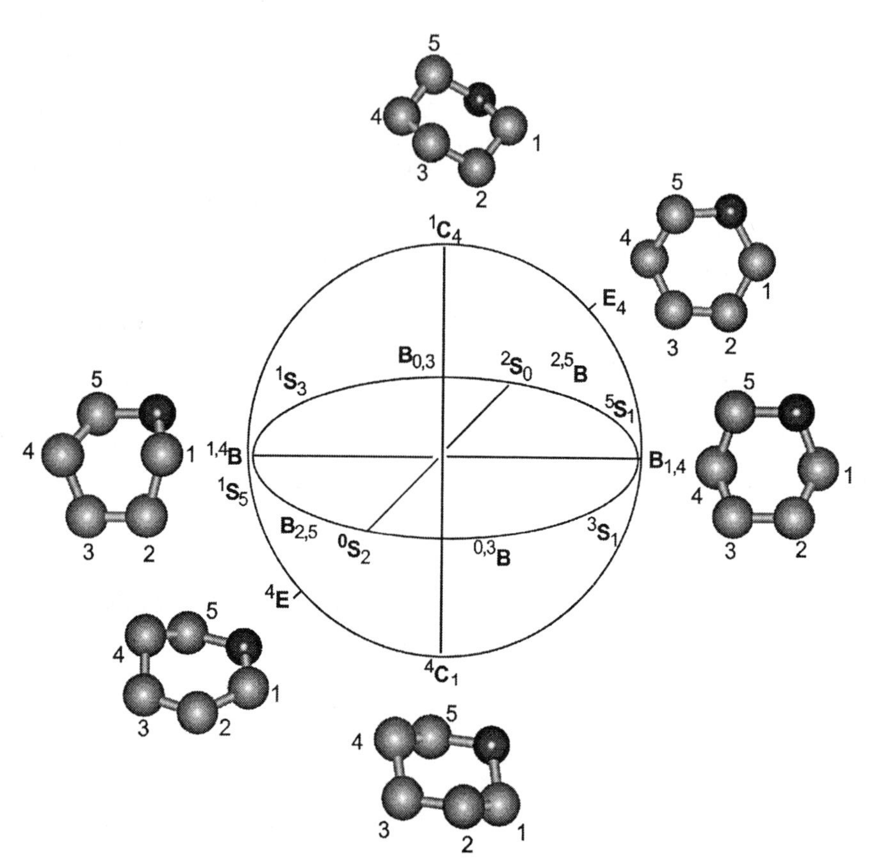

Figure 1. Spherical representation of pyranose conformations. The equatorial circle corresponds to pseudo-rotation.

pseudo-rotation plane consisting of the skew and boat conformers is situated at the mid-point between the two chairs. Any axis connected to the two chairs is orthogonal to the pseudo-rotation plane in this spherical representation. It is therefore easily realized that this canonical vector space could be represented with three basic canonical vectors.

By contrast, the number of components in the set are the six dihedral angles. Note that older qualitative characterizations use two molecular symmetry elements (σ-plane or C_2-axis). Such qualitative figures give us several hints for making appropriate canonical vectors. That is, each conformer's canonical vector should maintain an adequate symmetry. The presence of symmetry elements means that the sum of components in a canonical vector should be equal to 0. Based on these concepts, we define basic canonical vectors shown in Table 1. In order to precisely conform to the IUPAC convention we adopt 60° as a maximum torsion angle value according to Prelog and Cano (5).

For fitting to chemist's intuition, we define a transition conformer, a half-chair, that is situated at a middle point between chair and skewboat. In a similar manner, the envelope conformers are situated at a middle point between chair and boat, see E_4 and ^4E in Figure 1. By definition the canonical vector space presented by these vectors is isomorphic to 3 dimensional space. This means any point of canonical vector space can be represented adequately by three fundamental vectors. We can handle canonical vectors with the methods of three dimensional Euclidean space vectors. Any point in this space can be described as in equation **1**.

P is a point in the canonical space. **Ec** is a vector among chair, envelope, or half-chair basic canonical vectors.

Ea, **Eb**, (**Ea** $\neq$ **Eb**, **Ec**$\neq$ **Eb**, **Ec**$\neq$ **Ea**) are two arbitrary orthogonal basic canonical vectors.

fa, fb and fc are scalar coefficients.

$$\mathbf{P} = fa\cdot\mathbf{Ea} + fb\cdot\mathbf{Eb} + fc\cdot\mathbf{Ec} \qquad (1)$$

Since any vector in this canonical space can be presented by a linear combination of three basic canonical vectors this canonical space is a vector space. Therefore, we can define metric characters, inner product and absolute value in the same manner as those of a vector space.

$|\mathbf{E}|^2 = (\mathbf{E}, \mathbf{E})$, $|\mathbf{E}|>0$, an absolute value of vector **E**, (**E**,**E**) inner product.

We also use another metric character corresponding to an angle between two vectors of Euclidean space.

$$F = (\mathbf{E},\mathbf{P})/|\mathbf{E}|\cdot|\mathbf{P}|, \quad 0\leq F\leq 1 \qquad (2)$$

The scalar, F, in equation (2) is an indicator of closeness between two vectors. When we want to assign a conformation of a vector in a simple way, it is straightforward to calculate F between the vector and each basic canonical vector shown in Table 1. A basic canonical vector giving maximum F is closest to the target vector. When F is equal to 0, the two vectors are orthogonal to each other. For example, a chair vector is orthogonal to any vector, skew or boat, in the pseudo-rotational plane.

Table 1. Endocyclic torsion angles of 38 basic canonical conformations

Conformer	τ_1	τ_2	τ_3	τ_4	τ_5	τ_6
$^{1}C_4$	60.0	-60.0	60.0	-60.0	60.0	-60.0
$^{4}C_1$	-60.0	60.0	-60.0	60.0	-60.0	60.0
$^{1,4}B$	0.0	60.0	-60.0	0.0	60.0	-60.0
$B_{2,5}$	60.0	0.0	-60.0	60.0	0.0	-60.0
$^{O,3}B$	60.0	-60.0	0.0	60.0	-60.0	0.0
$B_{1,4}$	0.0	-60.0	60.0	0.0	-60.0	60.0
$^{2,5}B$	-60.0	0.0	60.0	-60.0	0.0	60.0
$B_{O,3}$	-60.0	60.0	0.0	-60.0	60.0	0.0
$^{1}S_5$	30.0	30.0	-60.0	30.0	30.0	-60.0
$^{O}S_2$	60.0	-30.0	-30.0	60.0	-30.0	-30.0
$^{3}S_1$	30.0	-60.0	30.0	30.0	-60.0	30.0
$^{5}S_1$	-30.0	-30.0	60.0	-30.0	-30.0	60.0
$^{2}S_O$	-60.0	30.0	30.0	-60.0	30.0	30.0
$^{1}S_3$	-30.0	60.0	-30.0	-30.0	60.0	-30.0
$^{1}H_2$	45.0	-15.0	0.0	-15.0	45.0	-60.0
$^{3}H_2$	60.0	-45.0	15.0	0.0	15.0	-45.0
$^{3}H_4$	45.0	-60.0	45.0	-15.0	0.0	-15.0
$^{5}H_4$	15.0	-45.0	60.0	-45.0	15.0	0.0
$^{5}H_O$	0.0	-15.0	45.0	-60.0	45.0	-15.0
$^{1}H_O$	15.0	0.0	15.0	-45.0	60.0	-45.0
$^{4}H_5$	-15.0	45.0	-60.0	45.0	-15.0	0.0
$^{O}H_5$	0.0	15.0	-45.0	60.0	-45.0	15.0
$^{O}H_1$	-15.0	0.0	-15.0	45.0	-60.0	45.0
$^{2}H_1$	-45.0	15.0	0.0	15.0	-45.0	60.0
$^{2}H_3$	-60.0	45.0	-15.0	0.0	-15.0	45.0
$^{4}H_3$	-45.0	60.0	-45.0	15.0	0.0	15.0
^{1}E	30.0	0.0	0.0	-30.0	60.0	-60.0
E_2	60.0	-30.0	0.0	0.0	30.0	-60.0
^{3}E	60.0	-60.0	30.0	0.0	0.0	-30.0
E_4	30.0	-60.0	60.0	-30.0	0.0	0.0
^{5}E	0.0	-30.0	60.0	-60.0	30.0	0.0
E_O	0.0	0.0	30.0	-60.0	60.0	-30.0
^{4}E	-30.0	60.0	-60.0	30.0	0.0	0.0
E_5	0.0	30.0	-60.0	60.0	-30.0	0.0
^{O}E	0.0	0.0	-30.0	60.0	-60.0	30.0
E_1	-30.0	0.0	0.0	30.0	-60.0	60.0
^{2}E	-60.0	30.0	0.0	0.0	-30.0	60.0
E_3	-60.0	60.0	-30.0	0.0	0.0	30.0

For quantitative definition of this canonical space, we defined three basic canonical vectors, namely, $^{1}\mathbf{C}_4$, $^{0}\mathbf{S}_2$, and $^{1,4}\mathbf{B}$ vectors as basis sets. These vectors are orthogonal to each other and can be used as reference ideal conformers in chemist's discussions. This basis set selection, one chair, one skewboat and one boat, from the pseudo-rotation plane is in line with the qualitative figure of Hendrickson. Of course, 11 other combinations are available for this purpose. The three orthogonal basis sets confirm a straightforward recognition because of being isomorphic to three dimensional Euclidean space.

We have discussed that for example an idealized pyranose conformation corresponding exactly to one of the IUPAC standard conformations should have a symmetry element. However, real molecules we encounter are distorted from such exact symmetric conformations. We define redundancy vectors, which are residual vectors out of a projection with the canonical vectors (three basis sets, chair, skew-boat, and boat). This redundancy is not significant in carbohydrate pyranose rings. It is, however, not negligible for some six-membered rings involving transition metals where bond lengths and bond angles are far from equal or tetrahedral respectively.

We have not precisely analyzed the details of these redundancy vectors. From our basic ideas about canonical vector space, we predict that during motions of pyranose rings, these molecules produce the redundancy in cases of symmetrical motions being energetically disfavored. This redundancy appears to originate from breaking the structural symmetry of the molecule. Most carbohydrate pyranose molecules are only slightly asymmetric and therefore they only produce small amounts of redundancy. We define 3 redundancy basis set vectors as in Table 2.

Table 2. Table 2. Redundancy vectors

Vector	τ_1	τ_2	τ_3	τ_4	τ_5	τ_6
Red1	60	60	60	60	60	60
Red2	0	60	60	0	60	60
Red3	60	30	-30	-60	-30	30

Each redundancy vector is orthogonal to any of the canonical vectors and is orthogonal to the other two redundancy vectors. With redundancy basis set vectors, six-membered rings could not be built. We consequently present a conformer with using a set of endocyclic torsion angles.

P a vector of endocyclic torsion angles on an arbitrary conformation.
$^{1}\boldsymbol{C}_4$ the canonical vector of conformation $^{1}C_4$ in Table 1.
$^{0}\boldsymbol{S}_2$ the canonical vector of conformation $^{O}S_2$ in Table 1.

$^{1,4}\boldsymbol{B}$ the canonical vector of conformation ^{1,4}B in Table 1.
Red1, **Red2**, **Red3**, the redundancy vectors in Table 2.
f1, f2, f3, f4, f5, f6 scalar coefficients.

$$\mathbf{P} = \mathrm{f1}\cdot{}^{1}\boldsymbol{C}_{4} + \mathrm{f2}\cdot{}^{0}\boldsymbol{S}_{2} + \mathrm{f3}\cdot{}^{1,4}\boldsymbol{B} + \mathrm{f4}\cdot\mathbf{Red1} + \mathrm{f5}\cdot\mathbf{Red2} + \mathrm{f6}\cdot\mathbf{Red3} \quad (3)$$

With equation (**3**), one conformer is mapped uniquely to the vector space which is built by three canonical basis set vectors and three redundancy basis set vectors. A convenient method of conformational characterization is to use only the first three terms, three canonical basis set vectors in the equation (**3**). So, a pyranose conformation and a conformational change can be represented with XYZ-coordinates or polar coordinates. As mentioned above, we make a new definition method which has advantages both of quantitative and qualitative methods.

Quantitative methods have advantages for illustrating pyranose conformational changes. Our new method inherits this strong point. Distortions from initial conformers can be visually and quantitatively understood with our method from the magnitude and type of distortion. For example we have used this canonical vector space for a constraint in molecular dynamic simulations of pyranose conformation change. Since we are interested in glycosylation reactions we need also to consider bond breaking and bond forming therefore with this quantitative formulation we can use such constraints in quantum molecular dynamics (6).

Section 2 - Permethylated Glycopyranosyl Oxacarbenium Ions - The Two Conformer Hypothesis

We confine our discussion in Sections two to four, to glycosylation reactions between pyranose sugars in which one sugar acts as an electrophilic donor reacting with a nucleophilic alcohol. This encompasses most, but not all experimental methods for glycosylation.

Early kinetic studies showed that, except for reactions of very strong nucleophiles where true S_N2 kinetics could be observed, most glycosylation reactions have considerable S_N1 character (7). This strongly implies that glycopyranosyl oxacarbenium ions are likely formed as intermediates or transitions states (TS). In all such cases the anomeric carbon must obtain appreciable sp^2 character. In most donors before activation the anomeric carbon is sp^3 hybridized and in the product glycosides it is also sp^3 hybridized. Therefore as a glycosylation reaction proceeds the anomeric carbon proceeds from sp^3 to sp^2 and back to sp^3. These changes have pronounced consequences for the conformation of the 6-membered pyranose ring.

It should be noted that the classical 5-coordinate *anti* S_N2 TS also requires the anomeric carbon to obtain a transient planar conformation. Although the details are clearly different, the concepts are similar. To describe these ring conformational changes we use the formalism developed in Section 1.

In a real glycosylation reaction the donor must be activated by a promoter or catalyst in a solvent. The reaction must also be kept free of adventitious water and any other nucleophiles that could compete with the alcoholic acceptor. As well, the proton of the acceptor must be transferred and some base must ultimately take up these protons. These realities make the reaction much more complicated than our models. However, it is our goal to assess the importance of the pyranose ring conformation especially in the intermediates and TS of the reaction.

The progress of glycosylation reactions is difficult to follow experimentally and usually the most one can measure is the disappearance of starting material and the appearance of products as a function of time. Importantly, it is difficult to detect transient intermediates if they are formed. In fact, in most cases all the data that is reported is the overall yield and the α:β ratio of the products. Typically, predominant or exclusive formation of one isomer is attributed to S_N2 attack on a reactive intermediate and approximately 50:50 α:β mixtures attributed to non-specific attack on a free oxacarbenium ion i.e. S_N1. Thus, it was a great surprise to us to discover that calculated intermediates between glycopyranosyl oxacarbenium ions and the achiral nucleophile methanol exhibited marked stereoselectivity.

Furthermore, in all cases studied to date the calculated intermediate with the lowest energy corresponds to the generally preferred anomer of the product. This unanticipated result led us to completely revise our thinking about the origin of the stereochemistry of glycosylation reactions.

How do we calculate these intermediates? First for these examples, we choose *O*-methyl substituted glycopyranosyl oxacarbenium ions[i] by suitable *in silico* modifications to optimized chair conformations of the parent per *O*-methyl glycopyranose. With these species there are no hydroxyls and hence no intramolecular hydrogen bonding. Conformations with intramolecular hydrogen bonding are found as the lowest energy ones by gas phase calculations of neutral monosaccharides and this factor may obscure other factors influencing reactivity (8, 9).

Methyl is also the smallest possible protecting group and begins to address the important issue of the effect of protecting groups on reactivity. In agreement with other calculations of permethylated sugars and experimental studies with model compounds, the preferred conformation of secondary *O*-methyl substituents is to have the methyl carbon *syn* to the sugar methane (10). These species with these *O*-methyl conformations are then subjected to minimization using Density Functional Theory (DFT) calculations. Visual analysis of graphical images of this first optimized structure is then performed to check for side chain orientations.

All appropriate side chain conformations are altered until a good structure is found which is then checked to be a minima by frequency calculations. These calculations include a continuum solvent (parameterized to CH_2Cl_2) correction and energies are corrected for Zero Point Energy (ZPE). Alternatively, different conformations may be generated by *in silico* modifications of one of the 38 standard IUPAC conformations of tetrahydropyran followed by the DFT optimization with side chain optimization process described above. These 38 conformations are shown in Table 1 and are available via the Internet at http://www.sao.nrc.ca/ibs/6ring.html.

The DFT calculations were carried out with the Amsterdam Density Functional (ADF) program system, ADF2000 (11). The atomic orbitals were described as an uncontracted double-ζ Slater function basis set with a single-ζ polarization function on all atoms which were taken from the ADF library. The 1s electrons on carbon and oxygen were assigned to the core and treated by the frozen core approximation. A set of s, p, d, f, and g Slater functions centered on all nuclei were used to fit the electron density, and to evaluate the Coulomb and exchange potentials accurately in each SCF cycle.

The local part of the V_{xc} potential (LDA) was described using the VWN parameterization (12), in combination with the gradient corrected (CGA) Becke's functional (13) for the exchange and Perdew's function for correlation (BP86) (14). The CGA approach was applied self-consistently in geometry optimizations. Second derivatives were evaluated numerically by a two point formula. The solvation parameters were dielectric constant ε =9.03, ball radius =2.4 angstroms, with atomic radii of C=1.7, O=1.4 and H=1.2 angstroms.

As a first example consider 2,3,4-tri-*O*-methyl-D-lyxopyranosyl cation, **1**. Two conformers were found that differ by ring inversion, see Figure 2. One conformer, **B0**, is the $^4H_3$ conformation, long considered a probable conformation for such glycopyranosyl oxacarbenium ions. In terms of the C, B and S terms (Section 1) it is $^4C_1$ 0.573, $B_{2,5}$ 0.059 and $^1S_3$ 0.389 i.e. approximately half a chair and half a twist boat therefore a half chair; $^4H_3$ 0.778. The second conformer, **B1**, has the $^3H_4$ conformation which is $^1C_4$ 0.580, $B_{2,5}$ 0.100 and $^3S_1$ 0.486 and therefore $^3H_4$ 0.972. In this conformation the substituents at O-3 and O-4 are pseudo-axial whereas the substituent at O-2 is pseudo-equatorial. The terms with small coefficients show not only how the dominant conformation is distorted but precisely measure the degree of distortion.

Considering that axial substituents are normally considered to be destabilizing it is surprisingly that this **B1** conformer is calculated to be 12.6 kJ mol^{-1} more stable than **B0**. This energy difference is sufficiently high that, if the barrier separating these conformers is low enough, and the lifetime of these ions is long enough, than **B1** should be the dominant species present. Specific solvation, ion-pairing, nucleophile preassociation and the effects of different protecting groups are all expected to modify this intrinsic selectivity. Note, for most synthetic purposes even if all ether protecting groups are used the stereoelectronic effects of protecting groups are expected to have an effect. This result, i.e. two conformers, does give the experimentalist a place to start their analysis from, which is our present goal.

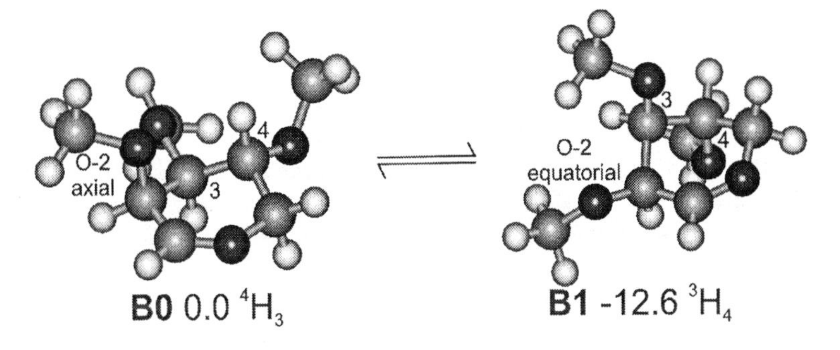

*Figure 2. Ball and Stick Representations of 2,3,4-tri-O-methyl-D-lyxopyranosyl cation, **1**, in its **B0** and **B1** Conformations, showing relative energies in kJ mol^{-1}*

Consideration of α or β-nucleophilic attack on **B0** and **B1** of **1** leads to four cases to consider. Using methanol as model nucleophile in order to avoid most issues of the stereoelectronic effects of protecting groups of the acceptor on stereoselectivity DFT calculations of the four possible cases were made (15). Values are: **1**(**B0**)+αMeOH -19.3 kJ mol^{-1}, **1**(**B0**)+βMeOH -48.9 (-20.9) kJ mol^{-1}, **1**(**B1**)+αMeOH -35.6 (-17.6) kJ mol^{-1}, **1**(**B1**)+βMeOH -64.6 (-36.6) kJ mol^{-1} (in brackets corrected for H-bonding, see below).

The results are best viewed by looking at the spherical representation of the conformations of 6-membered rings, see Figure 3. This shows the changes in conformation from the isolated cation to the ion:dipole complexes. Examination of a number of such complexes has led to the recognition of two important variables (16). One factor is intramolecular hydrogen bonding between the nucleophilic hydroxyl proton and the electronegative atoms of the oxacarbenium ions especially in cases where the species have appreciable hydronium ion character. The relative energetics of this effect are over estimated because intermolecular hydrogen bonding is not considered but it still is considered to be an important factor.

Secondly, the change in pyranosyl ring conformation from the isolated cation to the ion:dipole complex can be favorable or unfavorable. In all the isolated cations the C-5-O-5--C-1-C-2 (τ_5) torsion angle is nearly planar, for **1 B0** it is -10.6° and for **1 B1** it is 5.2°. For the D-sugars considered so far, this torsion angle increases for β-attack and decreases for α-attack.

For **1** α-attack ion:dipole complexes shift the pyranosyl ring conformation by lowering the C-5-O-5--C-1-C-2 torsion angle by 9° and 44.9° for **B0** and **B1** respectively. Whereas β-attack raises it by 31.7° and 45.2° for **B0** and **B1**. Both β-complexes allow for hydrogen bonding with the **B0** conformer exhibiting an acutely angled H-bond to the pseudo-axial O-2 and for the **B1** conformer a nearly linear H-bond to the pseudo-axial O-3 was found.

For α-attack on **B1** a H-bond to pseudo-axial O-4 was found, whereas for α-attack on **B0** no hydrogen bonds were found. Ball and stick representations of these structures are shown in Figure 3. The result of these effects is to put the **B0** + αMeOH in a $^4C_1$ conformer which is close to the starting $^4H_3$ whereas **B0** + βMeOH shifts to the less favourable ^2S_O conformer. This **B0** + αMeOH conformation is favourable for α-attack but the absence of hydrogen bonding probably underestimates the stability of this complex.

On the other hand the **B1** + αMeOH is in the unfavourable $^3S_1$ conformation whereas the **B1** + βMeOH is in the favourable $^1C_4$ chair as well as having a strong intramolecular hydrogen bond. Therefore it is not surprising that this is predicted to be the most stable complex and hence β-glycosides are predicted to be the preferred products. Indeed experimental results with a 2,3,4-tri-*O*-benzyl-D-lyxopyranosyl donor predominantly give β-glycosides (17).

D-lyxopyranose has the same configurations at C-2, C-3 and C-4 as D-mannopyranose and from the results with **1** could be expected to preferentially yield β-glycosides in glycosilation reactions. Since even ether protected D-mannopyranosyl donors usually give predominantly α-glycosides this suggests an important stereodirecting role for the substituent at C-5 in hexopyranoses. DFT calculations on 2,3,4,6-tetra-*O*-methyl- mannopyranosyl cation **2** provide some support for this hypothesis (18).

Again two conformers that differ by ring inversion were found for the *manno* cation. With **B0** characterized as $^4C_1$ 0.566, $B_{2,5}$ 0.043 and $^1S_3$ 0.397 therefore $^4H_3$ 0.793 and the inverted **B1** conformations characterized as $^1C_4$ 0.522, $^{O,3}B$ 0.432 and $^5S_1$ 0.048 therefore ^3E 0.864 (ΔE 1.9 kJ mol^{-1}). The ^3E conformation is near in configurational space to $^3H_4$ as visualized on the planar representation of the spherical representation, Figure 4a. The calculated C-5--C-6 torsion angle defined as ω_H (H-5C-5--C-6O-6) was found to be -178.7° (gg) for **B0** and -50.8° (gt) for **B1**. This shift for **B1** perhaps reflects the pseudo-axial orientation of O4.

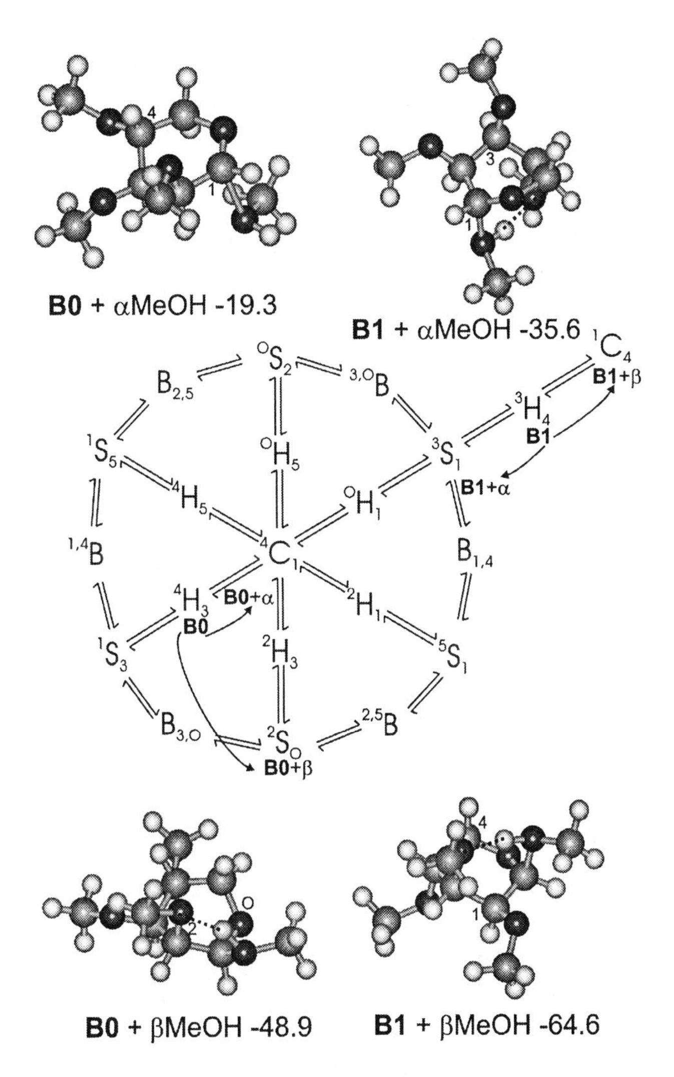

*Figure 3. Planar Representation of the Spherical Representation of Pyranose Ring Conformations Showing the Positions of the Parent **B0** and **B1** Conformers of **1** and the results of α- or β-Attack. Ball and Stick Representations of the Resulting Ion:Dipole Complexes of **1** and Methanol, showing relative energies in kJ mol^{-1}.*

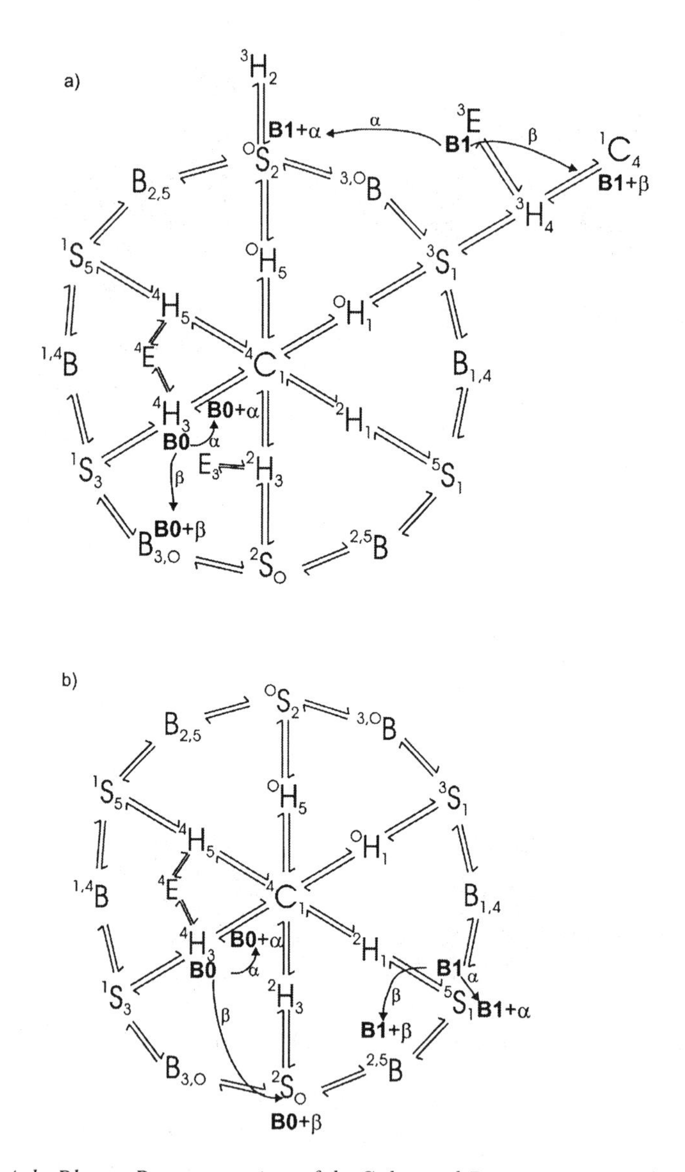

*Figure 4ab. Planar Representation of the Spherical Representation of Pyranose Ring Conformations Showing the Positions of the Parent **B0** and **B1** Conformers of **2** and **3** plus the Consequences of α- or β-Attack on Each. Top (a) Man **2** and Bottom (b) **3**.*

For **2** the β-face complexes are calculated to be more stable than those for α-face complexes. Values are: **2(B0)**+αMeOH -21.3 kJ mol^{-1}, **2(B0)**+βMeOH -51.3 (-23.3) kJ mol^{-1}, **2(B1)**+αMeOH -30.9 kJ mol^{-1}, **2(B1)**+βMeOH -49.5 (-21.5) kJ mol^{-1} (in brackets corrected for H-bonding). Most of this extra stabilization comes from hydrogen bonding to O-6 even though this forces the ring into a $B_{O,3}$ (**B0**) or an inverted $^1C_4$ (**B1**) conformations, respectively. The second consequence of hydrogen bonding is a shorter O_m--C-1 bond length. Values are: **2(B0)**+αMeOH 1.860 Å, **2(B0)**+βMeOH 1.498 Å, **2(B1)**+αMeOH 1.672 Å, **2(B1)**+βMeOH 1.510 Å.

Previously, we had estimated the energetic contribution of hydrogen bonding to be 28 kJ mol^{-1} by calculating for a deoxy analogue in the same conformation as the parent oxygen containing hydrogen bond accepting compound (16). These "corrected" values show the expected trend of α more favorable than β for mannopyranosyl donors.

Similar DFT calculations with 2,3,4,6-tetra-*O*-methyl-glucopyranosyl cation **3** test the effect of the configuration at C-2. Again two conformers were found with **3(B0)** characterized as $^4C_1$ 0.507, $B_{2,5}$ 0.015 and $^1S_3$ 0.509 therefore $^4H_3$ 1.013 and **3(B1)** characterized as $^1C_4$ 0.188, $B_{O,3}$ 0.117 and $^5S_1$ 0.779 (ΔE 4.8 kJ mol^{-1}). In this case $^5S_1$ is relatively far, see Figure 4b, from $^3H_4$. Examination of Figure 5h shows that this conformation allows O-2 to be pseudo-equatorial while O-3, O-4 and C-6 are pseudo-axial. This limited selection of results suggests that having O-2 in a pseudo-equatorial conformation stabilizes the oxacarbenium ion. This unanticipated result is a direction for future investigations.

Hydrogen bonding is found for **3(B0)**+βMeOH and **3(B1)**+βMeOH to O-6 and for **3(B1)**+αMeOH to O-4. The corresponding energies are: **3(B0)**+αMeOH -43.4 kJ mol^{-1}, **3(B0)**+βMeOH -56.2 (-28.2) kJ mol^{-1}, **3(B1)**+αMeOH -55.3 (-27.3) kJ mol^{-1} and **3(B1)**+βMeOH -43.0 (-15.0) kJ mol^{-1}. Thus, including H-bonding shows a preference for β-glycosides whereas without it a preference for α-glycosides. The torsion angle τ_5 is 0.3° for **B0** changing to -39.5° and 35.3° for α- and β-attack respectively whereas for **B1** τ_5 changes from -4.6° to -38.1° and 23.3° for α- and β-attack respectively.

For **1** to **3** and for several other sugars (unpublished and below) this trend of finding two conformers for the glycopyranosyl oxacarbenium ions which differ by ring inversion appears to be general. We have named this the Two Conformer Hypothesis. This hypothesis leads to four cases to consider when examining the α:β selectivity.

The resulting ion:dipole complexes with the model nucleophile methanol have as distinguishing characteristics intramolecular hydrogen bonding and ring conformational changes characterized by a decrease in C-5-O-5--C-1-C-2 (τ_5)

a)

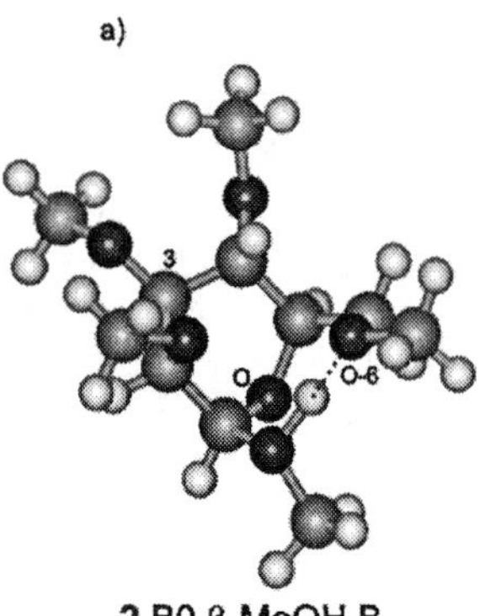

2 B0 β-MeOH $B_{3,O}$

b)

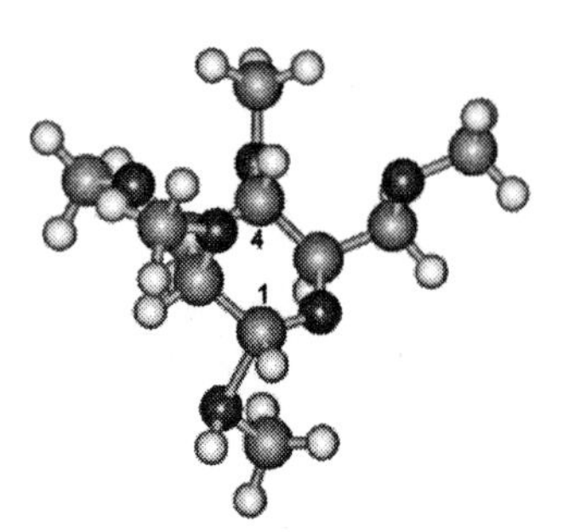

2 B0 α-MeOH $^{4}C_{1}$

c)

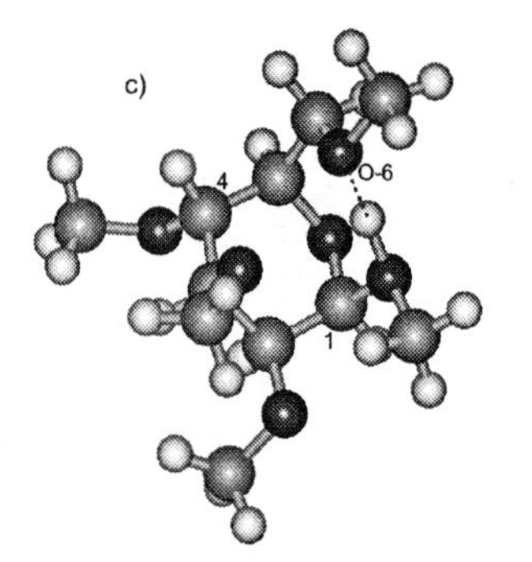

2 B1 β-MeOH $^{1}C_{4}$

d)

2 B1 α-MeOH $^{O}S_{2}$

e) f)

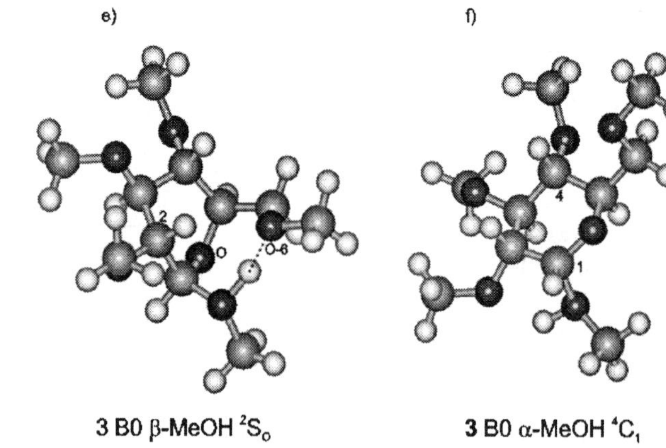

3 B0 β-MeOH $^{2}S_{O}$

3 B0 α-MeOH $^{4}C_{1}$

Figure 5. Ball and Stick Representations of the Ion:Dipole Complexes of ***2*** *(a-d) and* ***3*** *(e-h) plus Methanol.*

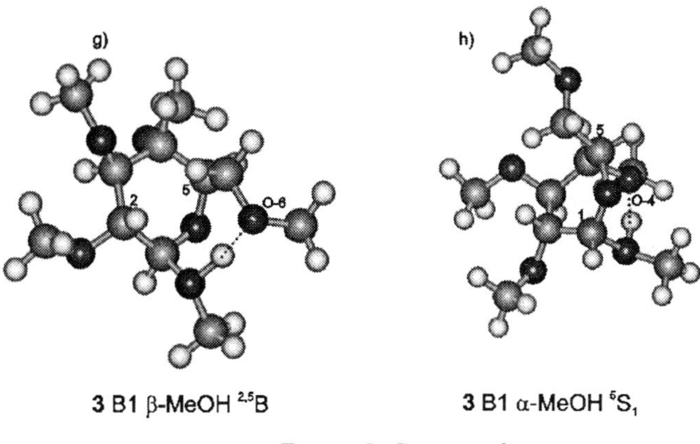

Figure 5. Continued.

for α-attack and an increase for β-attack on D-sugars from nearly planar configurations in the isolated cations. These conformational changes may be favorable or unfavorable depending on the conformations involved. These changes depend on the configuration of the sugar and the protecting groups.

Among other factors that could be studied, one variable that could be expected to be important is the variation of overall energy with the distance between the nucleophilic oxygen and the anomeric carbon, O_m--C-1. We have investigated this degree of freedom by systematically varying this "bond length" while allowing all other degrees of freedom to optimize by starting at the minima found in Figure 5 and varying this distance either forward or backward between 1.38 and 3.0 Å in small increments. Figure 6 shows one such study for the case of **3** plus βMeOH where each data point is a full constrained DFT optimization. We identified several technical problems with this study such as the discontinuities at long C-1--O_m bond lengths associated with changes in the ring conformation. However, in this case the complex does dissociate to the expected $^4H_3$ conformation at long C-1--O_m bond lengths. Several conclusions can be made from these studies.

In all cases including the one in Figure 6 the minima are shallow as in this case the region from about 1.42-1.68 Å is within 5 kJ mol^{-1} of the minimum. As mentioned above, this complex shows considerable hydronium ion character. Indeed at short separations the hydrogen bonded hydroxyl proton transfers to its oxygen acceptor, in this case O-6. Since many glycosylation reactions are done in the absence of bases this type of intramolecular proton transfer is highly plausible in real reactions.

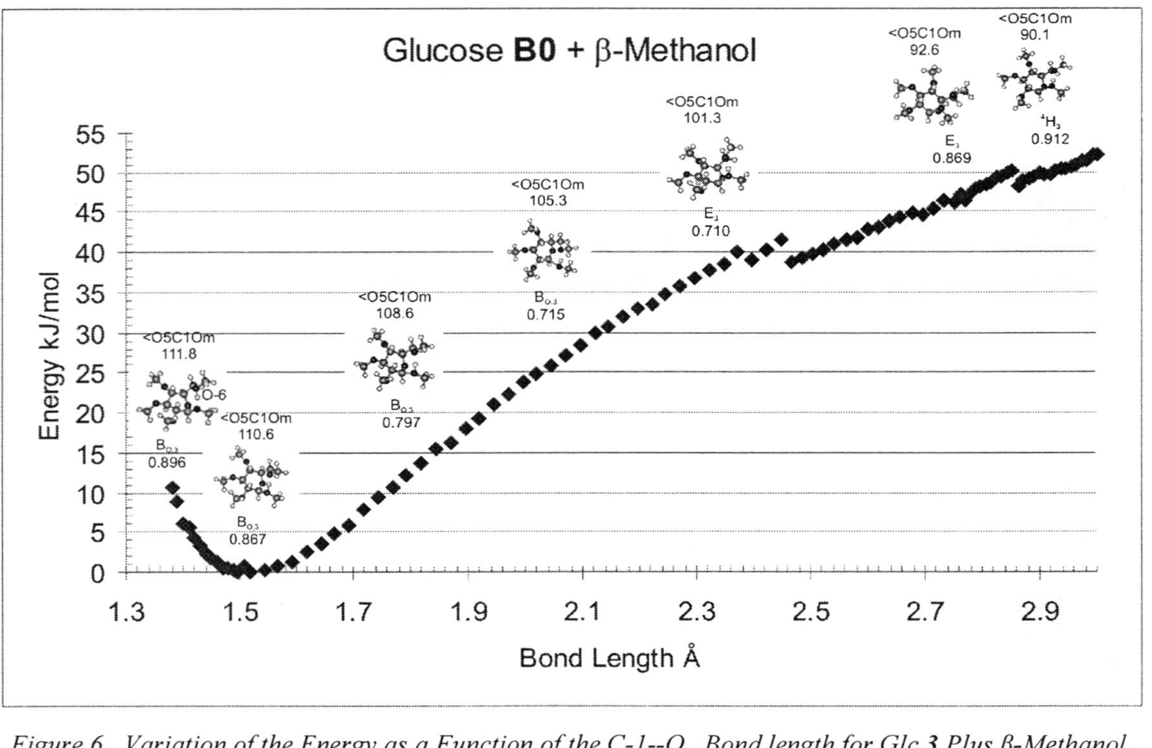

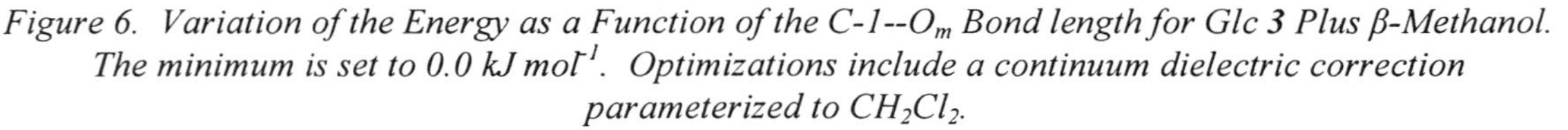

*Figure 6. Variation of the Energy as a Function of the C-1--O_m Bond length for Glc **3** Plus β-Methanol. The minimum is set to 0.0 kJ mol^{-1}. Optimizations include a continuum dielectric correction parameterized to CH_2Cl_2.*

Table 3. Conformational Descriptions, Selected Geometric Parameters and relative energies of Cations 4 and 5 and their methanol adducts (α and β).

Species	Chair	Boat	Skew -Boat	Half-Chair Envelope	C-1--O_m Å O_m °	<O-5 C-1 C-2 °	<C-5O -5--C-1 mol^{-1}	E kJ
Man								
4(B)	$^4C_1$ 0.052	**$B_{2,5}$ 0.834**	$^1S_3$ 0.090	-	-	-	2.2	0.0[a]
4 - β	$^1C_4$ 0.024	$^{O,3}B$ 0.251	**$^1S_5$ 0.970**	-	1.632	107.4	15.4	-51.5
4 - α	$^4C_1$ 0.111	**$B_{2,5}$ 0.825**	$^3S_1$ 0.018	-	2.298	96.3	-7.7	-13.5
Glc								
5(B)	$^4C_1$ 0.483	$^{1,4}B$ 0.495	$^OS_2$ 0.046	**^4E 0.967**	-	-	-0.4	0.0[a]
5 - β	$^4C_1$ 0.344	$^{1,4}B$ 0.629	$^OS_2$ 0.046	**^4E 0.688**	2.222	97.9	16	-18.9
5 - α	**$^4C_1$ 0.780**	$^{O,3}B$ 0.018	$^1S_5$ 0.266	-	1.645	108.1	-39.8	-67.2

a. The energy of solvated methanol is added to that of solvated **B** and then set to 0.0 kJ mol^{-1}.

Ultimately such species would transfer to the molecular sieves or other weakly basic components in the reaction mixtures. A third trend is apparent in all cases and is related to the O_m-C-1-O-5 bond angle. At large separations this angle is near 90° whereas at shorter separations it approaches the tetrahedral angle. We attribute this to a change from a π-complex to a σ-complex reflecting the hybridization changes at C-1 from sp^2 to sp^3.

The two factors, namely intramolecular hydrogen bonding and pyranose ring conformational changes, each suggest strategies for optimizing the stereoselectivity of glycosylation reactions. The possibility of hydrogen bonding could be stabilized or destabilized by the general principle of small electron donating (type 1) protecting groups favoring such interactions and large electron withdrawing (type 2) protecting groups disfavoring such interactions.

For *lyxo*-configured pyranoses β-glycosylation should be favoured by type 1 protecting groups at O-3 and perhaps O-2. Similarly for both **2** and **3**, type 1 groups should be at O-6 to favor β-glycosylation. Changing to type 2 groups should have the opposite effects. There is some experimental evidence for *gluco* configured donors that these observations are valid (19). Clearly more detailed studies are necessary. The possibility of fixing the ring conformation by for example fusing an additional ring to it could then lead to the stabilization of one of the two families of conformers.

Further the fused ring could lead to preferential glycosylation of one face over the other. Experimentally the 4,6-benzylidene derivatives of mannose and glucose are known under particular experimental conditions to give predominantly β- or α-glycosides respectively (20). To study this we have done DFT optimizations on 4,6-*O*-benzylidene-2,3-di-*O*-methyl-D-mannopyranosyl (**4**) and 4,6-*O*-benzylidene-2,3-di-*O*-methyl-D-glucopyranosyl (**5**) cations (16). The data presented in Table 2 show that these experimental observations are consistent with the ring conformational changes associated with changes in τ_5. A pictorial representation is shown in Figure 7. There are some related systems where similar control may be operative (21). The use of fused rings or other mechanisms to stabilize one oxacarbenium ion ring conformation appears to be an experimental strategy that could be further exploited.

Section 3 - Neighboring Group Participation from O-2

It has long been recognized that putting an acyl group adjacent to the anomeric carbon, in our examples on O-2, has a pronounced affect on reactions at the anomeric carbon. Lemieux showed that by successively substituting chlorine on the methyl carbon of a 2-*O*-acetyl substituted hexopyranosyl donor that the reactivity markedly decreased (22). Thus, the kinetic barrier to forming oxacarbenium ions is increased by the electron withdrawing groups.

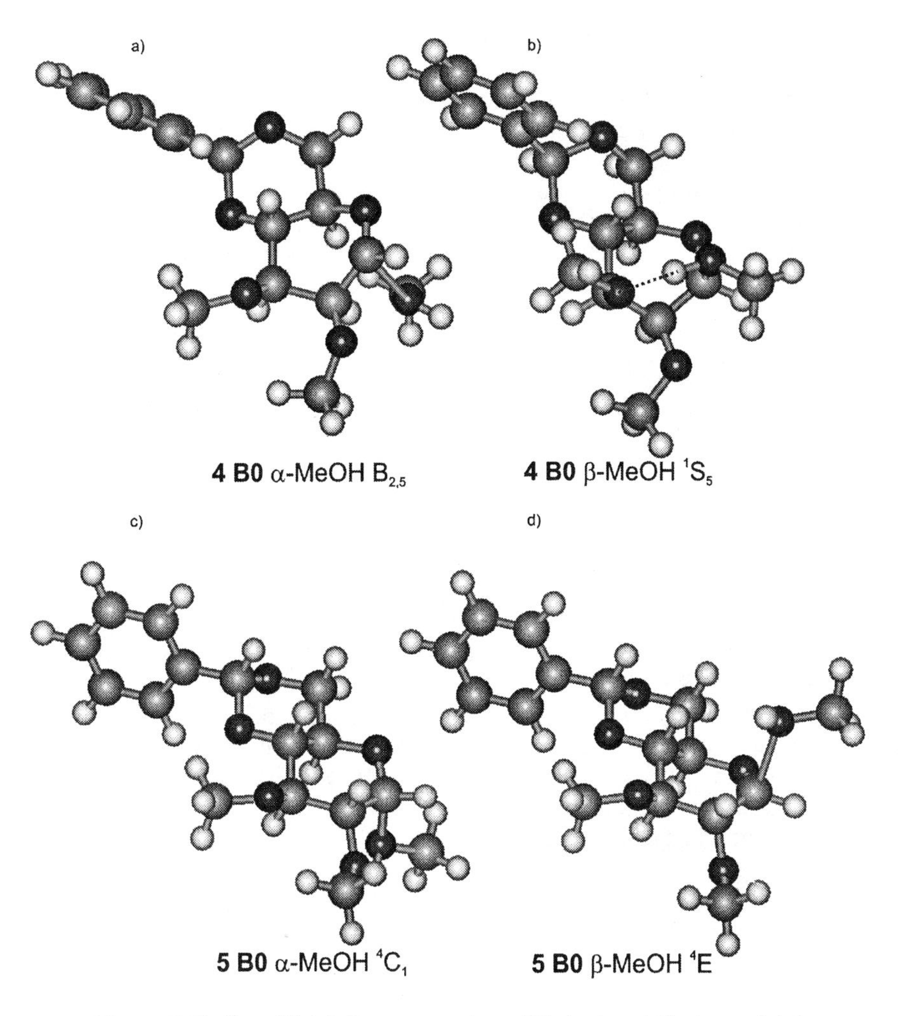

*Figure 7. Ball and Stick Representation of Calculated Cations of 4,6-Benzylidene Protected Man **4** and Glc **5** Showing the Favorable (β for **4** and α for **5**) and Unfavorable (α for **4** and β for **5**) Ring Changes Resulting from α- or β-attack of Methanol.*

The second major effect is that the carbonyl oxygen, O-7, can directly interact with C-1 to form a dioxolenium ion species (23). Such ions have been prepared and detected in solution and the solid state (24). One such 5-membered ring was found to be almost exactly planar in the solid state (25). Our DFT calculations of this species also found a planar geometry (26). In fact in all the cases we have calculated so far such rings are nearly planar. Accommodating this planarity has consequences for the fused pyranose ring.

When we started our studies of these neighboring groups systems we expected to be able to find a 5-coordinate *anti* S_N2 TS. Such TS are the conventional explanation for the 1,2-*trans* stereochemistry often but not always observed with O-2 acyl protecting groups (27). Our aim was to study the effect of various protecting groups on this TS such as studying different acyl groups in order to optimize the experimental choices. Unfortunately, in spite of numerous attempts we have never found this TS. It may be that our models are too simple and that inclusion of specific solvation and/or counterions may be necessary to stabilize this type of TS. However, our studies of these dioxolenium ions has led to some interesting observations that may be of interest.

One such observation is illustrated by the data in Table 4 where the results for DFT calculations on 2-*O*-acetyl-3,4,6-tri-*O*-methyl-D-glucopyranosyl **6**, mannopyranosyl **7** and galactopyranosyl **8** oxacarbenium ions are given. Species **6** and **7** are the 2-*O*-acetyl-analogues of **2** and **3**. As can be readily observed their respective **B0** and **B1** ring conformations ($^4H_3$ and $^5S_1$ for **6** and $^4H_3$ and ^3E for 7) are very similar to those for **2** and **3**. For **6** **B0** is more stable as for **3** but for **7** **B1** is calculated to be more stable in contrast to **2**.

Note that these **B0** and **B1** ring conformations have the O-2 acetyl group in its favoured conformation which is with the carbonyl oxygen *syn* to the methine at C-2, see Figure 8 (28). In all calculated pyranose structures with secondary esters this is found to be the minimum energy conformation. In either of these conformations if the ester rotates than the ring can close to form the dioxolenium ion. For *gluco* configured **6** this leads to two species **C0** and **C1** with conformations characterized as $^4H_5$ 0.813 and $^3S_1$ 0.654 respectively i.e. two conformers that differ by ring inversion.

Similarly for *manno* configured **7** two species characterized as $^4C_1$ 0.792 and $^1C_4$ 0.678, see Figure 9, and for *galacto* configured **8** two species characterized as $^4H_5$ 0.737 and $^1C_4$ 0.836 were found, see Figure 10. Intriguingly for **6** **C0** is energetically favored by about 10 kJ mol^{-1} and for **8** **C0** is still energetically favored but by about 40 kJ mol^{-1} whereas for **7** **C0** and **C1** are nearly isoenergetic (< 1 kJ mol^{-1}) i.e. all three configurational isomers are different.

We have previously studied both experimentally (29) and computationally 2-*O*-acyl-derivatives of galactopyranose. In particular we have made extensive DFT studies of 2,6-di-*O*-acetyl-3,4-isopropylidene-D-galactopyranosyl cation **9**. In these cases the same trends as for **8** were found but **C0** and **C1** have different conformations. Furthermore, the conversion of **B**'s to **C**'s for **9** was studied computationally using the formulations developed in Section 1 to enable DFT dynamic calculations. These results strongly supported a model where ring inversion preceded carbonyl ring closure and where the barrier to ring inversion is the main kinetic barrier (estimated barrier height 34 kJ mol^{-1}) (6).

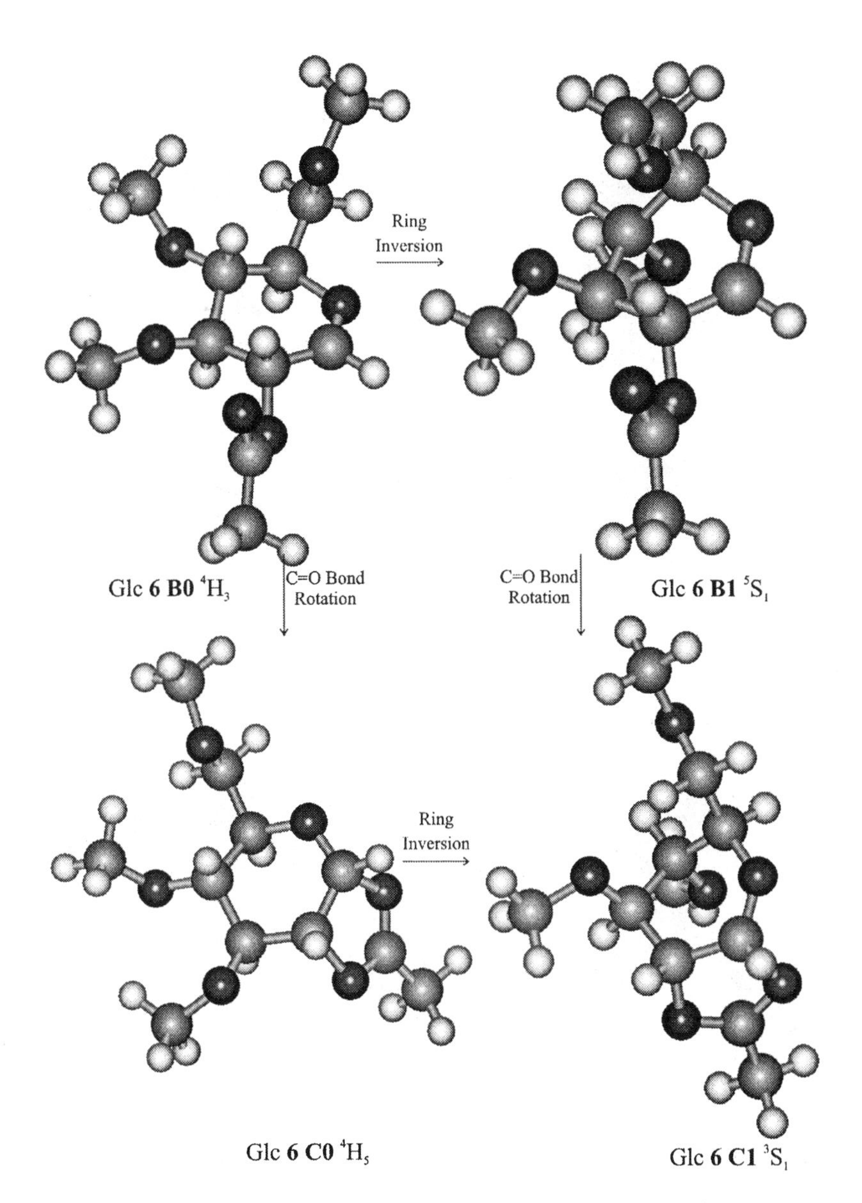

*Figure 8. Ball and Stick Representations of the **B0**, **B1**, **C0** and **C1** Species of Glc **6**.*

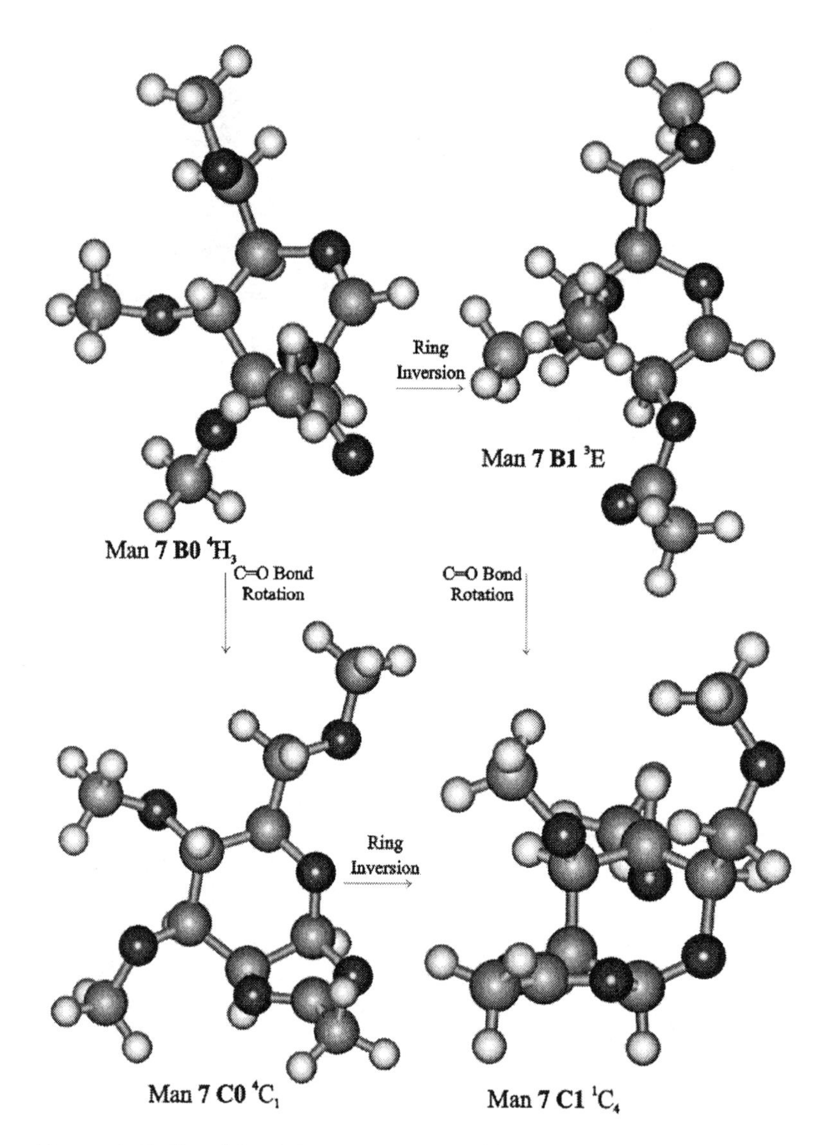

*Figure 9. Ball and Stick Representations of the **B0**, **B1**, **C0** and **C1** Species of Man 7.*

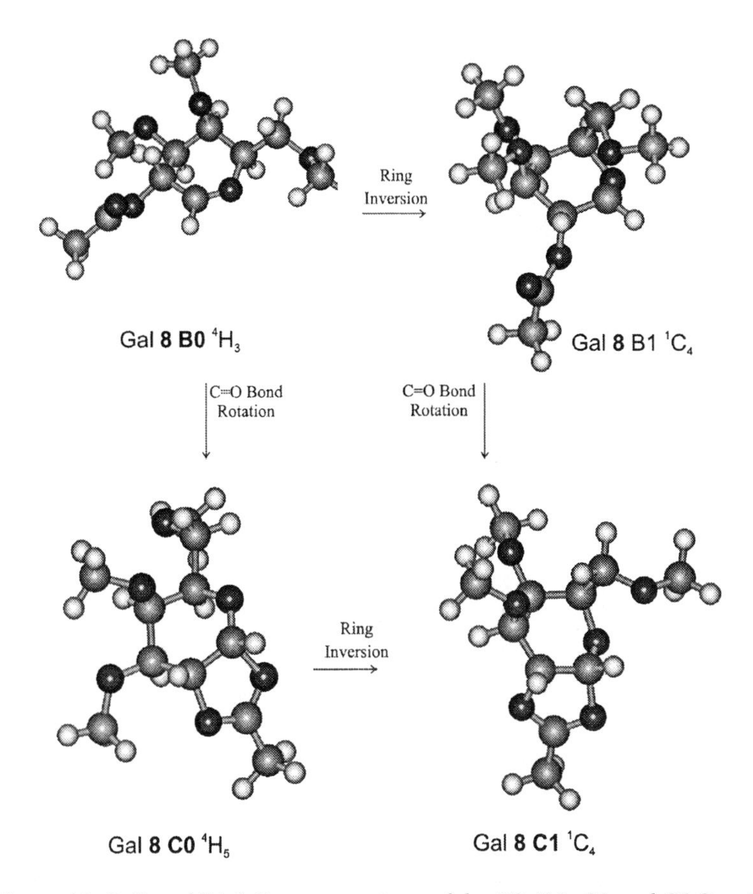

*Figure 10. Ball and Stick Representations of the **B0**, **B1**, **C0** and **C1** Species of Gal **8**.*

We have also studied the **C1** to **C0** interconversion to assess if this would shed some light on finding the *anti* S_N2 TS (30). In all conformers of the pyranose ring the C-1--O-7 did not appreciably lengthen. Therefore, from these studies we tentatively conclude that C-1--O-7 bond disassociation precedes ring inversion for the glycosylation process. In the extreme case this would be the same as nucleophilic attack on the isolated **B1**, not **B0** (for **9**). Since some degree of nucleophilic preassociation is probable and since this can only occur on the expected *trans* face this can partly explain the observed stereoselectivity. The cases where some *cis* (occasionally predominant) glycosylation occurs could be the result of attack on **B1**.

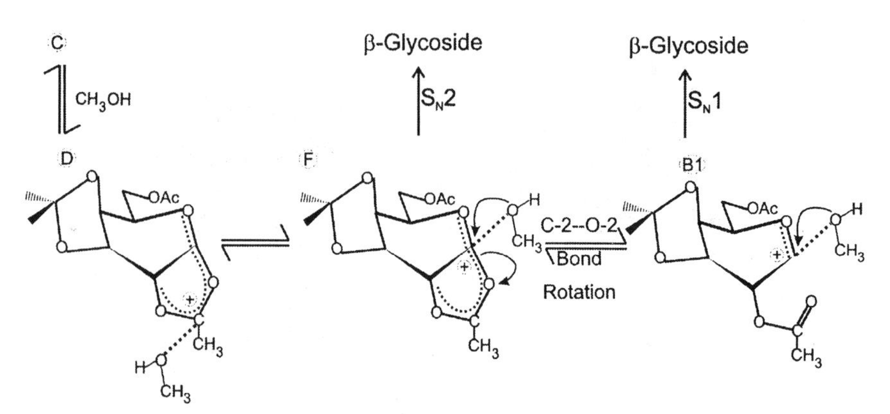

*Figure 11. Schematic Representation of Possible Intermediates in Neighboring Group Glycosylation Reactions of **9**.*

In reactions with very strong nucleophiles or cases where excess (e.g. solvolysis conditions) nucleophile is used attack on the first formed **B0** may out compete ring inversion and ring closure to **C** and therefore lead to anomeric mixtures. The order of reaction via **B** or **C** (i.e. ring inversion before or after carbonyl rotation) for derivatives other than *galacto* configured **9** is not known and is an active area of research for us.

Although entirely speculative at this point this observation of two conformers for dioxolenium ions and the possibility that at least in some cases ring inversion precedes 5-membered ring closure may assist in the comprehension of some cases in the literature. For example, it has been shown that n-pentenyl orthoesters can give rise to different glycosylation products than those derived from otherwise identically protected n-pentenyl glycoside donors (31). An interpretation following from our data would be that the dioxolenium ion formed by each process have different conformations and hence different reactivity.

Similar explanations may apply to cases where oligosaccharide donors with 2-acyl groups show different stereoselectivities depending on the anomericity of the sugar linkage attached to the reducing end sugar (32).

Again, it would seem possible that different conformations of the reducing end sugar could be formed after activation that have different reactivities. A detailed study of these possibilities appears warranted.

Table 4. Ring Conformations of 2-*O*-Acetyl-3,4,6-tri-*O*-methyl-D- Glucopyranosyl 6, Mannopyranosyl 7 and Galactopyranosyl 8 and 9 Oxacarbenium Ions

Species	τ_1 °	τ_2 °	τ_3 °	τ_4 °	τ_5 °	τ_6 °	Chair	Boat	Skew-Boat	Half-Chair Envelope	Energy kJ mol^{-1}
6B0	-42.9	61.0	-50.2	21.3	-3.9	15.6	$^{4}C_1$ 0.541	$B_{2,5}$ 0.055	$^{1}S_3$ 0.473	**$^{4}H_3$ 0.946**	0.0[a]
6B1	-9.5	-34.3	59.5	-40.0	-4.6	31.8	$^{1}C_4$ 0.244	$B_{O,3}$ 0.044	**$^{5}S_1$ 0.752**	-	7.2
6C0	-26.9	51.3	-65.3	56.2	-31.6	16.7	$^{4}C_1$ 0.689	$^{O,3}B$ 0.040	$^{1}S_5$ 0.406	**$^{4}H_5$ 0.813**	-32.5
6C1	24.5	-52.1	43.0	-3.0	-26.3	14.2	$^{1}C_4$ 0.228	$^{2,5}B$ 0.149	**$^{3}S_1$ 0.654**	^{5}E 0.298	-23.2
7B0	-43.6	60.1	-46.7	17.0	-0.7	14.7	$^{4}C_1$ 0.508	$B_{2,5}$ 0.023	$^{1}S_3$ 0.493	**$^{4}H_3$ 0.986**	0.0[a]
7B1	54.9	-60.6	37.9	-7.6	3.2	-28.7	$^{1}C_4$ 0.536	$^{O,3}B$ 0.436	$^{5}S_1$ 0.079	**^{3}E 0.873**	-12.5
7C0	-26.3	37.6	-57.3	70.0	-58.0	35.8	$^{4}C_1$ 0.792	$^{1,4}B$ 0.004	$^{O}S_2$ 0.359	-	-30.0
7C1	46.4	-57.6	50.7	-33.0	24.4	-32.0	$^{1}C_4$ 0.678	$^{2,5}B$ 0.022	$^{3}S_1$ 0.273	-	-29.1
8B0	-52.3	65.2	-45.9	12.9	-0.8	22.1	$^{4}C_1$ 0.553	$^{2,5}B$ 0.065	$^{1}S_3$ 0.533	**$^{4}H_3$ 1.067**	0.0[a]
8B1	35.9	-47.6	66.8	-69.8	61.9	-46.9	**$^{1}C_4$ 0.913**	$B_{1,4}$ 0.023	$^{2}S_O$ 0.284	-	5.5
8C0	-37.1	59.6	-65.7	48.9	-27.0	21.6	$^{4}C_1$ 0.722	$B_{O,3}$ 0.087	$^{1}S_5$ 0.369	**$^{4}H_5$ 0.737**	-38.8
8C1	31.8	-43.4	59.2	-67.3	56.9	-38.2	**$^{1}C_4$ 0.824**	$B_{1,4}$ 0.031	$^{2}S_O$ 0.293	-	0.3
9B0	-59.6	35.4	7.6	-33.4	7.7	40.7	$^{4}C_1$ 0.242	$B_{1,4}$ 0.022	**$^{2}S_O$ 0.771**	-	0.0[a]
9C0	-51.6	33.1	17.5	-58.3	40.9	15.7	$^{1}C_4$ 0.045	$^{1,4}B$ 0.170	**$^{2}S_O$ 0.908**	-	-20.1
9C1	22.1	17.6	-58.3	62.4	-21.4	-22.3	$^{4}C_1$ 0.320	**$B_{2,5}$ 0.688**	$^{3}S_1$ 0.032	-	-55.2

a. Energy of this species set to 0.0 kJ mol^{-1}. Total energies **6B0** 18 442.1, **7B0** 18 425.9 and **8B0** 18430.3.

Section 4- Acyl Transfer as a Side Reaction to Neighboring Group Participation

Our study of **9** was prompted by a glycosylation of the polymer linker combination, monomethyl ether of polyethylene glycol-dioxyxylene (MPEGDOXOH), with the donor ethyl 2,6-di-*O*-benzoyl-3,4-isopropylidene-1-thio-β-D-galactopyranoside, **10**. This glycosylation under a large variety of conditions led to a complex mixture of products. For example, under conventional NIS/triflic acid conditions the desired polymer bound glycoside **11** was obtained in <40 % yield. The remaining products can be attributed to the side reaction know as acyl transfer. These include: the benzoylated acceptor MPEGDOXOBz, the 2-OH glycoside analogue of **11** namely **12** and β1,2-linked oligomers, **13**, of the donor, see Figure 12. This side reaction is long known but little recognized. It is obvious that for successful polymer-supported synthesis where high yields and high purity are mandatory this side reaction is very unproductive. Since polymer-supported oligosaccharide synthesis is a major endeavour in our research groups (33) we devoted considerable effort to discovering the mechanism of this reaction.

After considering a large number of possibilities including most of the suggested mechanisms in the literature we found a mechanism that starts from ion:dipole

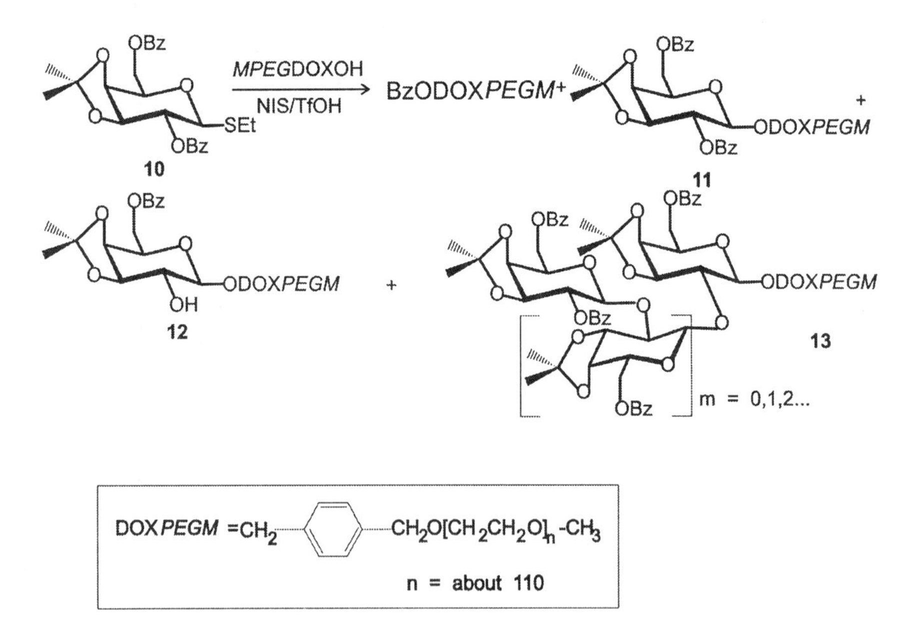

*Figure 12. Schematic Representation Showing Polymer-supported Glycosylation Reaction with Donor **10** Resulting in >60% Benzoyl Transfer Related Products.*

complex **D** and is triggered by proton transfer to O-2 followed by several bond breaking and bond forming reactions, see Figure 13. Consideration of the LUMO's of **B**'s and **C**'s suggests that for **B**'s nucleophilic attack should occur at C-1 where as for **C**'s it should occur at C-7, the former carbonyl carbon of the dioxolenium ion. Ion:dipole complexes **D** have long bonds to this carbon from the nucleophilic oxygen, O-8. In order to model this reaction we developed a procedure that uses complex constraints. This complex constraint is a linear combination of the bond lengths that are formed or broken during the reaction, see equation **4**:

$$Q = r_1 - r_2 + r_3 - r_4 + r_5 - r_6 \qquad (4)$$

where, r_1 through r_6 are defined in Figure 14.

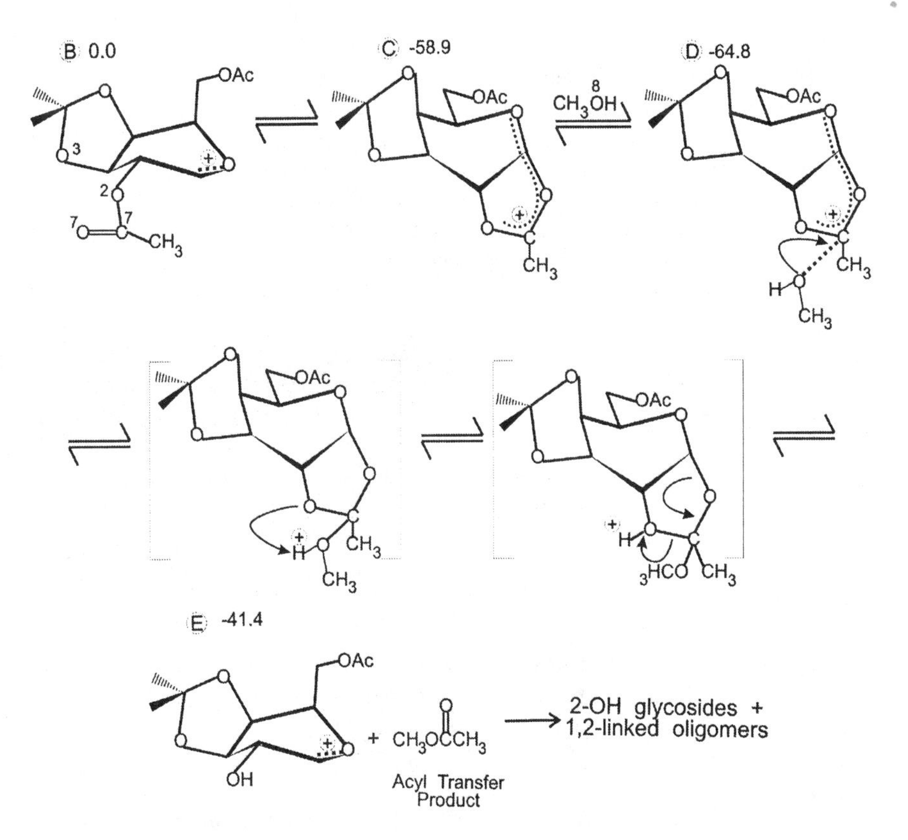

Figure 13. Schematic Representation of Mechanism for Acyl Transfer Triggered by Proton

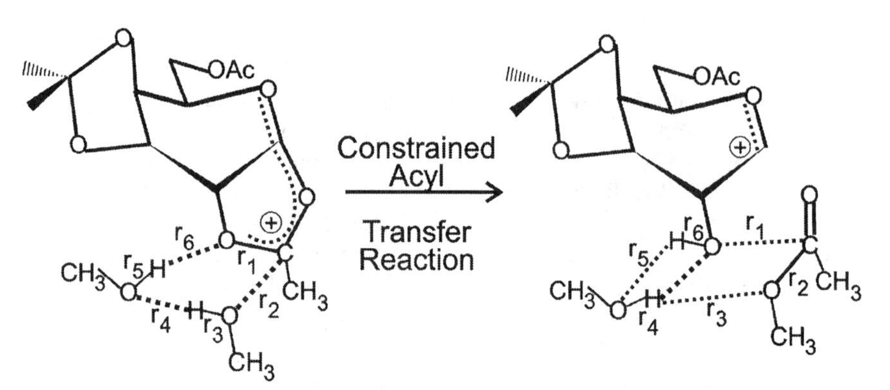

Figure 14. Bond Lengths (r_1 to r_6) used for the Constrained Acyl Transfer Reaction

The mechanism that results starts by having the O-8--C-7 bond length shorten. This shortening has the result that the complex develops appreciable hydronium ion character and hence most of the positive charge moves to the nucleophilic hydroxyl. In the presence of base this species could undergo proton transfer to form a stable orthoester. In the absence of added base this proton can transfer to another oxygen. In order to find a low enough barrier to proton transfer we had to introduce a relay molecule which in our case is another methanol but in the real reaction could be any species with an exchangeable hydrogen.

The second step in our mechanism is the transfer of the hydroxylic proton to the relay molecule followed by transfer to O-2. The 1st proton transfer is concurrent with O-8--C-7 covalent bond formation. The first TS is associated with the 2nd proton transfer. Once transferred the O-2--C-7 bond breaks with the positive charge now mostly on C-7. This process is associated with a second TS. Finally the O-7--C-1 breaks to lead to an ion:dipole complex of the newly formed ester and a 2-OH analogue of **B** i.e. **E**. The 2-OH glycosides and β1,2-linked oligomers are assumed to arise from **E** but this has not been studied in detail.

We studied this mechanism using the same constrained methodology with four different acyl groups namely formyl, acetyl, benzoyl and pivaloyl. These were chosen since experimental evidence from our collaborators has shown that acyl transfer decreases in the order formyl > acetyl > benzoyl > pivaloyl (34). Our studies found exactly the same mechanism for all 4 analogues including the location of the two TS's and the order of bond forming and bond breaking. This was highly surprising considering that the constraint in no way influenced this outcome and therefore we regard these results as quite convincing.

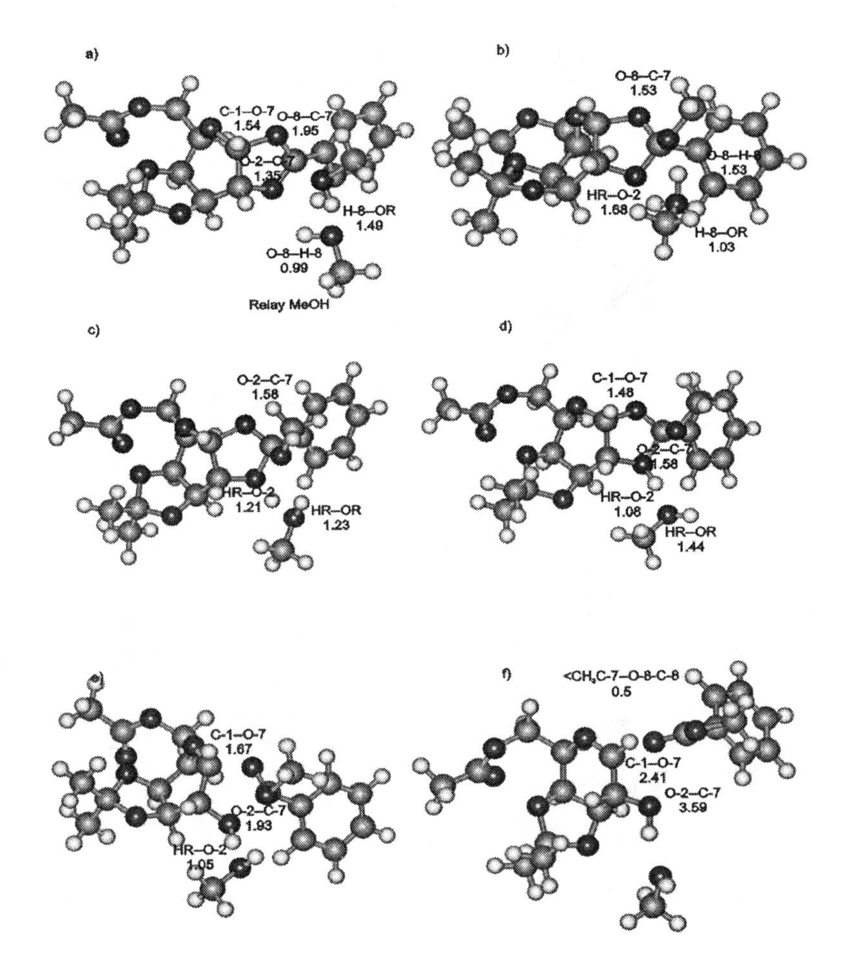

Figure 15a-f. Ball and Stick Representations of Snapshots of the Calculated Mechanism of Acyl Transfer. The Acyl Group is Benzoyl. Starting Structure Showing Relay MeOH (a) Proton Transfer to Relay O and Formation of O-7--C-7 bond (b) 1st TS with Proton being Transferred to O-2 (c) Proton Transferred to O-2 (d) 2nd TS and breaking of O-2--C-7 bond (e) Final structure with bond C-1--O7 broken too, note cis conformation of ester (f).

Detailed examination of the various steps and intermediates led to two interesting observations. The first is shown in Figure 16 that shows a portion of the calculated structures of ion dipole complexes **D**. In these structures one can see that the O-7--C-1 bond length correlates with the ease of acyl transfer. The O-7--C-1 is longest for formyl and shortest for benzoyl and pivaloyl. This is not an intuitive result and demonstrates the power of calculations to find novel correlations. Correlations only suggest causation and do not prove it. We come back to this correlation after examining the second point.

Graphical examination of the second TS for the benzoyl case, see Figure 15e, suggested to us that 2,6-substitution of the benzoyl ring might further restrict

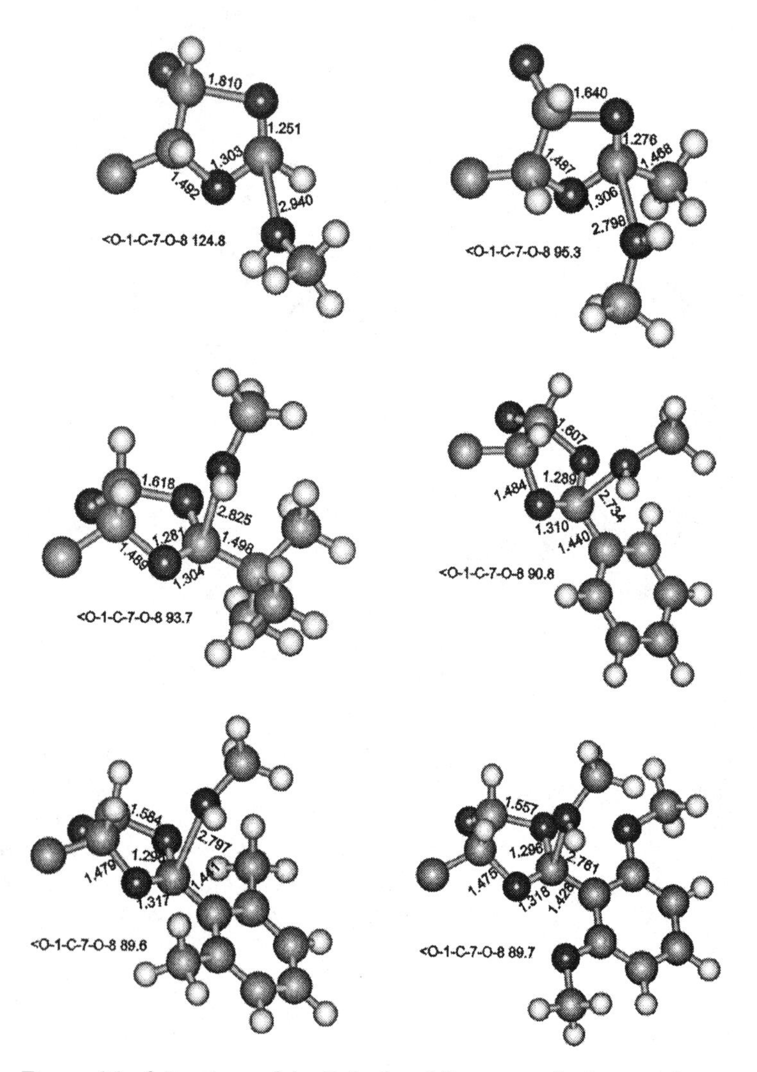

Figure 16a-f. Portions of the Calculated Structures for Intermediates ***D***: *Formyl (a) Acetyl (b) Benzoyl (c) Pivaloyl (d) 2,6-dimethylbenzoyl (e) and 2,6-dimethoxybenzoyl (f).*

rotation about the C-7--C(substituent bond C_{ipso} for benzoyl). This idea prompted us to first calculate and then prepare 2,6-disubstituted analogues **14** and **15** of benzoate **10**. In their calculated **D** structures the O-7-- C-1 is even shorter than for unsubstituted benzoyl suggesting an electronic component to this effect, see Figures 16e and f. More importantly under the same reactionconditions that resulted in >60% benzoyl transfer the 2,6-dimethylbenzoyl donor **14** exhibited no detectable transfer to give glycoside **16** in nearly quantitative yield and the 2,6-dimethoxybenzoyl donor **15** exhibited 8% transfer and >90% glycoside **17**.

The 2,6-dimethoxybenzoates are easier to remove than 2,6-dimethylbenzoate to yield the diol **18** and so are the protecting group of choice, see Figure 17. Further improvements can likely be made by adding an electron withdrawing group in the 4-position. Thus we have gone from a calculated TS to a modified protecting group that eliminates this side reaction having first done the calculations before the experiment.

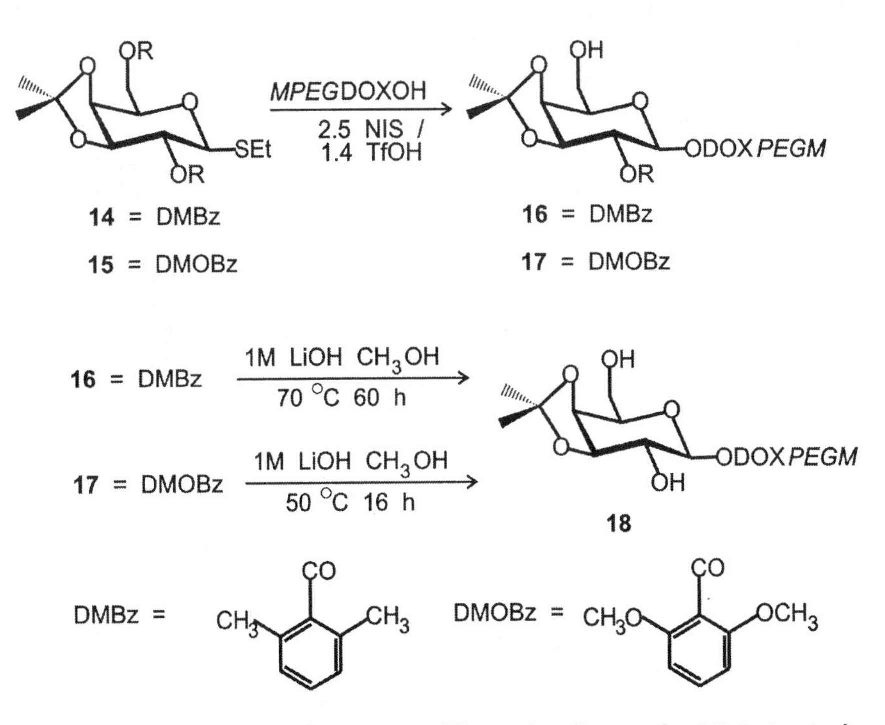

Figure 17. Structures and Reactions of Protecting Groups that Minimize Acyl Transfer Developed from a Calculated TS.

Conclusion

It is our hope that organic chemists will use our quantitative method of describing pyranose ring conformations. The examples shown here show how this aids in defining the probable conformations of the common glycopyranosyl oxacarbenium ions. Our calculations support a model where such ions are found in at least two different conformations that depend on the protecting groups and the configuration. This ability to exist in two different conformations is a probable explanation for some of the experimental observations on glycosylation reactions. Furthermore, it is anticipated that experimentalists should be able to take advantage of this idea to design stereoselective glycosylation reactions.

Acknowledgements

The authors gratefully acknowledge the use of the VPP770 Fujitsu parallel computer situated in the RIKEN Computer Center. This work was partly supported by the iHPC multiscale modeling initiative of the NRC.

References

1. Bérces, A.; Whitfield, D.M.; Nukada, T. *Tetrahedron* **2001**, *57*, 477.
2. Stoddart, J. F. In *Stereochemistry of Carbohydrates*; Wily-Interscience: New York, **1971**.
3. There are many methods for example:.(a) Cremer, D.; Pople, J.A. *J. Am. Chem. Soc.* **1975,** *97*, 1354. (b) Haasnot, C.G.A. *J. Am. Chem. Soc.* **1992,** *114,* 882.
4. Hendrickson, J.B. *J. Am. Chem. Soc.* **1967,** *89*, 7047.
5. (a) Klyne, W.; Prelog, V. *Experimentia* **1960**, *16*, 521. (b) Cano, F. H.; Foces-Foces, C.; Garcia-Binoco, S. *Tetrahedron* **1977**, *35*, 797.
6. Bérces, A.; Nukada, T.; Whitfield, D.M. *J. Am. Chem. Soc.* **2001** *123*, 5460.
7. Rhind-Tutt, A.J.; Vernon, C.A. *J. Chem. Soc. Part 2*, **1960**, 4637.
8. Bennet, A.J. *J. Chem. Soc. Perkin Trans 2* **2002**, 1207.
9. (a) Barrows, S.E.; Dulles, F.J.; Cramer, C.J.; French, A.D.; Truhlar, D.G. *Carbohydr. Res.* **1995**, *276*, 219. (b)Whitfield, D.M. *J. Molec. Struct.*(*Theochem*), **1997**, *395*, 53. (c) Polavarapu, P.L.; Ewig, C.S. *J. Comput. Chem.* **1992**, *13*, 1255. (d) Jeffrey, G.A. *J. Molec. Struct.* **1990**, *237*, 75.
10. Mendonca, A.; Johnson, G.P; French, A.D.; Laine, R.A. *J. Phys. Chem. A* **2002**, *106*, 4115.
11. Baerends, E. J.; Ellis, D.E.; Ros, P. *Chem. Phys.* **1973**, *2*, 41. (b) te Velde, G.; Baerends, E. J.; *J. Comput. Phys.* **1992**, *99*, 84. (c) Fonseca Guerra, C.; Snijders, J.G.; te Velde, G.; Baerends, E.J. *Theor. Chim. Acta* **1998**, *99*,

391. (d) Versiuis, L. Ziegler, T. *J. Chem. Phys.* **1988,** *88*, 322. (e) Fan, L.; Ziegler, T. *J. Chem. Phys.* **1992**, *96*, 9005.
12. Vosko, S.H.; Wilk, L.; Nusair, M. *Can. J. Phys.* **1980**, *58*, 1200.
13. Becke, A.D. *Phys. Rev. A* **1988**, *38*, 3098.
14. Perdew, J.P. *Phys. Rev. B* **1986**, *34*, 7506.
15. (a) Spijker, N.M.; Basten, J.E.M.; van Boeckel, C.A.A. *Recl. Trav. Chim. Pays-Bas.* **1993**, *112*, 611. (b) Spijker, N.M.; van Boeckel, C.A.A. *Angew, Chem. Int. Ed. Engl.* **1991**, *30*, 180. (c) L.G. Green, S.V. Ley, in *Carbohydrates in Chemistry and Biology Vol. 1* (Eds.: B. Ernst, G.W. Hart, P. Sinaÿ) Wiley-VCH Weinheim, **2000**, pp 427.
16. Nukada, T.; Bérces, A.; Whitfield, D.M. *Carbohydr. Res.* **2002**, *337*, 765.
17. Gurdial, S.; Patila, S. G.; Whitfield, D.M. unpublished observations.
18. Nukada, T.; Bérces, A.; Wang, L.; Zgierski, M.Z.; Whitfield, D.M. *Carbohydr. Res.* **2005**, *340*, 841.
19. (a) Maroušek, V.; Lucas, T.J.; Wheat, P.E.; Schuerch, C. *Carbohydr. Res.* **1978**, *60*, 85. (b) Houdier, S.; Vottero, P.J.A. *Carbohydr. Res.* **1992**, *232*, 349. (c) Fréchet, J.M.; Schuerch, C. *J. Am. Chem. Soc.* **1972**, *94*, 604.
20. (a) Abdel-Rahman, A.A.-H.; Jonke, S.; El Ashry, E.S.H.; Schmidt, R.R. *Angewandte. Chem. Int. Ed. Engl.* **2002**, *41*, 2972. (b) Crich, D.; Cai, W. *J. Org. Chem.* **1999**, *64*, 4926. (c) Crich, D.; Li, H. *J. Org. Chem.* **2000**, *65*, 801. (d) Weingart, R.; Schmidt, R.R. *Tetrahedron Lett.* **2000**, *41*, 8753. (e) Yun, M.; Shin, Y.; Chun, K.H.; Nam Shin, J.E. *Bull. Korean Chem. Soc.* **2000**, *21*, 562. (f) Crich, D.; Dudkin, V. *Org. Lett.* **2000**, *2*, 3941. (g) Ito, Y.; Ohnishi, Y.; Ogawa, T.; Nakahara, Y. *SYNLETT*, **1998**, 1102.
21. Tamura, S.; Abe, H.; Matsuda, A. Shuto, S. *Angewandte. Chem. Int. Ed. Engl.* **2003**, *42*, 1021. (b) Yule, J.E.; Wong, T.C.; Gandhi, S.S.; Qiu, D.; Riopel, M.A. ; Koganty, R.R. *Tetrahedron, Lett.* **1995**, *36*, 6839. (c) Toshima, K.; Nozaki, Y.; Tatsuta, K. *Tetrahedron, Lett.* **1991**, *32*, 6887.
22. Lemieux, R.U. *Adv. Carbohydr. Chem.* **1954**, *9*, 1.
23. a) Lemieux, R. U. *Chem. Canada* **1964**, *16*, 14. (b) Wulff, G.; Röhle, G. *Angew. Chem. Int. Ed.* **1974**, *13*, 157.
24. (a) Wallace, J.E.; Schroeder, L.R. *J. Chem. Soc. Perkin Trans 2*, **1977**, 795. (b) Paulsen, H.; Herold, C.P. *Chem. Ber.* **1970**, *103*, 2450. (c) Crich, D.; Dai, Z.; Gastaldo, S. *J. Org. Chem.* **1999**, *64*, 5224.
25. Paulsen, H.; Dammeyer, R. *Chem. Ber.* **1973**, *106*, 2324.
26. Nukada, T.; Bérces, A.; Zgierski, M.Z.; Whitfield, D.M. *J. Am. Chem. Soc.* **1998**, *120*, 13291.
27. (a) Toshima, K.; Tatsuta, K. *Chem. Rev.* **1993**, *93*, 1503. (b) Sinaÿ, P. *Pure & Appl. Chem.* **1991**, *63*, 519. (c) Schmidt, R.R. *Pure & Appl. Chem.* **1989**, *61*, 1257. (d) Paulsen, H. *Chem. Soc. Rev.* **1984**, *13*, 15. (e) Ogawa, T. *Chem. Soc. Rev.* **1994**, *23*, 397. (f) Kochetkov, N. E. in *Studies in Natural Products Chemistry* (Rahman, A. Ed.) **1994**, *14*, 201. (g) Demchenko, A.V. *Curr. Org. Chem.*, **2003**, *7*, 35.
28. Schweizer, W.B.; Dunitz, J.D. *Helv. Chim. Acta* **1982**, *65*, 1547.
29. Nukada, T.; Bérces, A.; Whitfield, D.M. *J. Org. Chem.* **1999**, *64*, 9030.

30. Bérces, A.; Nukada, T.; Whitfield, D.M. unpublished observations.
31. (a) Fraser-Reid, B.; Grimme, S.; Piacenza, M.; Mach, M; Schlueter, U. *Chem. Eur. J.* **2003**, *9*, 4687. (b) Fraser-Reid, B.; López, J.L.; Radhakrishnan, K.V.; Nandakumar, M.V.; Gómez, A.M.; Uriel, C. *J. Chem. Soc. Chem. Commun.* **2002**, 2104. (c) Mach, M.; Schlueter, U.; Mathew, F.; Fraser-Reid, B.; Hazen, K.C. *Tetrahedron*, **2002**, *58*, 7345.
32. (a) Zeng, Y.; Ning, J.; Kong, F. *Tetrahedron Lett.* **2002**, *43*, 3729. (b) Zeng, Y.; Ning, J.; Kong, F. *Carbohydr. Res.* **2003**, *338*, 307.
33. (a) Hanashima, S.; Manabe, S.; Ito, Y. *Synlett*, **2003**, 979. (b) Yan, F.; Gilbert, M.; Wakarchuk, W.W.; Brisson, J.R.; Whitfield, D.M. *Org. Lett.* **2001**, *3*, 3265.
34. Bérces, A.; Whitfield, D.M.; Nukada, T.; do Santos Z.I.; Obuchowska, A.; Krepinsky, J.J. *Can. J. Chem.* **2004**, *82*, 1157.

Indexes

Author Index

Subject Index

B

C

E

F

G

H

N

T

U

V

Z